Forensic Fire Scene Reconstruction

BRADY

Forensic Fire Scene Reconstruction

David J. Icove
The University of Tennessee

John D. DeHaan
Fire-Ex Forensics, Inc.

PEARSON

Prentice Hall

Upper Saddle River, New Jersey 07458

Library of Congress Cataloging-in-Publishing Data

Icove, David J.
Forensic fire scene reconstruction/David J. Icove and John D.
   DeHaan.—1st ed.
           p. cm.
Includes bibliographical references and index.
   ISBN 0-13-094205-7
   1. Fire investigation.   2. Arson investigation.   I. DeHaan, John D. (John David),
1948- II.   Title.
   TH9180.I26 2004
   363.37'65—dc21   2003012949

Publisher: Julie Levin Alexander
Publisher's Assistant: Regina Bruno
Senior Acquisitions Editor: Katrin Beacom
Assistant Editor: Kierra Kashickey
Senior Marketing Manager: Katrin Beacom
Channel Marketing Manager: Rachele Strober
Director of Production and Manufacturing: Bruce Johnson
Managing Production Editor: Patrick Walsh
Manufacturing Buyer: Pat Brown
Production Liaison: Julie Li
Production Editor: Wendy Druck, *The GTS Companies*/York, PA Campus
Creative Director: Cheryl Asherman
Cover Design Coordinator: Christopher Weigand
Cover Designer: Christopher Weigand
Cover Photographer: Eddie Sperling
Compositor: *The GTS Companies*/York, PA Campus
Printer/Binder: Courier Westford
Cover Printer: Phoenix Color Corporation

Credits and acknowledgments borrowed from other sources and reproduced, with permission, in this textbook appear on appropriate pages within text.

Note: A portion of this book was prepared by David J. Icove, who is currently an employee of the Tennessee Valley Authority (TVA). The views expressed in this book are the views of Dr. Icove and his coauthor. They are not necessarily the views of TVA or the United States Government. Reference in this book to any product, process, or service by trade name, trademark, manufacturer, or otherwise does not necessarily constitute its endorsement or recommendation by TVA or the United States Government.

Pearson Education LTD.
Pearson Education Singapore Pte. Ltd
Pearson Education, Canada, Ltd
Pearson Education–Japan

Pearson Education Australia PTY, Limited
Pearson Education North Asia Ltd
Pearson Educación de Mexico, S.A. de C.V.
Pearson Education Malaysia, Pte. Ltd

PEARSON
Prentice Hall

10 9 8 7 6 5 4 3 2 1
ISBN 0-13-094205-7

To Bud Nelson, who started it all.

Contents

List of Figures

List of Tables

Preface

For centuries the investigation of fires was limited to inspecting the ruins, asking questions of witnesses, and applying common sense to conclude how it all began. It is a particularly challenging endeavor because of fire's transient power and destructiveness. It is tempting to apply "common knowledge" because fire has been part of man's experience for thousands of years and everyone "knows" about fires. But it turns out that some of that "knowledge" was not well based and it sometimes led to erroneous conclusions. Unlike the situation in many other disciplines, there was usually no way to do controlled tests of the conclusions to determine whether or not they were reliable. There were too many variables and no way to duplicate all the necessary conditions. Over the years more systematic knowledge of science and engineering was applied as chemists and fire scientists became involved. Forensic scientists like Paul Kirk and fire scientists like Howard Emmons and Philip Thomas contributed greatly to improving the accuracy of investigations by demonstrating the usefulness of scientific analysis, but concerned professional fire investigators realized that even more science was needed.

The confluence of events that dramatically influenced fire investigation really began in the 1980s with the publication of several keynote textbooks: the National Bureau of Standards' *Fire Investigation Handbook* in 1980, the first edition of *Combating Arson-for-Profit* in 1980, the first revision of *Kirk's classic Fire Investigation* in 1983, Cooke and Ide's *Techniques of Fire Investigation* in 1984, and Drysdale's epic *Introduction to Fire Dynamics* in 1985.

A marked change in the investigation of fires occurred when Harold E. "Bud" Nelson, of the National Institute of Standards and Technology (NIST), assisted federal agencies in the investigation of the tragic fire at the Dupont Plaza Hotel in Puerto Rico in 1986. Bud offered to study various scenarios to help explain the growth of the fire using computer modeling in exchange for data and information from the scene. That tentative collaboration was very successful and led to the confirmation of not only the area of origin, but also the manner of fire and smoke spread that led to the terrible loss of life.

From that time on, the interaction among fire engineers, fire scientists, and investigators grew rapidly. By 1992, the National Fire Protection Association (NFPA) had published its first collaboration between those disparate groups with *NFPA 921*, and the Bureau of Alcohol, Tobacco and Firearms (ATF) had made fire engineering an essential part of its training for Certified Fire Investigators (CFIs) with the development of Quintiere's *Principles of Fire Behavior* in 1998. Each successive edition of *NFPA 921* and *Kirk's Fire Investigation* has strengthened the bonds, and it is hoped that this book will extend the capacity of investigators to apply engineering and criminalistics to the accurate investigation of fire.

Fire scene reconstruction has been part of good fire investigation for many years, but the concept was generally limited to the physical scene. This meant identifying fuel packages by their postfire remains and replacing them in their positions during the fire. This reconstruction could be confirmed by interviews with residents, tenants, visitors, or customers and by reviewing prefire photos or videos.

Forensic fire scene reconstruction goes well beyond establishing what furnishings were present and where they were placed. Reconstruction involves identification and documentation of all relevant features of the fire scene—dimensions, materials, location, and physical evidence—that help establish human activities and contacts. This information is then placed in context with principles of fire engineering and human behavior and is used to evaluate various scenarios of the origin, cause, and development of the fire and the interaction of people with it.

Identification of the cause of an incendiary fire is often much more straightforward than the elimination of possible accidental ignition sources, but this is a necessary element in all arson cases. Systematically employed in accidental fires, forensic reconstructions using the scientific method can eliminate all but one conflicting hypothesis. Such a systematic approach can be used to test for expected consequences if each possible hypothesis were correct. For instance, if a transmission failure in a car being driven caused it to catch fire, one would expect there to be physical evidence of such a failure—burned transmission fluid, melted plastic, metal particles, and so forth. If examination reveals no such defects, another cause or mechanism must be sought out. It is the intent of this text to explore and explain the process of applying the full spectrum of forensic reconstruction to fire events. Today, fire scene reconstruction is based primarily upon physical evidence of burn patterns on the remnants of structures and vehicles. What is not always taken into consideration when looking at the fire's pattern of damage is the physical evidence of pre- and postfire human activities and the behaviors that controlled these activities.

This textbook is intended for use by fire and law enforcement investigators, prosecutors, forensic scientists, and fire protection professionals. A thorough understanding of fire dynamics is valuable in applying these forensic engineering techniques. This book describes and illustrates a totally new systematic approach for reconstructing fire scenes. The approach applies the principles of fire protection engineering along with forensic science and behavioral science.

Using historical fire cases, the authors provide new lessons and insight into the ignition, growth, development, and outcome of those fires. All documentation in the case examples follows or exceeds the methodology set forth by the NFPA in *NFPA 921—Guide for Fire and Explosion Investigations*. The best features of this work include the use of real-world case examples to illustrate the concepts discussed herein. These examples shed new light on the forensic science, fire engineering, and human factor issues. Each example is illustrated using the guidelines from *NFPA 921*. In cases where fire engineering analysis or fire modeling is applicable, these techniques are explored.

Chapters 1–9 also conclude with a set of short self-study problems. The references in Appendix A include the latest public-domain publications on major fire investigations, U.S. government major fire reports, fire modeling software, and other documents that may be of interest to the reader.

AUTHOR BACKGROUNDS

This textbook is coauthored by two of the United States' most experienced fire scientists. Their combined talents total nearly 65 years of experience in the fields of behavioral science, fire protection engineering, fire behavior, investigation, criminalistics, and crime scene reconstruction.

David J. Icove, Ph.D., P.E.

An internationally recognized forensic fire engineering expert with over 30 years of experience, Dr. Icove is coauthor of *Combating Arson-for-Profit,* the leading textbook on the crime of economic arson. He also serves as a principal member of the *NFPA 921 Technical Committee on Fire Investigations.* A career law enforcement officer, Dr. Icove has experience as a criminal investigator on the federal, state, and local levels.

He presently serves as an Inspector in the Criminal Investigations Division of the U.S. Tennessee Valley Authority (TVA) Police, Knoxville, Tennessee, and is presently assigned to the Federal Bureau of Investigation (FBI) Joint Terrorism Task Force (JTTF). In addition to conducting major case investigations, Dr. Icove oversees the development of advanced fire investigation training and technology programs in cooperation with various agencies, including the Federal Emergency Management Agency's U.S. Fire Administration.

Before transferring to the U.S. TVA Police, he served 9 years as a program manager in the elite behavioral science and criminal profiling units at the FBI, Quantico, Virginia. At the FBI, he implemented and became the first supervisor of the Arson and Bombing Investigative Support (ABIS) Program, staffed by FBI and ATF criminal profilers. Prior to his work at the FBI, Dr. Icove served as a criminal investigator at arson bureaus of the Knoxville Police Department, the Ohio State Fire Marshal's Office, and the Tennessee State Fire Marshal's Office.

His expertise in forensic fire scene reconstruction is based on a blend of onscene experience, conduction of fire tests and experiments, and participation in prison interviews of convicted arsonists and bombers. He has testified as an expert witness in civil and criminal trials, as well as before U.S. Congressional Committees seeking guidance on key arson investigation and legislative initiatives.

Dr. Icove holds B.S. and M.S. degrees in Electrical Engineering and a Ph.D. in Engineering Science and Mechanics from The University of Tennessee. He also holds a B.S. degree in Fire Protection Engineering from the University of Maryland—College Park. He is presently Adjunct Assistant Professor in the Department of Electrical and Computer Engineering at The University of Tennessee, Knoxville, and is a Registered Professional Engineer in Tennessee and Virginia.

John D. DeHaan, Ph.D., FABC, CFI, FSSDip

An internationally recognized forensic science expert, Dr. DeHaan is the author of *Kirk's Fire Investigation,* the leading textbook in the field of fire and arson investigation. He is also a former principal member of the *NFPA 921 Technical Committee on Fire Investigations.*

Dr. DeHaan has been a criminalist for 33 years and has gained considerable expertise in fire and explosion evidence as well as human hair, shoe print, and instrumental analysis and crime scene reconstruction. He has been employed as a criminalist by the Alameda County Sheriff's Office, the U.S. Treasury Department, and the California Department of Justice.

His research into forensic fire scene reconstruction is based on first-hand fire experiments on fire behavior involving over 500 observed full-scale structure, and 100 vehicle fires under controlled conditions, as well as laboratory-scale studies. Dr. DeHaan has testified as an expert witness in civil and criminal trials across the United States and overseas.

Dr. DeHaan graduated from the University of Illinois—Chicago Circle in 1969 with a B.S. degree in Physics and a minor in Criminalistics. He was awarded a Ph.D. in Pure and Applied Chemistry (Forensic Science) by Strathclyde University in Glasgow,

Scotland, in 1995. Dr. DeHaan is a Fellow, American Board of Criminalistics (Fire Debris), and holds Diplomas in Fire Investigation from the Forensic Science Society and the Institution of Fire Engineers and a Certified Fire Investigator certification from the International Association of Arson Investigators.

ABOUT THIS BOOK

This is the first edition of a book that started as a series of lectures for members of fire and police agencies and the insurance industry, fire protection engineering professionals, and academicians. These lectures were used over time in college classrooms, at law enforcement training academies, at in-service fire investigation seminars, and in forensic laboratories.

The scientific approach that the authors presented during these lectures is consistent with present-day expert witness guidelines in federal and state courts. Based upon inquiries from both our colleagues and our students, we decided that this textbook would take the form of a stand-alone volume capable of furthering the expertise of any individual or group actively involved in the pursuit of fire and arson investigation. This target audience includes:

- public safety officials charged with the responsibility of investigating fires;
- prosecutors of arson and fire-related crimes who seek to add to their capabilities of evaluating evidence and presenting technical details to a nontechnical judge and jury;
- judicial officials seeking to comprehend better the technical details of cases over which they preside;
- private-sector investigators, adjusters, and attorneys representing the insurance industry who have the responsibility of processing claims or otherwise have a vested interest in determining responsibility for starting or causing a fire;
- citizens and civic community service organizations committed to conducting public awareness programs designed to reduce the threat of fire and its devastating effect on the economy; and
- scientists, engineers, academicians, and students engaged in the education process pertaining to forensic fire scene reconstruction.

SCOPE OF THE BOOK

Forensic Fire Scene Reconstruction is divided into the following chapters.

Chapter 1, **"Principles of Reconstruction,"** describes a systematic approach to reconstructing fire scenes in which investigators rely on the combined principles of fire protection engineering along with forensic and behavioral science. Using this approach, the investigator can more accurately document a structural fire's origin, intensity, growth, direction of travel, and duration as well as the behavior of the occupants.

Chapter 2, **"Basic Fire Dynamics,"** provides the investigator with a firm understanding of the phenomenon of fire, heat release rates of common materials, heat transfer, growth, and development, fire plumes, and enclosure fires.

Chapter 3, **"Fire Pattern Analysis,"** describes the underpinnings of how fire patterns are used by investigators in assessing fire damage and determining a fire's origin. Fire patterns are often the only remaining visible evidence left after a fire is extinguished. The ability to document and interpret fire pattern damage accurately is a skill paramount to investigators when they are reconstructing fire scenes.

Chapter 4, **"Fire Scene Documentation,"** details a systematic approach needed to support forensic reports. The purpose of forensic fire scene documentation includes recording visual observations, emphasizing fire development characteristics, and authenticating and protecting physical evidence. The underlying theme is that documentation produces sound investigations and courtroom presentations.

Chapter 5, **"Arson Crime Scene Analysis,"** reviews the techniques used in the analysis of arsonists' motives and intents. It presents nationally accepted motive-based classification guidelines along with case examples of the crimes of vandalism, excitement, revenge, crime concealment, and arson-for-profit. The geography of serial arson is also examined, along with techniques to profile the targets selected by arsonists.

Chapter 6, **"Fire Modeling,"** discusses the use of various mathematical and computer-assisted techniques for modeling fires, explosions, and the movement of people. Numerous models are explored, along with their strengths and weaknesses. Several case examples are also presented.

Chapter 7, **"Fire Death and Injuries,"** provides an in-depth examination of the impact and tenability of fires on humans. The chapter examines what kills people in fires, particularly their exposure to by-products of combustion, toxic gases, and heat. It also examines the predictable fire burn pattern damage inflicted on human bodies and summarizes postmortem tests and criminalistic examinations desirable in comprehensive death investigations.

Chapter 8, **"Fire Testing,"** reviews the applicable standard fire testing methods that are important to forensic fire scene analysis and reconstruction.

Chapter 9, **"Case Studies,"** profiles many of the concepts used in forensic fire scene investigations using real-world sanitized cases.

Chapter 10, **"Future Tools for the Fire Investigator,"** describes various concepts recently introduced in fire investigation and presents proactive strategies being used to increase investigative effectiveness and strengthen successful case documentation in the future.

Appendix A contains the References. Appendix B is a Glossary covering common terminology used in the fire investigation field. Appendix C is a Mathematics Refresher.

PEER REVIEWERS

Peer review is an important concept for ensuring that a textbook is well balanced, useful, authoritative, and accurate. The following agencies, institutions, companies, and individuals provided invaluable support during the peer-review process.

Dr. Vytenis (Vyto) Babrauskas, Fire Science and Technology, Inc., Issaquah, Washington.

Guy E. "Sandy" Burnette, Jr., Attorney, Tallahassee, Florida.

Robert F. Duval, National Fire Protection Association, Quincy, Massachusetts.

Daniel Madrzykowski, National Institute of Standards and Technology, Gaithersburg, Maryland.

John E. "Jack" Malooly, Bureau of Alcohol, Tobacco and Firearms, Chicago, Illinois.

Michael Marquardt, Bureau of Alcohol, Tobacco and Firearms, Grand Rapids, Michigan.

Lamont "Monty" McGill, McGill Investigations, Gardnerville, Nevada.

Robert R. Rielage, Ohio Division of State Fire Marshal, Reynoldsburg, Ohio.

Robert K. Toth, Iris Fire, LLC, Denver, Colorado.

ACKNOWLEDGMENTS

We have many people to thank for both their help on and their inspiration for the first edition of this book, including the many present and past employees at the following institutions and agencies.

Bureau of Alcohol, Tobacco and Firearms (ATF): Steve Carman, Gerald A. Haynes, Steve Avato, Dennis C. Kennamer, Jack Malooly, Wayne Miller, Michael Marquardt, Luis Velaszco and John Mirocha.

J. H. Burgoyne & Partners (U.K.): Roger Berretts, Robin Holleyhead, and Roy Cooke.

California State Fire Marshal's Office: Lamont "Monty" McGill (retired), Jim Allen (retired), and Joe Konefal.

Eastern Kentucky University: Ronald L. Hopkins.

Federal Bureau of Investigation (FBI): Richard L. Ault (retired), Steve Band, S. Annette Bartlett, John Henry Campbell (retired), R. Joe Clark, Roger L. Depue (retired), Joseph A. Harpold (retired), Timothy G. Huff (retired), Sharon A. Kelly, John L. Larsen, James A. O'Connor (retired), William L. Tafoya (retired), and Arthur E. Westveer.

Fire Safety Institute: John M. Watts, Jr.

Fire Science and Technology: Vyto Babrauskas.

Gardiner Associates: Mick Gardiner, Jim Munday.

Kent Archaeological Field School: Paul Wilkinson.

Knox County (Tennessee) Sheriff's Office: Michael K. Dalton.

McKinney (Texas) Fire Department: Chief Mark Wallace.

National Association of Fire Investigators: Patrick M. Kennedy.

New South Wales Fire Brigade: Ross Brogan.

Ohio State Fire Marshal's Office, Reynoldsburg: Eugene Jewell (deceased), Charles G. McGrath (retired), Mohamed M. Gohar (retired), J. David Schroeder, Jack Pyle (deceased), Harry Barber, Lee Bethune, Joseph Boban, Kenneth Crawford, Dennis Cummings, Dennis Cupp, Robert Davis, Robert Dunn, Donald Eifler, Ralph Ford, James Harting (retired), Robert Lawless, Keith Loreno, Mike McCarroll, Matthew J. Hartnett, Brian Peterman, Mike Simmons, Rick Smith, Stephen W. Southard, and David Whitaker.

Richland Washington Fire Department: Glenn Johnson and Grant Baynes.

Rodger H. Ide.

Sacramento County (California) Fire Department: Jeff Campbell (retired).

Saint Paul (Minnesota) Fire Department: Jamie Novak.

Santa Ana (California) Fire Department: Jim Albers (retired) and Bob Eggleston (retired).

Tennessee State Fire Marshal's Office: Richard L. Garner, Robert Pollard, Eugene Hartsook (deceased), and Jesse L. Hodge (retired).

Tennessee Valley Authority (TVA): Carolyn M. Blocher, James E. Carver, W. Chris McRae, R. Douglas Norman, Larry W. Ridinger, Sidney G. Whitehurst, and Norman Zigrossi (retired).

University of Arkansas, Department of Anthropology: Elayne J. Pope and O'Brien C. Smith.

University of Edinburgh, Department of Civil Engineering: Professor Dougal Drysdale.

University of Maryland, Department of Fire Protection Engineering: John L. Bryan (Emeritus), James A. Milke, Frederick W. Mowrer, James G. Quintiere, Marino di Marzo, and Steven M. Spivak (Emeritus).

The University of Tennessee, College of Engineering: A. J. Baker, J. Douglas Birdwell, Samir M. El-Ghazaly, Rafael C. Gonzalez, M. Osama Soliman, Jerry Stoneking (deceased), Tse-Wei Wang, and the many students in the ECE 599 course on Computer Fire Modeling.

The University of Tennessee, Medical Group, Memphis: O'Brien C. Smith.

U.S. Consumer Products Safety Commission: Gerard Naylis and Carol Cave.

U.S. Fire Administration, Federal Emergency Management Agency: Edward J. Kaplan, Kenneth J. Kuntz, and Denis Onieal.

Our sincere, heartfelt thanks go to Carolyn M. Blocher, Angi M. Christensen, Vivian McLaughlin, Edith De Lay, Wendy Druck, and Larry Harding, who reviewed the early manuscripts, addressed technical questions, and made many beneficial suggestions as to the format and content of this textbook.

A very special mention goes to our engineering technical editor, Laura Morris Edwards.

Also, special thanks go to our families and close friends over the years for their patient support.

SPECIAL ACKNOWLEDGMENT

The authors specifically acknowledge the contributions and encouragement of Harold E. "Bud" Nelson, P.E., one of the most influential fire protection engineers of the 20th century. Through all his innovative work at the GSA, NIST, and, later, Hughes Associates, Inc., he advanced the science of fire scene analysis through his development and application of fire modeling and scientific methods to fire investigation.

An active member of the NFPA Technical Committee on Fire Investigation, Bud could always be counted upon to educate its members on the latest findings, interpretations, and rationale for explaining fire dynamics. His thorough yet matter-of-fact investigations into the Dupont Plaza Hotel fire in Puerto Rico and other incidents, both large and small, have provided benchmarks for the fire investigation community and serve as the guides for many of the fire scene assessment techniques discussed in this textbook.

Forensic Fire Scene Reconstruction

Principles of Reconstruction

1 CHAPTER

It is a capital mistake to theorize before one has data. Insensibly one begins to twist facts to suit theories, instead of theories to suit facts.

—Sir Arthur Conan Doyle,
"A Scandal in Bohemia"

The goal of this chapter is to introduce a new look on the principles and concepts of science-based fire scene reconstruction. The use of the scientific method in forming an expert opinion as to the fire's origin and cause, development, and impact upon human lives is discussed from several aspects.

The process of fire scene reconstruction is used to determine the most likely development of a fire using a scientifically based methodology. Reconstruction follows the fire from ignition to extinguishment; and it explains aspects of the fire and smoke development, the role of fuels, effects of ventilation, the impact of manual and automatic extinguishment, the performance of the building, life safety features, and manner of injuries or death.

Underlying principles of forensic fire scene reconstruction rely firmly upon a comprehensive review of the fire pattern damage, sound fire protection engineering principles, human factors, physical evidence, and an appropriate application of the scientific method. These factors often form the basis for an expert opinion as to the most probable origin and cause of the fire or explosion. The expert opinion may be part of a written report or the basis for oral testimony in depositions or courtrooms.

In order to be effective, these expert opinions must be able to pass the eventual scrutiny of cross-examination during sworn depositions, peer review, and courtroom testimony. Recent court decisions place more weight on expert forensic testimony based upon scientific, rather than merely experience-based, knowledge (Lentini 2001).

When determining the origin and cause of a fire, a comprehensive reconstruction often involves a fire engineering analysis that tests various scenarios. This analysis may use fire modeling to compare actual events with predicted outcomes by varying causes and growth scenarios. This engineering analysis adds value, understanding, and clarity to an already complex fire scene investigation.

◆ 1.1 NEED FOR SCIENCE IN FIRE SCENE RECONSTRUCTION

A federal conference in November 1997 assessed the current state of the art and identified technical gaps in fire investigation (Nelson and Tontarski 1998). The International Conference on Fire Research for Fire Investigation concluded that many scientific gaps existed in the methodology and principles used to reconstruct fire scenes.

This federal conference cited research, development, training, and education needs including the following.

- **Fire incident reconstruction** — laboratory methods for testing ignition source hypotheses
- **Burn pattern analysis** — methods for validation and training for evaluating patterns on walls and ceilings and patterns resulting from liquids on floors
- **Burning rates** — determination of burning rates for different items, development of a burning rate database
- **Electrical ignition** — validation of means to identify electrical faults as an ignition source
- **Flashover** — impacts of flashover on fire patterns and other indicators
- **Ventilation** — effects of ventilation on fire growth and origin determination
- **Fire models** — methods for validation of, training in, and education on the use of fire models
- **Health and safety** — methods to evaluate fire investigator occupational health and safety
- **Certification** — development of training and certification programs for investigators and laboratory personnel
- **Train-the-trainers** — methods to pass along training objectives and expertise to groups of qualified trainers

The primary goal of the conference was to encourage the development and use of scientific principles and methodologies by fire investigators and determine what facilities could improve investigation by public-sector investigators. The conference issued a white paper under the auspices of the Fire Protection Research Foundation (FPRF 2002). That study concluded that there was a basic lack of a scientific foundation for many methods now in use to identify the area of origin and the cause of fires. Complex types and forms of burning materials, building geometry, ventilation, and fire fighting actions were listed as contributing effects that complicate these determinations.

◆ 1.2 THE SCIENTIFIC METHOD

The underpinning of forensic fire scene reconstruction is the use and application of relevant scientific principles and research in conjunction with a systematic examination of the scene. This is particularly true in cases that later require expert witness opinions.

The scientific method, which embraces sound fire protection engineering principles combined with peer-reviewed research and testing, is the best approach for conducting fire scene analysis and reconstruction. *NFPA 921—Guide for Fire and*

Explosion Investigations authoritatively defines the scientific method as

> the systematic pursuit of knowledge involving the recognition and formulation of a problem, the collection of data through observation and experiment, and the formulation and testing of a hypothesis. (National Fire Protection Association [NFPA] 2001, part 1.3.105)

The process of the scientific method continuously refines and explores a working hypothesis until arriving at a final expert conclusion or opinion, as illustrated in Figure 1.1. An important concept in the application of the scientific method to fire scene investigations is that all fires should be approached by the investigator without a presumption of the cause. Until sufficient data have been collected by the examination of the scene

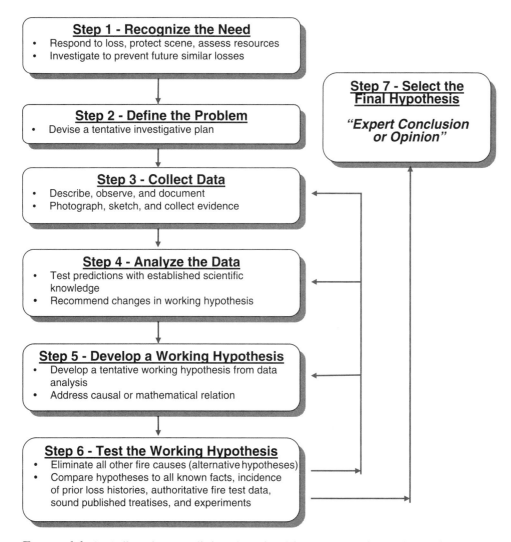

FIGURE 1.1 ◆ A flowchart outlining the scientific method as it applies to fire scene investigation and reconstruction. *Derived from guidelines, NFPA 921 [NFPA 2001, part 2.3].*

(assuming that it is still available), no specific hypothesis should be reasonably formed or treated (NFPA 2001, part 2.3.6.7).

Listed below is an annotated seven-step systematic process, based upon *NFPA 921* (NFPA 2001, part 2.3) for applying the scientific method to fire investigation and reconstruction.

1. **Recognize the Need.** Paramount is the recognition by first responders that a scene needs to be protected until a full investigation can be started. After initial notification, the investigator should proceed to the scene as soon as possible and determine the resources needed to conduct the investigation thoroughly. This includes not only the origin and cause of this event but recognizing the responsibility to determine if future fires, explosions, or loss of life can be prevented through new designs, codes, or enforcement strategies.

2. **Define the Problem.** Devise a tentative investigative plan to preserve and protect the scene, determine the cause and nature of the loss, conduct a needs assessment, formulate and implement a strategic plan, and prepare a report. This step also includes determining who has the primary responsibility and authority to interview witnesses, protect evidence, and review preliminary findings and documents describing the loss.

3. **Collect Data.** Collect facts and information about the incident through direct observations, measurements, photography, evidence collection, testing, experimentation, historical case histories, and witness interviews. All collected data should be subject to verification of how it was legally obtained, its chain of custody, and notation as to its reliability and authoritative nature.

4. **Analyze the Data** (Inductive Reasoning). Using inductive reasoning, analyze all data collected. The investigator relies upon his or her knowledge, training, and experience in evaluating the totality of the data. This subjective approach to the analysis may include knowledge of similar loss histories (observed or obtained from references), training and understanding of fire dynamics, fire testing experience, and experimental data.

5. **Develop a Working Hypothesis.** Based upon the data analysis, develop a tentative working hypothesis to explain the fire's origin, cause, and development that is consistent with on-scene observations, physical evidence, and testimony from witnesses. The hypothesis may address a causal mechanism or a mathematical relation (e.g., plume flame height).

6. **Test the Working Hypothesis** (Deductive Reasoning). Compare the working hypothesis with all other known facts, the incidence of prior loss histories, authoritative fire test data, sound published treatises, and experiments. Use the working hypothesis to eliminate all other reasonable origins and causes for the fire or explosion. Recommend the collection and analysis of additional data, seek new information from witnesses, and develop or modify the working hypothesis. This may involve reviewing the analysis with other investigators with relevant experience and training. Interactively repeat Steps 4, 5, and 6 until there are no discrepancies with the working hypothesis. By testing the working hypotheses rigorously against the data, those that cannot be conclusively eliminated should still be considered viable.

7. **Select Final Hypothesis** (Conclusion or Opinion). When the working hypothesis is thoroughly consistent with evidence and research, it becomes a final hypothesis and can be authoritatively presented as a conclusion or opinion of the investigation. If a final hypothesis cannot be determined, the cause should be reported as "undetermined."

THE WORKING HYPOTHESIS

The concept of the *working hypothesis* is central within the framework of the scientific method. In the case of fire scene reconstruction, the working hypothesis is based upon how an investigator describes or explains the fire's origin, cause, and subsequent development.

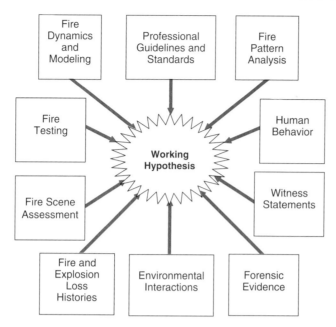

FIGURE 1.2 ◆ Sources of information that may contribute to a working hypothesis. *Courtesy of D. J. Icove.*

As shown in Figure 1.2, the investigator molds a working hypothesis and then refines it by drawing upon and examining many sources of information, including past investigative experiences. The working hypothesis may include peer reviews of the fact sequences by other qualified investigators, who themselves bring institutional knowledge and experiences. Alternative hypotheses may be created and tested.

New hypotheses are often built upon their predecessors, particularly when the knowledge base of a discipline broadens. When an accepted hypothesis cannot explain new data, we strive to construct a new hypothesis and in some cases generate completely new insight into the problem. An historical example of the application of the hypothesis dealt with proof of the geocentric versus solar system concept, simply "Does the earth revolve around the sun—or vice versa?" This is where the early hypothesis that the sun revolved around the earth was replaced with a new model proposing that the earth revolved around the sun that better fitted new data.

Even when a working hypothesis agrees with all known experimental evidence, it may, nevertheless, be disproved through subsequent experimentation or an emergent discovery. One area of recent experimental evidence is in the interpretation of fire patterns (FEMA 1997). Certain classes of fire pattern damage once thought to be produced solely by accelerants at fire scenes have been disproved by new principles, practice, and testing (DeHaan 2002; NFPA 2001). These include, for example, spalling of concrete, collapsed furniture springs, and crazed glass.

The knowledge of the application of ***fire dynamics and modeling*** is a significant advantage, which can clearly set novices apart from expert fire investigators. Since investigators are applying fire science to their profession, the use of the same terminology, descriptions of the scene, and interpretations are important in developing and communicating the working hypothesis. This is also important during peer review.

Professional guidelines and standards are often relied upon in developing the hypotheses, particularly those that are authoritative in nature. Engineering handbooks in wide use include the NFPA (2002) *Fire Protection Handbook* and *The SFPE Handbook of Fire Protection Engineering* (SFPE 2002a). Guides having widespread distribution include the NFPA's (2001) *NFPA 921—Guide for Fire and Explosion Investigations,* the National Institute of Justice's (NIJ 2000) *Fire and Arson Scene Evidence: A Guide for Public Safety Personnel, Kirk's Fire Investigation,* 5th ed. (DeHaan 2002), and *Combating Arson-for-Profit: Advanced Techniques for Investigators* (Icove, Wherry, and Schroeder 1998). Other NIJ national guidelines cover topics such as general crime scenes, death investigations, eyewitness identification, bombings, digital images, and electronic evidence.

Fire pattern analysis is an important aspect to interpreting the impact of the fire plume on the structure and contents of the building, vehicle, forest land, victim, or other property. With the frequent use of liquid accelerants by arsonists, authoritative fire testing continues to extend the body of knowledge into fire pattern analysis, particularly those involving flammable and combustible spill and burn patterns (DeHaan, 1995; Putorti 2001).

An understanding of *human behavior* is paramount when evaluating the movement and tenability of persons involved in fires. Not all people respond to perceptions of fire danger in the same way. The investigator must be able to understand and describe how occupants may respond to smoke and fire that changes in time and location, resulting in either their successful escape or their death from exposure to various by-products of combustion.

Witness statements giving the accounts of actions taken prior to, during, and after the fire are important to include when developing the working hypothesis. Often, witnesses may need to be reinterviewed to bring out additional details left out of the initial statements given to authorities. This information may include observations seen by a witness prior to the fire and as the fire was discovered and progressed.

Forensic science can add additional information to a working hypothesis. Examples include the use of traditional forensic examinations of fire debris, the analysis of impression and trace evidence, and, lately, the increased use of DNA in cases of death investigations.

The knowledge of *environmental interactions* of the initial temperature, humidity, wind speed, and direction on the developing fire may play an important role in developing the working hypothesis. For example, in a high-rise building fire, the environmental factors of wind and temperature along with the information of whether the fire occurred above or below the neutral plane may impact the initial discovery of odors, smoke, and fire development (SFPE 2002a, 4-275). Lightning strikes are an environmental interaction and potential cause for ignition that should be considered.

Fire and explosion loss histories are significant to the development of any working hypothesis. The knowledge of similar historical case histories may shed new knowledge in developing the hypotheses that should be tested before selecting the final hypothesis. Note that such statistical data should not be relied upon to prove the cause of a particular fire under investigation.

Many public and private organizations collect fire data for statistically profiling occupancies to develop risk assessment and fire prevention efforts. Listed in Table 1.1 is a profile of fire losses occurring in offices listed by cause based upon the combined loss histories of Allendale, Arkwright, and Protection Mutual Insurance Companies for the years 1991 through 1995 (Factory Mutual 1996).

These loss histories may also be included in the investigator's vast experience of *fire scene assessments* while working, reviewing, or reading about similar cases. The investigator's heuristic knowledge (rules of thumb), training, and experience, combined

TABLE 1.1 ◆ Profile of Office Fire Losses by Cause for 1991 to 1995

Fire Cause	*No. of Losses*	*Amount of Losses (Thousands of U.S. $)*
Electricity	84	34,947
Arson	47	11,563
Smoking	17	3,400
Exposure	16	1,472
Hot work	12	2,375
Hot surface	12	1,998
Overheating	6	1,472
Spontaneous ignition	2	4,957
Other	14	2,316
Unknown	119	13,446

Source: Factory Mutual (1996).

with professional investigative guidelines and standards, are also drawn upon during the development of a working hypothesis.

The combination of generally accepted scientific principles and research, when supported by empirical *fire testing,* can yield valuable insight into fire scene reconstruction. Scientifically conducted test results can normally be reproduced and their error rates validated. When this is true, the results are considered to be "objective"— neither prejudiced nor biased by the researcher. Accepted fire testing procedures are developed and maintained by the Committee on Fire Standards (E05) of the American Society for Testing and Materials (ASTM 1999a).

APPLYING THE WORKING HYPOTHESIS

It is vital for an experienced investigator to become proficient in the development of the working hypothesis. This can be accomplished through consistently evaluating, documenting, and protecting evidence from the fire scene, recording observations and behaviors of witnesses to the fire, comparing similar case histories, reading the fire literature, actively participating in authoritative fire testing, and writing and publishing peer-reviewed research.

The scientific method, as it applies to forensic fire scene reconstruction, provides fundamental principles for developing a hypothesis about a fire, testing the hypothesis, and deriving a sound theory as to its origin, development, damage, and cause. Through an iterative process, the hypothesis is revisited, revised, and reformulated until it is reconciled with all the available data. At this point, the hypothesis becomes a final theory and can be presented as a conclusive opinion of the investigation.

Struggling with the concept of the working hypothesis is not new in fire investigations. Example 1.1 relates the many hypotheses that arose during the 1942 Cocoanut Grove Night Club fire. Some of the theories appeared far-fetched, but investigators examined each one during the case. Clearly, the authorities pursued all theories that surfaced in this case. This is an example of how the scientific method can be used to discount those theories that could not be supported by analytical data, fire protection principles, and witness observations.

EXAMPLE 1.1 The Working Hypothesis

The Cocoanut Grove Night Club fire of November 28, 1942, cost the lives of 489 people, with many others seriously injured, including 131 treated at nearby hospitals. Fire was first discovered in the basement area called the Melody Lounge.

While the investigating authorities were probing the fire, they were flooded with publicly expressed opinions explaining how the fire started and progressed. Even though the exact source of ignition is still disputed today, many of the publicly expressed opinions had to be addressed in the report. The major issues were the use of combustible decorations, the rapid rate of fire spread, the lack of adequate exits, and the cause of so many postfire deaths.

What is known is that witnesses first reported the fire originating in the lounge, igniting adjacent combustible decorations and cloth-covered ceilings and walls, and progressing up the stairs—blocking the only visible means of exiting the building.

Many of the theories on the fire's origin, cause, and rapid spread were reported in the official NFPA report issued January 11, 1943. Even though the published report does not go into sufficient technical detail on these theories, it still serves as an example of a working hypothesis in fire scene reconstruction and analysis. The initial NFPA report listed the following theories and eliminated some based upon their improbability.

Ordinary Fire Theory. The ordinary fire theory stated that the decorations, cloth on the ceiling and walls, and other available combustible materials generated an abundance of smoke and carbon monoxide. The report notes the possibility of another unexplained element that could have accelerated the growth of the fire.

The report did not further discuss the unexplained element. In retrospect, the high number of deaths of "survivors" hours after the fire led to suspicion of unusual factors in the toxicity of smoke.

Alcohol Theory. The alcohol theory suggested that vapors from the mouths of heavy drinkers contributed to the rapid fire spread. This included the proposal that warm alcohol evaporating from drinks on tables could also have contributed to the fire. This theory was discounted as a possibility, as the concentration of such vapors would be far too low to enhance flame spread.

Pyroxylin Theory. The pyroxylin (nitrocellulose) theory suggested that the extensive use of artificial leather used on the walls of the building caused the rapid fire spread and production of nitrous by-products of combustion, accounting for the numerous fatalities. These materials contained only a small percentage of pyroxylin, which was insufficient to contribute to the number of fatalities. A scientific assessment led to the elimination of this theory.

Motion Picture Film Theory. The motion picture film theory was raised when it was determined that the building had previously been used as a motion picture film exchange. The theory suggested that a quantity of decomposing scrap nitrocellulose film may have remained in a hidden storage area and was ignited by a source of ignition. The observations of witnesses did not support this theory.

Refrigerant Gas Theory. The air-conditioning supplying the Melody Lounge was located in a false corner wall. Even though refrigerant gases may be toxic, a theory suggesting that ignition of these refrigerant gases occurred was discounted. Furthermore, the refrigerant gas ran in copper tubing, which would have needed to be thermally or mechanically breached to release the gas during the initial stages of the fire. Historically, air-conditioning systems have used ammonia or sulfur dioxide as refrigerant.

Flameproofing Theory. The flameproofing chemicals theory was based on the suggestion that additives to the combustible decorations may have given off ammonia and other gases when heated. The investigation failed to determine if any flame retardants were ever applied to the decorations, and the surviving victims did not report any symptoms associated with this type of exposure.

Fire Extinguisher Theory.   The fire extinguisher theory came about when it was learned that fire extinguishers were used on an incipient fire in an artificial palm tree and that toxic gases may have been emitted by the extinguishing liquid. The investigation found no supporting evidence for this theory.

Insecticide Theory.   The insecticide theory emerged when it was learned that insecticides were used in the basement kitchen. It was suggested that if flammable vapors had collected in concealed wall spaces, they might have accounted for the reported initial flash fire. There was no information as to whether the insecticide contained an ignitable liquid, and no indications of excess quantities of any insecticide.

Gasoline Theory.   The gasoline theory was suggested when it was determined that the building had previously been used as a garage and that gasoline tanks still existed under the basement floor. The theory that gasoline vapor emerged from these tanks was discounted since the fumes were heavier than air and would have remained at floor level. Furthermore, the physical scene examination proved that there was not an adequate concentration of vapors to support the flame spread seen in this fire. Vapors would have been detected by occupants at levels far below ignitable concentrations.

Electric Wiring Theory.   The electric wiring theory came about when it was learned that an unlicensed electrician installed part of the wiring in the building. The theory suggested that the heated insulation on overloaded wiring was responsible for forming flammable and toxic vapors. No evidence of overloaded wiring was found that could support this theory.

Smoldering Fire Theory.   The smoldering fire theory was suggested when it was learned that witnesses reported that several walls were hot to the touch and that a smoldering fire went undetected. Investigators physically examining the fire scene found no physical evidence to support this theory.

As a postscript to these theories, over half a century later the cause of the fire is still listed as "undetermined." Hypotheses are still being tested as to the cause of the Cocoanut Grove fire. Recent reviews of the investigation by the NFPA beginning in 1996 suggest that fire modeling combined with the use of the scientific method from *NFPA 921* (NFPA 2001) may shed new insights into this case (Beller and Sapochetti 2000).

BENEFITS OF USING THE SCIENTIFIC METHOD

The International Conference on Fire Research for Fire Investigation concluded that many scientific gaps existed in the methodologies and principles used to reconstruct fire scenes (Nelson and Tontarski 1998). Major conference goals included an assessment of the use of scientific principles and methodologies in fire investigation and identification of fire investigation needs and education. Numerous benefits exist for using the scientific method to examine fire and explosion cases. Three major benefits are as follows.

- ◆ Acceptance of the methodology in the scientific community
- ◆ Use of a uniform, peer-reviewed protocol of practice, such as *NFPA 921* (NFPA 2001)
- ◆ Improved reliability of testimony from opinions formed using the scientific method

First, the use of the scientific method is well received in both the technical and the research communities. An investigation conducted using this approach is more likely to be embraced by those who would tend to doubt a less thoroughly conducted probe.

Second, the scientific method is an accepted protocol of practice for *NFPA 921*. Those persons ignoring or deviating from *NFPA 921* practices would bring closer scrutiny to their reports and opinions.

Another guide that repeatedly cites *NFPA 921* is the NIJ's (2000) *Fire and Arson Scene Evidence: A Guide for Public Safety Personnel*. The NIJ guide notes that actions taken at the scene of a fire or arson investigation can play a pivotal role in the resolution of a case. A thorough investigation is key to ensuring that potential physical evidence is neither tainted nor destroyed, nor potential witnesses overlooked.

Third, and most important, expert testimony in fire and explosion cases will rely more heavily on opinions formed using the scientific method. Recent U.S. Supreme Court decisions underscore these principles, with many state courts following the trend. The following section discusses at length many issues surrounding expert testimony.

◆ 1.3 FOUNDATIONS OF EXPERT TESTIMONY

When applying the scientific method in order to state an expert opinion on the origin and cause of a fire or explosion, the investigator sets a level for the degree of confidence in forming the opinion at a prescribed threshold. Specifically, when this degree of confidence in the data or hypothesis is at the "possible" or "suspected" level, "undetermined" should be cited as the cause of the fire or explosion. An opinion should be expected to withstand the challenges of a reasonable examination by peer review or a thorough cross-examination in the courtroom (NFPA 2001, part 16.7).

FEDERAL RULES OF EVIDENCE

For the purposes of this textbook, emphasis is placed on testimony primarily before the federal court system. The Federal Rules of Evidence (FRE) clearly state who may offer testimony in federal court (FRCP 2000). Many state courts model their guidelines on the federal rules.

Recent amendments to the FRE impacted the admissibility of evidence and opinion testimony under Rules 701, 702, and 703. These changes state the following.

Rule 701. Opinion Testimony by Lay Witnesses

If the witness is not testifying as an expert, the witness' testimony in the form of opinions or inferences is limited to those opinions or inferences which are (a) rationally based on the perception of the witness, (b) helpful to a clear understanding of the witness' testimony or the determination of a fact in issue, and (c) not based on scientific, technical, or other specialized knowledge within the scope of Rule 702.

Rule 702. Testimony by Experts

If scientific, technical, or other specialized knowledge will assist the trier of fact to understand the evidence or to determine a fact in issue, a witness qualified as an expert by knowledge, skill, experience, training, or education may testify thereto in the form of an opinion or otherwise, if (1) the testimony is sufficiently based upon reliable facts or data, (2) the testimony is the product of reliable principles and methods, and (3) the witness has applied the principles and methods reliably to the facts of the case.

Rule 703. Bases of Opinion Testimony by Experts

The facts or data in the particular case upon which an expert bases an opinion or inference may be those perceived by or made known to the expert at or before the hearing. If of a type reasonably relied upon by experts in the particular field in forming opinions or inferences upon the subject, the facts or data need not be admissible in evidence in order for the opinion or inference to be admitted. Facts or data that are otherwise inadmissible shall not be disclosed to the jury by the proponent of the opinion or inference unless the court determines that their probative value in assisting the jury to evaluate the expert's opinion substantially outweighs their prejudicial effect.

Note that these federal guidelines do not apply to all state-by-state or international situations, and the reader is encouraged to seek guidance from an appropriate legal advisor.

SOURCES OF INFORMATION FOR EXPERT TESTIMONY

In general, experts will normally use three major sources of information on which to base their opinions (Kolczynski 2000):

- first-hand observations
- facts presented to experts prior to trial
- facts supplied in court

Examples of ***first-hand observations*** include those observed at a fire scene, in a laboratory examination or fire test, or from evaluation of evidence under FRE 703. It is important that experts try to actually visit the fire scene first-hand and not merely rely upon sketches and photographic evidence taken by another party.

Although important, it is not always required or possible to actually visit the scene, particularly if the scene has been considerably altered or destroyed. If photographic documentation is sufficiently comprehensive and can be documented or validated by other means, the expert can rely on it (see *United States of America v Ruby Gardner* 2000).

Experts usually critically review the case and gain insight and knowledge of ***facts prior to trial.*** This knowledge may be based upon facts gleaned from scientific manuals, learned treatises, or results of historical testing. Examples include information learned through expert treatises such as *NFPA 921, Kirk's Fire Investigation,* and *The SFPE Handbook.*

Rule 701 allows for lay witnesses to offer nonexpert, opinion evidence testimony. These opinions might include the state of intoxication, a vehicle's speed, and the memory of an identifiable odor of gasoline. For example, firefighters who are not experts in fire investigation may be allowed to testify that they detected a strong odor of gasoline when entering a certain room of a burning structure.

A lay witness might also offer expert testimony under Rule 702 when the opinion is based upon an accepted evaluation method, as long as the witness is qualified to make this evaluation. Examples of accepted evaluation methods include the guidelines in *NFPA 921* and various ASTM standards.

Rule 703 historically (since its enactment in 1975) allowed for experts to form an opinion based upon facts, whether or not in evidence, as long as these facts were necessary to form professional judgments from a nonlitigation point of view. This recent change makes the clarification that just because an expert witness uses information to form an opinion, its use does not automatically make that information admissible in court. Under FRE 703, experts may also rely upon hearsay through statements of other witnesses. For example, the expert may use information from the persons who witnessed the fire to help form an opinion, testimony of other witnesses, documents and reports, and related written reports submitted in the case.

Facts supplied in court may also be a basis for expert witness testimony. These facts may include information from testimony of other witnesses and evidence presented. The expert witness may also be asked a hypothetical question. For example, an expert may be asked, "Assume that three separate gasoline-filled plastic jugs with partially burned wicks were found in separate rooms of the house. What would be your conclusions based upon these facts and your 25 years of experience as a fire investigator?"

DISCLOSURE OF EXPERT TESTIMONY

The following is an excerpt regarding guidelines on the requirement to disclose expert testimony in civil cases from Rule 26(a)(2)(B) of the Federal Rules of Civil Procedure (FRCP 2000) effective December 1, 2000, as amended.

(2) Disclosure of Expert Testimony.

(A) In addition to the disclosures required by paragraph (1), a party shall disclose to other parties the identity of any person who may be used at trial to present evidence under Rules 702, 703, or 705 of the *Federal Rules of Evidence.*

(B) Except as otherwise stipulated or directed by the court, this disclosure shall, with respect to a witness who is retained or specially employed to provide expert testimony in the case or whose duties as an employee of the party regularly involve giving expert testimony, be accompanied by a written report prepared and signed by the witness. The report shall contain a complete statement of all opinions to be expressed and the basis and reasons therefore; the data or other information considered by the witness in forming the opinions; any exhibits to be used as a summary of or support for the opinions; the qualifications of the witness, including a list of all publications authored by the witness within the preceding ten years; the compensation to be paid for the study and testimony; and a listing of any other cases in which the witness has testified as an expert at trial or by deposition within the preceding four years.

(C) These disclosures shall be made at the times and in the sequence directed by the court. In the absence of other directions from the court or stipulation by the parties, the disclosures shall be made at least 90 days before the trial date or the date the case is to be ready for trial or, if the evidence is intended solely to contradict or rebut evidence on the same subject matter identified by another party under paragraph (2)(b), within 30 days after the disclosure made by the other party. The parties shall supplement these disclosures when required under subdivision (e)(1).

FRCP Rule 705 addresses the disclosure of facts or underlying data in which an expert forms his/her expert opinion. The expert may testify as to an opinion without first introducing the underlying facts or data used to develop that opinion, unless otherwise directed by the court. However, the expert may be required to later disclose the underlying facts or data they used during a cross-examination.

Under FRCP Rule 26(a)(2)(B), experts are required to provide a written expert witness disclosure report. This report, which must be prepared and signed by the expert witness, must contain the following.

- ◆ Complete statement of all opinions to be expressed along with the basis for these opinions
- ◆ Data or information considered by the expert used in forming the opinions
- ◆ Exhibits planned to be used in summarizing or supporting the opinions
- ◆ Qualifications of the expert witness, including a list of all publications authored within the preceding 10 years
- ◆ Compensation to be paid for the report and testimony
- ◆ Listing of all cases in which the expert has testified in trial or deposition within the preceding 4 years

The outlined format of this written expert disclosure report varies. Shown in Table 1.2 is an example of an outline for a summary disclosure letter, sent by an expert to an attorney. Table 1.3 lists areas to cover when qualifying as a fire expert (Beering 1996).

TABLE 1.2 ◆ Example Expert Disclosure

I. Qualifications

I, _____, P.E., am a consulting engineer and have been retained as an expert witness by the Defendant. A copy of my resume is attached as Exhibit A. In the course of my professional practice, I have over XX years of experience as a specialist in fire scene investigation and reconstruction. In the past XX years, I have testified as an expert witness in XX cases. Information about those cases is listed in Exhibit B. In the past XX years, I have coauthored XX textbooks, XX chapters in textbooks, and XX articles; and the titles are listed in my vitae in Exhibit C.

II. Scope of Engagement

I am retained by the Defendant in this case to evaluate the initial growth and development of a fire on XXXXXX XX, XXXX, at a XXXX located at XXXX Street, XXXXX, XXXXX County, XXXXXX; the impact of smoke detectors to notify the victims of the incipient fire; and the impact of conditions and human behavior during the fire which claimed the life of XXX X. XXX. The analysis performed for this case conforms to the generally accepted fire protection engineering methods of fire scene investigation and reconstruction.

III. Information Reviewed

To prepare my opinion, I reviewed the following documents: (LIST)

IV. Summary of Opinion

The analysis and opinions expressed in this report are based on my knowledge of the facts and information made available to date. If additional information becomes available which has a bearing on these opinions as expressed below, I will amend or supplement these opinions appropriately. Based on my understanding of the issues in the complaint and on the scope of this engagement, it is my opinion that: (LIST)

V. Compensation

I have billed $X,XXX to the attorneys representing the Defendant for my services through the date of this report. Billings for future services, including testifying at depositions and trial, will be at $XXX.XX per hour.

TABLE 1.3 ◆ Areas to Cover When Qualifying as a Fire Expert	
Category	*Example*
Identification	Name
	Title, department
	Employment
Education and training	Formal education
	Fire training academies
	National Fire Academy
	FBI Academy
	Federal Law Enforcement Training Center
	Annual international and state seminars
	Specialized training
Certifications	Fire investigator
	Police officer
	Fire protection specialist
	Professional engineer
	Instructor
Experience	Fire suppression
	Fire investigations
	Arrests and convictions
	Fire testing
	Laboratory
Prior testimony	Criminal, civil, and administrative
	Lay and expert witness
	Jurisdictions (state, federal, international)
Professional associations	Professional fire investigation
	Fire and forensic engineering
Teaching experience	Local, state, federal, and international
	National academies
	Colleges and universities
Professional publications	Articles
	Books
	National standards
Awards and honors	Local, state, federal, and international

Source: Summarized from Beering (1996).

◆ 1.4 RECOGNITION OF FIRE INVESTIGATION AS A SCIENCE

The recent trend is that U.S. Supreme Court decisions now, more than ever, continue to define the admissibility of expert scientific and technical opinions, particularly as they relate to fire scene investigations. These decisions impact how expert testimony is accepted and interpreted.

The bottom line is that although much controversy exists over these court decisions, fire and explosion investigation is emerging more as a "science" and is less thought of as an "art." This is particularly true when combining the use of the scientific method with relevant engineering principles and research in providing expert witness testimony (Chesbro 1994; Ogle 2000).

A judge has the discretion to exclude testimony that is speculative or based upon unreliable information. In the case, *Daubert v Merrell Dow Pharmaceuticals,* 509 U.S. 579 1993 (Daubert 1993), the Court placed the responsibility upon a trial judge to ensure that expert testimony was not only relevant but also reliable. The judge's role is to serve as a "gatekeeper" to determine the reliability of a particular scientific theory or technology. The Court defined four criteria to be used by the "gatekeeper" to determine whether the expert's theory or underlying technology should be admitted. *Daubert* allows the Court to gauge whether the expert testimony aligns with the facts of the case as presented.

In a recent decision, *Kumho Tire Co. Ltd. v Carmichael* (1999), the Court applied the four-guideline criteria to expert testimony to determine whether it was based upon science or experience. The four-guideline *Daubert* criteria consist of testing, peer review and publication, error rates and professional standards, and general acceptability as shown in Table 1.4, which lists the pertinent issues examined by each criterion. In short, expert testimony must rely on a balance of valid peer-reviewed literature, testing, and acceptable practices if it is to be considered credible by the courts.

A recent federal case citing *Daubert* and Federal Rules 702 and 703 allowed the expert testimony of Dr. John DeHaan pertaining to the origin and cause of an arson fire based upon his review of case materials relied upon by other experts (*United States of America v Ruby Gardner* 2000). Dr. DeHaan relied upon a review of reports, photographs, and third-party observations, some of which may not have been directly admissible as evidence.

The court determined that this procedure was reliable and appropriate, confirming the conviction. Furthermore, the court drew a parallel to other testimony, including hearsay and third-party observations used by an arson investigation expert in forming an opinion or the ability of a psychiatrist to testify as an expert by relying only upon staff reports, interviews with other doctors, and background information.

TABLE 1.4 ◆ Criteria and Pertinent Issues Examined by *Daubert*

Criteria	*Pertinent Issues Examined*
Testing	Has the methodology, theory, or technique been tested?
Peer review and publication	Has the methodology, theory, or technique been subject to peer review and publication?
Error rates and professional standards	What are the known or potential error rates of this methodology, theory, or technique? Does the methodology, theory, or technique comply with controlling standards, and how are those standards maintained?
General acceptance	Is the methodology generally accepted in the scientific community?

IMPACT OF *NFPA 921* ON SCIENCE-BASED EXPERT TESTIMONY

The importance of *NFPA 921* has also been cited along with *Daubert* as an interlinking element of expert testimony since it establishes guidelines for the reliable and systematic investigation or analysis of fire and explosion incidents (Campagnolo 1999). Several clusters of recent federal court opinions and rules fall into the following areas.

- ◆ Use of investigative protocols, guidelines, and peer-reviewed citations
- ◆ Methodological explanations for burn patterns
- ◆ Qualifications to testify

Professional education is paramount in any profession. It is incumbent on all professional fire investigators continuously to read and keep abreast of all relevant fire, engineering, and legal publications and critically evaluate their conclusions with this ever-changing knowledge. As with science, today's knowledge may change with new developments and impact upon the reviews of an established hypothesis.

References to *NFPA 921*. In 1997, a federal judge upheld a motion by a defendant insurance company in a civil case barring the plaintiff's expert fire investigator from mentioning *NFPA 921* during his testimony. *NFPA 921* was first published in 1992 after the November 16, 1988, fire in question. The judge requested that the plaintiffs provide him a copy of *NFPA 921*. The judge found that, under Rule 703 of the *Federal Rules of Evidence,* any reference to *NFPA 921* would have little probative value and would confuse the jury (*LaSalle National Bank et al. v Massachusetts Bay Insurance Company et al.* 1997).

Investigative Protocols. In a 1999 federal criminal case, the defendant moved that the court exclude a state fire marshal's testimony on the incendiary nature of a fire, arguing that the testimony did not meet the standards for admissibility under *Daubert* (*U.S. v Lawrence R. Black Wolf, Jr.* 1999). The defendant argued that the testimony should be excluded since the investigator did not arrive at the scene until 10 days after the fire occurred and did not take samples for laboratory analysis, the theory of the fire's origin and cause had not been subject to accurate and reliable peer review, and no scientific methods or procedures were used to reach the conclusions. The defendant's Exhibit A was the 1998 edition of *NFPA 921*.

In this case, the U.S. Magistrate determined that the investigator's proffered testimony was reliable since he indeed used an investigative protocol consistent with the basic methodologies and procedures recommended by *NFPA 921,* and he passed the requisite scrutiny demanded of fire origin and cause experts (Figure 1.3). The magistrate ruled that the proffered testimony was reliable according to the first part of the Rule 702 *Daubert* test.

Regarding the relevancy test, the magistrate determined that the investigator's testimony would help a jury understand the circumstances surrounding the fire's origin and cause. The magistrate further found that Rule 703 had been met since the investigator's proffered testimony relied upon facts and data normally collected by experts when forming opinions as to the fire's origin and cause. The magistrate concluded that the defendant would have ample opportunity to address the expert's opinion during cross-examination, through presentation of contrary evidence, and in instructions to the jury.

Use of Guidelines and Peer-Reviewed Citations. In a case involving a November 16, 1994, residence fire, a number of independent investigators attempted to determine the exact origin and cause of the fire. One investigator, an electrical engineer, offered the opinion that a television set located in the basement family room caused

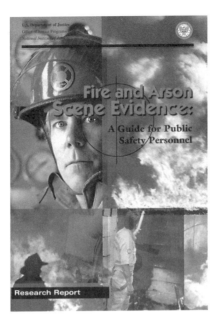

FIGURE 1.3 ◆ An investigator's proffered testimony may be deemed reliable if in accordance with an investigative protocol, such as the one published by the U.S. Department of Justice, that cites methodologies and procedures recommended by *NFPA 921. Courtesy of U.S. Department of Justice; www.usdoj.gov.*

the fire. The plaintiffs sued the television's manufacturer, claiming product liability, negligence, and breach of warranties (*Andronic Pappas et al. v Sony Electronics, Inc. et al.* 2000).

A federal judge hearing this case held a 2-day *Daubert* hearing and concluded that the one investigator's causation testimony was inadmissible. The judge cited that the issue in this case was the second Rule 702 factor, that an expert's opinion should be based upon a reliable methodology. The judge noted that the investigator did not use a fixed set of guidelines in determining the cause of the fire. Notably, the investigator did confirm that, even though he was aware of the existence of established guidelines in *NFPA 921* and *Kirk's Fire Investigation,* he relied upon his own experience and knowledge.

During the hearing, the plaintiffs did not submit any books, articles, or witnesses on proper fire causation techniques other than the one investigator. The judge specifically commented on the fact that the one investigator made reference to his reliance on a burn pattern inside the television set. No citations were made to peer reviewed sources to support this opinion.

Peer Review of Theory.   In a case involving a residence fire that occurred around December 22, 1996, an insurance company hired an investigator to determine the cause of the fire, which all parties agreed had originated in an area to the right of a fireplace firebox. In dispute by the defendants was the cause of the fire, and they filed a motion to exclude the testimony of the plaintiff's expert (*Allstate Insurance Company v Hugh Cole et al.* 2001).

Even though the plaintiff filed his complaint on December 21, 1998 (prior to the December 1, 2000, effective date of the amended Rule 702), the judge applied the amended Rule 702 to the admissibility of the testimony (FRCP 2000). The plaintiff asserted that in his trial testimony, the investigator would rely upon data from *NFPA 921* to support his opinion.

The opinion of the plaintiff's investigator was that heat from a metal pipe ignited adjacent combustibles. The judge noted that the theory, based upon *NFPA 921,* had already been subject to peer review, was generally accepted in the scientific community, and met the sufficient reliability guidelines to satisfy the admissibility requirements of Rule 702. Peer review can also include consultation with other qualified fire experts, both scientific and investigative.

Methodology Needed. On September 13, 1996, a fire in a residence started in the corner of a kitchen containing a dishwasher, toaster oven, and microwave oven. The municipal fire marshal concluded that the fire was caused by the microwave oven, while in a federal civil case, the plaintiff's expert asserted that a defective toaster oven caused the fire (*Jacob J. Booth and Kathleen Booth v Black and Decker, Inc.* 2001).

A senior federal judge held a *Daubert* hearing in which evidentiary and testimonial records were reviewed and ruled that the opinion of the plaintiff's investigator was not admissible under Rule 702. The judge concluded that the plaintiff's investigator did not provide sufficient reliable evidence to support the methodology for investigating the cause of the fire. The judge also noted that the comprehensive nature of *NFPA 921* contained a methodology that could have supported the opinion and tested the hypothesis for the fire's cause.

In a federal civil case opinion on March 28, 2002, involving automobile ignition switch fires, a U.S. District Court judge examined the proposed use of statistical fire loss database of vehicle fires in litigation. Also examined in detail was the importance for reliance upon *NFPA 921* in fire investigations when forming a hypothesis using the scientific method (*Snodgrass et al. v Ford Motor Company and United Technologies Automotive, Inc.* 2002).

In a federal civil case opinion on July 10, 2002, an insurance company sued to recover subrogation paid by the company for fire damage to the property of its subrogors. The judge denied the defendant's motion to exclude the trial testimony of the plaintiff's origin and cause expert witness. The judge noted that the expert's testimony is the product of reliable principles based upon the scientific method as outlined in *NFPA 921* and that the expert applied these principles and methods in a reliable and relevant manner to satisfy the admissibility requirements of *Daubert* and the Federal Rules of Evidence 702 (*Royal v Daniel Construction* 2002).

Methodological Explanations for Burn Patterns. In the case of an April 23, 1997, building fire, a federal U.S. magistrate denied the plaintiff's motion to bar opinion testimony as to the origin and cause of the fire at the time of trial (*Eid Abu-Hashish and Sheam Abu-Hashish v Scottsdale Insurance Company* 2000). In this case, both a municipal fire department investigator and a private insurance investigator concluded from their examinations of the scene that the fire was incendiary in origin. No physical evidence was taken for laboratory examination.

The plaintiff argued that these investigators were not reliable and their testimonies were inadmissible under Rule 702 because they did not use the scientific method as outlined in *NFPA 921* and relied upon only physical evidence observed at the scene. The case also contained parallels with the *Benfield* case (*Michigan Miller's Mutual Insurance Company v Benfield* 1998). In denying the plaintiff's motion, the federal magistrate noted in his decision that the investigators were able to provide an adequate methodological explanation for the analysis of burn patterns that led to how they reached their conclusion as to the fire's incendiary origin (Figure 1.4).

FIGURE 1.4 ◆ Investigators should be able to provide adequate and methodical explanations of the fire pattern analysis methods they use to reach conclusions as to the origin of a fire. *Photo by D. J. Icove.*

Methodology and Qualifications. In the case of a July 1, 1998, residential fire, which resulted in a federal lawsuit for products liability, the judge granted the defendant's motion to exclude the testimony of the electrical engineer citing *Daubert,* Rule 702, and Rule 704 (*American Family Insurance Group v JVC America Corp.* 2001). In this case, an insurance company investigator called upon an electrical engineer to remove and examine the charred remains of a bathroom exhaust fan, clock, lamp, timer, compact disc player, computer with printer and monitor, ceiling fan, power receptacle, and power strip. The engineer came to an opinion that a defect in the compact disc player caused the fire. His observations were based on burn patterns in the room, on appliance remains, and on his experience, education, and training.

In his ruling, the judge noted that the training and experience of the engineer did not qualify him to offer an analysis of burn patterns and theory of fire origin. Furthermore, the judge noted that the engineer did not use the scientific methodology recommended by *NFPA 921* to form a hypothesis from the analysis of the data, nor did he satisfy the requirements for expert testimony under *Daubert.*

AUTHORITATIVE SCIENTIFIC TESTING

The *Daubert* criteria lists testing as one of the primary considerations in evaluating and demonstrating the reliability of a scientific theory or technique. In fire investigations, the scientific theory relies heavily upon established and proven aspects of nature, such as those demonstrated in the science of fire dynamics.

The combination of generally accepted scientific principles and research, when supported by empirical fire testing, can yield valuable insight into fire scene reconstruction.

Much of the accumulated knowledge of fire scene reconstruction is impossible for individual investigators to gain from personal experience alone, regardless of the skill and diligence with which their analyses are conducted.

Scientifically conducted test results can normally be reproduced and their error rates validated or estimated. When this is true, the results are considered to be "objective," neither prejudiced nor biased by the researcher. An example of peer-reviewed published guidelines is incorporated in the accepted fire testing procedures developed and maintained by the Committees on Fire Standards (E05) and Forensic Sciences (E30) of the ASTM (1999a). Table 1.5 lists many of the ASTM testing standards applicable to fire investigations.

PEER REVIEW AND PUBLICATIONS

A credible, reliable theory must take into account the body of research that has been compiled, verified, and published by experts in the field. The necessity for credibility underscores advantages of using the scientific method.

Recently, there has been a tendency for courts to hold experts to the same standards that scientists use in evaluating each other's work, sometimes referred to as ***peer review*** (Ayala and Black 1993). Peer review is common among respected journals that publish the results of theories, testing, and methodology. The common approach in literary peer review is to use reviewers who are quite familiar with the published scientific data and methodologies and determine whether the author's views are of the quality of these publications. Peer-reviewed journals useful in the field of forensic fire investigation include *Fire Technology, Journal of Fire Protection Engineering, Fire and Materials, Fire Safety Journal, Combustion and Flame,* and *Journal of Forensic Sciences.*

Investigators can also participate in peer review when their cases are submitted to supervisors for review. In law enforcement, this is the primary function of supervisory review to assure that all questions, logical investigative leads, laboratory examinations, and plausible theories are addressed.

Some experts develop opinions based solely on the results of tests conducted specifically to support expert testimony. The *Daubert II* (1995) Court places greater weight on testimony based upon preexisting research that uses the scientific method, as it is considered more reliable. (Clifford 2000).

NEGATIVE CORPUS

The Eleventh Circuit Court of Appeals applied *Daubert* and excluded the testimony of a fire investigator in the *Benfield* case (*Michigan Miller's Mutual Insurance Company v Benfield* 1998). The Court held that the investigation of fires is science-based and that the *Daubert* criterion applies.

The use of the "negative corpus" or "arson by default" approach (the process of ruling out all accidental causes for a fire without sufficient scientific and factual basis to determine what did cause the fire) is rarely an acceptable methodology for determining that a fire was intentionally set. In this case, the investigator was not able to articulate the methodology of how he eliminated possible ignition sources and had no scientific basis for his opinion.

For reference purposes, negative corpus does not appear in *NFPA 921;* rather the concept is handled based upon the existence rather than the absence of evidence in determining the cause of a fire (NFPA 2001, part 16.2.5). The concept of elimination of all other potential causes uses the scientific method of testing and rejecting alternative hypotheses.

TABLE 1.5 ◆ Testing Standards Applicable to Fire Investigations

Technique	*Instruction*	*Standard*
Laboratory testing	Examining and Testing Items (Litigation)	ASTM E 860
	Debris Samples by Steam Distillation	ASTM E 1385
	Debris Samples by Solvent Extraction	ASTM E 1389
	Debris Samples by Gas Chromatography	ASTM E 1387
	Debris Samples by Headspace Vapors	ASTM E 1388
	Debris Samples by Passive Headspace	ASTM E 1412
	Debris Samples by GC-MS	ASTM E 1618
	Practice for Evidence in Forensic Science Lab	ASTM E 1492
Flash and fire point	Tag Closed Tester	ASTM D 56
	Cleveland Open Cup	ASTM D 92
	Pensky Martens Closed Tester	ASTM D 93
	Tag Open Cup Apparatus	ASTM D 1310
	Setaflash Closed Tester	ASTM D 3278
	Plastic Tests using a Hot Air Furnace	ASTM D 1929
Autoignition temperature	Liquid Chemicals	ASTM E 659
Heat of combustion	Hydrocarbon Fuels by Bomb Calorimeter	ASTM D 2382
Flammability	Apparel Textiles	ASTM D 1230
	Apparel Fabrics by Semi-restraint Method	ASTM D 3659
	Finished Textile Floor Covering	ASTM D 2859
	Aerosol Products	ASTM D 3065
	Chemical Concentration Limits	ASTM E 681
	Field Flame Test for Textiles and Films	NFPA 705
Cigarette ignition	Mockup Upholstered Furniture Assemblies	ASTM E 1352
	Upholstered Furniture	ASTM E 1353
Surface burning	Building Materials	ASTM E 84
Fire tests and experiments	Roof Coverings	ASTM E 108
	Room Fire Experiments	ASTM E 603
	Measurement of Gases Present or Generated	ASTM E 800
Critical radiant flux	Floor Covering Systems	ASTM E 648
Release rates	Heat and Visible Smoke	ASTM E 906
	Heat and Visible Smoke using an Oxygen Consumption Calorimeter	ASTM E 1354
Rate of pressure rise	Combustible Dusts	ASTM E 1226
Electrical	Dielectric Withstand Voltage	MILSTD202F
	Insulation Resistance	MILSTD202F
Self-heating	Spontaneous Heating Values of Liquids and Solids (Differential MacKey Test)	ASTM D3523

NFPA 921 specifically cautions that "the elimination of all accidental causes to reach a conclusion that a fire was incendiary is a finding that can rarely be justified scientifically." If pursued vigorously, however, the scientific method can be used to demonstrate successfully that the only mechanism for ignition had to be deliberate by demonstrating that all relevant accidental mechanisms had been specifically evaluated, tested, and eliminated.

ERROR RATES, PROFESSIONAL STANDARDS, AND ACCEPTABILITY

Under *Daubert,* the court considers the known or potential rate of error and the existence of acceptable professional standards on the techniques used by the expert. Error rates from repeated tests form the basis for many equations, relationships, and models used to describe fire and explosion dynamics. These error rates are produced during fire testing, such as listed in Table 1.4.

A particular practice that has a broad impact on investigators is the *Standard Practice for Examining and Testing Items That Are or May Become Involved in Litigation (ASTM E860* [ASTM 1999b]). This practice covers evidence (actual items or systems) that may have future potential for testing or disassembly and are involved in litigation.

This practice sets forth the following guidance:

- documentation of evidence prior to removal and/or disassembly, testing, or alteration
- notification of all parties involved
- proper preservation of evidence after testing

This practice also stresses the importance of safety concerns associated with testing and disassembly of evidence. This is particularly important when dealing with energized equipment or evidence containing potentially hazardous chemicals.

Other ASTM professional standards include *E620—Practice for Reporting Opinions of Technical Experts, E678—Practice for Evaluation of Technical Data, E1020—Practice for Reporting Incidents,* and *D1188—Practice for Collection and Preservation of Information and Physical Items by a Technical Investigator.* These practices are a part of the ASTM E30 committee's responsibility and are presently being revised to make them suitable in all litigation, both civil and criminal.

◆ 1.5 FORENSIC FIRE SCENE RECONSTRUCTION

Forensic fire scene reconstruction extends beyond the systematic evaluation of the scene by using the scientific method to conduct a comprehensive review of the fire pattern damage, witness accounts, and evidence supporting human activities. According to *NFPA 921,* fire scene reconstruction is defined as

> the process of recreating the physical scene during fire scene analysis through the removal of debris and the replacement of contents of structural elements in their pre-fire positions. (NFPA 2001, part 1.3.51)

Specifically, the forensic fire scene reconstruction approach is more comprehensive than the *NFPA 921* approach. It takes into account identifiable fire pattern damage (changes to exposed surfaces by heat transfer), human factors (witness accounts, detection, behavior, tenability, and escape paths), physical forensic evidence of human activity (burn injuries, wounds, latent fingerprints, shoeprints, and blood spatters), and the application of the scientific method based upon relevant scientific principles and research (fire testing, dynamics, suppression, modeling, pattern analysis, and historical cases).

The applicability of using modern fire dynamics relationships to fire scene investigations has been advanced over the years. The impact is documented in key textbooks and articles by Nelson (1987), Drysdale (1999), Babrauskas (1997), Lilley (1995), and Grosselin (1998).

Because these and many other sources of data must be taken into consideration when a comprehensive fire analysis is undertaken, the authors recommend the following steps when performing a forensic fire scene reconstruction.

- Evaluate the heat transfer damage by fire plumes and gases
- Correlate the human observations and factors
- Document the fire growth, burn pattern indicators, and processing of the scene
- Establish the starting conditions—fuels, heat sources, environment, ventilation
- Conduct a fire engineering analysis
- Formulate, evaluate, and test conclusions

These steps form the basis for the remaining theme of this textbook and are illustrated by example and references. In cases where ample fire patterns or damage exists, investigators can often accurately determine the area and point of origin of the fire after extinguishment. In this process, the investigator traces the fire's area, growth, and direction of travel taking into account the identifiable fire patterns of smoke deposit, damage due to heat transfer, human factors, forensic physical evidence, effects of fire suppression, and a broad knowledge of case histories of similar scenarios.

Reconstruction may point to obvious areas to collect fire debris evidence, particularly when flammable or combustible liquids are suspected to have been used to start the fire (DeHaan 2002; Stickevers 1986). In evidence collection, reconstruction should be combined with an extensive knowledge of fire dynamics. Armed with this knowledge, an investigator who suspects that an ignitable liquid was used to start a fire will look for an area at floor level close to the edge of a fire plume's fuel source that may have remained cooler than the surrounding area during the fire. These cooler areas are where residues of liquids sometimes survive, especially when absorbed and protected by a floor covering such as carpet. These areas are usually evidenced by a demarcation or loss of material. The analysis of fire patterns left by fire plumes can also provide valuable information, especially in complicated fires where multiple points of origin may exist, as is often the case in arson fires.

Reconstruction involves establishing what the scene looked like at the start of the fire, not only physically (what was present and where was it) but also environmental (temperatures, weather, etc.). A very useful concept for testing accidental scenarios is for the investigator to consider, "What was different the day of (or just before) the fire?" A change in environmental conditions, equipment operation, time of use, or source of material can sometimes be the factor that shifts a marginally "stable" system into a finally unstable state, triggering a fire by accidental means. Reconstruction may also indicate specific evidence of accidental or natural fires.

STEP 1—EVALUATE THE HEAT TRANSFER DAMAGE BY FIRE PLUMES

Heat is transferred by three fundamental methods: ***conduction, convection,*** and ***radiation.*** The relative importance of these three methods that dominate a fire depends upon not only the intensity and size of the fire, but also the physical environment. The

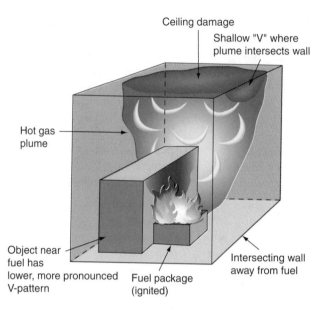

Ceiling damage

Shallow "V" where plume intersects wall

Hot gas plume

Object near fuel has lower, more pronounced V-pattern

Fuel package (ignited)

Intersecting wall away from fuel

FIGURE 1.5 ◆ Typical fire plume damage may produce a circular burn pattern on the ceiling and "V" or "inverted-V" patterns on intersecting vertical surfaces. The area of most intense damage on the ceiling is centered over the highest-temperature region of the plume's flame. *From Kirk's Fire Investigation, 5th/ed., by DeHaan, 2002. Reprinted by permission of Pearson Education, Inc., Upper Saddle River, N.J.*

heat transferred increases the surface temperature of the target, which can cause observable and measurable effects. Heat transfer is covered in more detail in Chapter 2.

Rapid detection and suppression normally account for fires being contained within their rooms or compartments of origin. In these simple fires of limited extension, the heat transfer relationships of conduction, convection, and radiation can be used to explain further the fire plume burn pattern damage (Quintiere 1994). Understanding plume damage is a key to reconstruction, and a simplified diagram showing a typical plume-damaged room is shown in Figure 1.5. These processes are examined in more detail in Chapter 3.

STEP 2—CORRELATE THE HUMAN OBSERVATIONS AND FACTORS

It is extremely important to document and correlate the observation of witnesses and victims to the fire scene reconstruction. Investigators' fire analysis should include eyewitness testimony, news videos, and fire fighters' descriptions as to the fire's growth during their presence. This documentation can be used to develop and test hypotheses correlating the actions of persons prior to, during, and after the fire.

Early research work evaluated the decision-making behaviors of the occupants and the various factors that affected their decisions in large fires involving high-rise buildings (Paulsen 1994). An investigator who is involved with forensic fire scene reconstructions should become knowledgeable of the existing research into the wide range of human behavior in fire situations. Unlike decisions made during violent crimes (fight, hide, or run), decisions made by people in fire environments can have numerous possibilities—ignore, investigate, approach and observe, run away and call for help, run away and return for property, fight fire, etc.

These human factors may include acts or omissions with respect to the ignition of fires either deliberately or by acts of carelessness, improper extinguishment of cigarettes, impaired judgment while under the influence of drugs or alcohol, or reactions to the environment. Other factors include the reactions and decisions, either proper or improper, to trigger alarms and escape to areas of refuge or remaining to fight the fire.

There are also times in criminal cases that the confession of an arsonist comes into question, particularly when investigators attempt to determine the veracity of the statement. For example, the type, quantity, and distribution of an accelerant used to set the fire or its identification can arise from the admissions of someone involved in the ignition of the fire and comparison to the physical evidence.

The location and intensity of the fire origin can sometimes be established by combining witness observations with an understanding of fire dynamics and physical evidence. This strategy is particularly important when the fire has caused extensive damage to the building or compartment or when the damage extends beyond the compartment of origin, making the determination of the location of the initial fire plume difficult.

Thermal burns on a fire victim's body or carboxyhemoglobin (COHb) levels in the blood may be indicative of their location at the time of the incipient fire, direction of travel, activities, and escape route. For example, an arsonist running away from a premature and rapidly developing flash fire (often characteristic of a flammable liquid) may suffer radiant heat burns on exposed skin and arms or to the backsides of legs and clothing.

STEP 3—DOCUMENT THE FIRE SCENE PROCESSING

A combination of photographs, sketches, and fire scene analysis is necessary to record accurately the information obtained during the fire scene reconstruction. This step is explored in depth in Chapter 4.

NFPA 921 and *Kirk's Fire Investigation* provide numerous examples of how to graphically record and document photographic, witness, and interpretive fire scene analysis. This documentation should also identify the operation of fire protection equipment, such as heat and smoke detectors, sprinklers, and audible alarms. A single photograph may not accurately represent the scene, as illustrated in Figure 1.6.

The systematic documentation of a fire scene from its initial stages can play an essential role in a professional effort to record the events for all parties involved. This approach captures all available information that is needed for later use in criminal, civil, or administrative matters. The bottom line is that systematic documentation, when it is correctly conducted, will be better suited to pass the *Daubert* challenges for courtroom admissibility and will ensure that an independent investigator will arrive at the same opinions given the total facts and circumstances.

Fire scenes often contain complex information that must be documented, a task of paramount importance to the investigator. A single photograph and scene diagram are not often sufficient to capture information on fire dynamics, building construction, evidence collection, and avenues of escape for its occupants.

This task is properly accomplished through a comprehensive effort of forensic photography, sketches, drawings, and analysis. In summary, the purposes of forensic fire scene documentation include documenting visual observations, emphasizing development

FIGURE 1.6 ◆ A single photograph may not be sufficient to document the fire scene processing. *Photo by D. J. Icove.*

characteristics, and authenticating physical evidence found at the scene. On fire scene sketches, it is important to record the investigator's observations as to the significant fire pattern damage, location of the fire plumes, and the area and point of fire origin. The evidence and explanations contained in this documentation should be sufficiently clear and precise so that an independent and qualified investigator who is unfamiliar with the fire scene can independently arrive at the same conclusions as to its origin and cause.

STEP 4—ESTABLISH STARTING CONDITIONS

Nearly all rooms contain some sort of fuel. Location, distribution, and type of fuel are critical in evaluating the ignition and spread of a fire. These fuels may be in the form of furniture (live load) but also in the form of carpet and pad, draperies and wall and ceiling coverings, as well as the building materials themselves (e.g., wood floors, walls and ceilings, fiberboard walls). It is vital to know what these materials were, where they were located in the room, and what their ignition and combustion properties were.

It is also important to know the initial conditions before the fire. For example, the investigation may need to document the prefire conditions of environment (temperature, wind, humidity), ventilation (doors, windows, furnace, HVAC), and operability of potential ignition sources. The first fuel ignited is a vital piece of information in a reconstruction. The assessment of this fuel should be done to determine what the fuel was, how it could be ignited, how long ignition took, and how large a fire would be produced.

STEP 5—CONDUCT A FIRE ENGINEERING ANALYSIS

The behavior of many fires can be analyzed by applying fundamental fire protection engineering calculations to estimate a fire's size, growth, and damage effects from heat and smoke. Even though these calculations are never 100% accurate, they can assist in the interpretation and explanation of fire behavior (Quintiere 1997). Common factors used in fire engineering calculations useful to investigators include heat release rates, flame height of a fire plume, location of the virtual source, heat flux, and flashover.

Fire modeling can also be an important tool in fire engineering analysis. A fire model is generally a computer program used to predict the environment in single- and multicompartment structures subjected to a fire. Models usually calculate temperatures associated with the distribution of smoke and fire gases throughout a building. Fire calculations and models are discussed in Chapter 6.

◆ 1.6 BENEFITS OF FIRE ENGINEERING ANALYSIS

The primary goal of this text is to extend and explain scientific approaches to fire analysis for use by the fire investigator. However, investigators need a reasonable degree of training in fire protection engineering principles to construct and interpret the results in their cases correctly.

FIRE ENGINEERING ANALYSIS

Fire engineering analysis, which can range from a basic "back-of the-envelope" first-level calculation to a sophisticated fire model, is an important step in fire scene reconstruction that provides numerous benefits. Historical investigations using fire engineering analysis have accurately assessed fire development, measured the performance of fire protection features and systems, and predicted the survivability and behavior of people during the incident.

Several significant fire engineering analysis studies have been undertaken by the NIST over the years. These detailed reports consist of the Dupont Plaza fire (Nelson 1987), First Interstate Bank Building fire (Nelson 1989), Pulaski Building fire (Nelson 1990), Hillhaven Nursing Home fire (Nelson 1991), Happyland Social Club fire (Bukowski 1992), 62 Watts Street fire (Bukowski 1996), and Cherry Road fire (Madrzykowski and Vettori 2000).

Some form of fire engineering analysis is now considered a necessary step in comprehensive fire scene analysis and reconstruction. Fire engineering analysis combined with modeling, which is explored in detail later in this text, offers the following benefits.

- ◆ Establishes the basis for the collection of data needed to construct event time lines, ignition sequences, and failure modes and effects analysis (FMEA)
- ◆ Invites application of the scientific method when testing hypotheses and validating fire scene reconstruction
- ◆ Provides a viable alternative to full-scale fire testing and can extend full-scale test results to differing ranges of conditions (sensitivity analysis)
- ◆ Provides answers to many important questions raised on human factors, ignition sequences, equipment failure, and fire protection systems (detectors, alarms, and sprinklers)
- ◆ Identifies important future research areas in fire investigation

Historically, the use of engineering analysis and modeling in fire scene reconstruction was conducted on a case-by-case basis, due mainly to the complexity of the process. A vast amount of knowledge and time were required to collect and analyze the information on the development of the fire.

Engineering analysis and modeling are becoming more commonplace. Despite the complexity of the processes, fire engineering analysis and modeling should be applied to cases involving multiple deaths or cases where code deficiencies contributed to the fire or in the anticipation of extensive civil or criminal litigation.

FIRE MODELING

A fire engineering analysis often uses fire modeling to compare actual events with predicted outcomes using varying fire causes and growth scenarios, as illustrated in Figure 1.7, involving a chair fire in an apartment. Surface temperatures on the walls are calculated by the fire model and displayed using a color-coded gradient scale. Results from the surface temperatures can be compared with burn patterns to confirm the sequence of the fire. This analysis often adds value, understanding, and clarity to complex fire scene investigation. Fire modeling, a less expensive and more environmentally sensitive approach, can also serve as an alternative to full-scale fire testing to explain burn patterns and the fire dynamics.

Results can also be extended to predict what effects different conditions will have on the fire, a process called *sensitivity analysis.* For example, a model based upon heat release rates, plume height, fuel configuration, and placement of the origin within the room

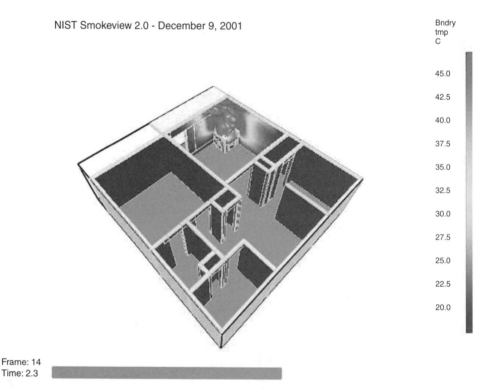

FIGURE 1.7 ◆ A fire engineering analysis often uses fire modeling to compare actual events with predicted outcomes. Heat transfer contact areas from a fire initiated in a chair in the upper-right corner of the living room are shown at 2.3 min of the computer simulation. This pattern can be compared with witness accounts. *Courtesy of D. J. Icove.*

(middle, wall, or corner) could potentially resolve whether the fire was the result of an accidental trash can fire or a fire caused by a pool of intentionally spread flammable liquid.

The model can also predict if the actions of an open door or window could impact the fire. In addition, the comparison of a time line is a beneficial product of the fire model. Example 1.2 shows the successful application of fire modeling as a tool in litigation.

EXAMPLE 1.2   Fire Model Helps Defend U.S. Government Against Lawsuit

At 6:45 am in a three-bedroom townhouse on a military reservation, just before the winter holidays, a 5-year-old was involved in a storage closet fire. The smoke alarm notified the parents, and the father tried unsuccessfully to extinguish the fire. Their telephone being out of order, the husband told his wife to awaken their other three children while he went next door to call the fire department.

The husband's wife and two youngest children perished in the fire. Three older children survived. The responding fire department was slightly delayed, by 2.5 min, when they had to yield to troops on the road. This delay formed the basis for a $36 million wrongful death lawsuit against the U.S. Government. The U.S. Department of Justice asked the National Institute for Standards and Technology to model the fire to determine what, if any, role the reported delay played in the deaths. A NIST fire protection engineer used HAZARD I (which at the time contained the earlier version 2.0 of CFAST) to model the fire. He relied upon construction plans, fire incident reports, radio logs, and witness statements to input the variables into the model.

The fire model accurately predicted areas of fire and smoke damage, the safe egress routes of the survivors, and the impact that attempts by neighbors to gain access by opening a bedroom window had on the resulting carbon monoxide and heat intensity levels. These predictions agreed with the autopsy findings, blood toxicity tests, and thermal injuries to the victims.

A further time-line analysis showed that the delayed response played no significant role in the fatalities. After depositions were taken from the NIST engineer, the lawsuit was settled out of court for less than $200,000 (Bukowski 1991).

Fire engineering analysis and modeling may also provide answers to important questions raised during fire investigations. These questions might include the following:

- What was the most probable cause of the fire (i.e., can several possible causes be eliminated)?
- How long did it take to activate fire or smoke detectors and water sprinkler heads?
- What were the smoke and carbon monoxide levels in each room after 10 min?
- Why didn't the occupants of the building survive the fire?
- Was an accelerant used in the fire and, if so, what type?
- How much time elapsed from when the owner of the structure left the building until the fire reached flashover?
- Did a negligent building design or failed fire detection and suppression system contribute to the growth of the fire?
- What changes to policy, building, or fire codes are necessary to ensure that similar fires will not occur in the future?

Discussion and cooperative research between fire investigators and fire protection engineers underscore the need to include fire modeling as an integral and required step in fire scene reconstruction. Fire investigators and engineering professionals need to collaborate further in order to expand their knowledge and validate the methods shared in common by both groups.

◆ 1.7 CASE STUDY 1—*DAUBERT* AND THE FIRE INVESTIGATOR

The debate over the application of *Daubert* to fire scene investigations has intensified at the point of deciding whether origin and cause determination is to be considered scientific evidence or nonscientific technical evidence. The advocates of the strict scientific approach bristle at the suggestion that fire scene investigation is in any way "nonscientific," pointing to the many misconceptions previously used by fire investigators (spalling, V-patterns, etc.), which were exposed by fire scientists only in recent years. They advocate the use of *Daubert* in fire scene analysis as the only means of preventing a return to the improper fire scene methodologies employed by unqualified investigators lacking proper scientific training. In contrast, the "technicians" argue that fire scene investigation has never been a pure science like chemistry or physics, as it employs elements of both disciplines. The term "nonscientific" in the context of *Daubert* is a legal distinction, rather than a scientific one. It is not to say that fire investigation is "unscientific" or devoid of any application of scientific principles. Instead, it is a recognition of the objective and subjective components that form a part of every fire scene investigation, more precisely, the human component in examining, analyzing, and, ultimately, interpreting fire scene evidence to reach a conclusion about the fire's origin and cause.

Early on in *Daubert*'s history, the Tenth Circuit directly addressed the testimony of a fire investigator in an arson case. In *United States v Markum,* 4 F. 3d 891 (10th Cir. 1993), the Court found the admission of a fire chief's testimony that a fire was the result of arson to be proper based primarily on his extensive experience in fire investigations. There the Court said,

> . . . Experience alone can qualify a witness to give expert testimony. See *Farner v Paccar, Inc.,* 562 F. 2d 518, 528-29 (8th Cir. 1977); *Cunningham v Gans,* 501 F. 2d 496, 500 (2nd Cir. 1974).

> Chief Pearson worked as a firefighter and Fire Chief for 29 years. In addition to observing and extinguishing fires throughout that period, he attended arson schools and received arson investigation training. The trial court found that Chief Pearson possessed the experience and training necessary to testify as an expert on the issue whether the second fire was a natural rekindling of the first fire or was deliberately set. That finding was not clearly erroneous. *Markum* at 896.

Another case that directly addressed the application of *Daubert* to fire investigation was *Polizzi Meats, Inc.* (PMI) *v Aetna Life and Casualty,* 931 F. Supp. 328 (D.N.J. 1996). In that case, the Federal District Court of New Jersey said,

> PMI's counsel argues that because of a lack of "scientific proof" of the fire's causation, none of Aetna's witnesses may testify at trial. This astounding contention is based on a seriously flawed reading of the United States Supreme Court's decision in *Daubert v Merrell Dow Pharmaceuticals, Inc. Daubert* addresses the standards to be applied by a trial judge when faced with a proffer of expert scientific testimony based upon a novel theory or methodology. Nothing in *Daubert* suggests that trial judges should exclude otherwise relevant testimony of police and fire investigators on the issues of the origins and causes of fires. *Polizzi* at 336–37. (Citations omitted)

These two decisions were the only reported cases considering *Daubert* in the specific context of fire investigation, until the Eleventh Circuit announced its decision in a case that has taken an entirely different view on the process of fire scene investigation. The *Markum* case analysis was provided courtesy of Guy E. Burnette, Jr., Tallahassee, Florida (Burnette 2003).

◆ 1.8 CASE STUDY 2—THE *JOINER* CASE: A CLARIFICATION OF *DAUBERT*

As courts from various jurisdictions are still trying to shed light on the full meaning of *Daubert,* the United States Supreme Court recently took up the issue again and provided some guidance and insights. In *General Electric Company v Joiner,* 66 U.S.L.W. 4036 (1997) the Supreme court reviewed a case where the trial judge had entered summary judgment in favor of the defendant in a lawsuit alleging that the plaintiff had contracted cancer as the result of exposure to PCB chemicals. The scientific evidence in support of the plaintiff's claim was derived from laboratory studies of mice that had been injected with massive doses of PCB chemicals and certain limited epidemiological studies suggesting a causal connection between PCB chemicals and cancer in humans. The trial judge ruled that the evidence offered by the plaintiff failed to satisfy the requirements of *Daubert,* describing the evidence offered by the plaintiff's experts as "subjective belief or unsupported speculation." It was noted that *Joiner* failed to present any credible scientific evidence of a direct causal connection between exposure to PCB chemicals and cancer.

On appeal, the ruling was reversed by the Eleventh Circuit, which held that the evidence should have been presented to the jury for a decision. The appellate court observed that the Federal Rules of Evidence favor the admissibility of expert testimony as a general rule. Further, the appellate court applied a more stringent standard of review of the trial court's ruling, since the ruling was "outcome determinative" (i.e., resolved the entire case).

The United States Supreme Court overturned the decision of the Eleventh Circuit and reinstated the ruling of the trial court. In doing so, the Supreme Court reiterated and clarified some of the points made in the *Daubert* decision. First, the role of the trial judge as "gatekeeper" was reaffirmed. In particular, the trial judge not only was allowed to draw his own conclusions about the weight of evidence offered by an expert witness, but was expected to do so. It was noted that this had been a function of the trial judge long before the *Daubert* decision itself. Since this was a proper role of the trial judge, the decision to accept or reject expert testimony would not be subjected to a more stringent standard of review on appeal. The decision of the trial judge would be given deference on appeal and it would require showing an "abuse of discretion" for the decision of the trial judge to be overturned.

The Supreme Court held that the application of *Daubert* to expert testimony is not merely a review and approval of the methodology employed. It includes scrutiny of the ultimate conclusions reached by the expert witness based upon the methodologies and data employed to reach those conclusions.

Notably, the Supreme Court did not clarify the controversy over scientific evidence versus technical evidence. The Supreme Court did not address the issue in *Joiner* because it was clearly a "scientific evidence" case. That remains a major part of the controversy in construing Daubert and the admissibility of expert testimony. The *Benfield* decision did directly address this issue and demonstrated a new perspective on this critical aspect of fire investigation.

This case history was provided courtesy of Guy E. Burnette, Jr., Tallahassee, Florida (Burnette 2003).

In *Michigan Miller's Mutual Insurance Company v Janelle R. Benfield,* Case Number 93-1283, United States Court of Appeals for the Eleventh Circuit, the *Daubert* analysis was applied to a fire scene investigation. This case has attracted great attention within the fire investigation community and has become a focal point of the *Daubert* controversy.

In January 1996 the *Benfield* case was tried in federal district court in Tampa, Florida. The case involved a house fire in which the insurance company, Michigan Miller's Mutual, refused to pay on the policy based, in part, upon the fire being incendiary and the involvement of an insured party in setting the fire. As a part of the insurer's case, a fire investigator with over 30 years of experience in fire investigations was called as an expert witness to present his opinion of the origin and cause of the fire. He testified that the fire was started on top of the dining room table where some clothing, papers, and ordinary combustibles had been piled together. He examined the fire scene primarily by visual observation and concluded that the fire was incendiary based upon the absence of any evidence of an accidental cause, along with other evidence and factors noted at the scene. After cross-examining the investigator, the plaintiff moved to exclude the testimony under *Daubert* and the trial court agreed. In the trial court's ruling striking the expert's testimony, the judge specifically found that the witness

> . . . cites no scientific theory, applies no scientific method. He relies on his experience. He makes no scientific tests or analyses. He does not list the possible causes, including arson, and then using scientific methods exclude all except arson. He says no source or origin can be found on his personal visual examination and, therefore, the source and origin must be arson. There is no question but that the conclusion is one to which *Daubert* applies, a conclusion based on the absence of accepted scientific method. . . .

> And finally, it must be noted that [his] conclusion was not based on a scientific examination of the remains, but only on his failure to be able to determine a cause and origin from his unscientific examination. This testimony is woefully inadequate under *Daubert* principles and pre-*Daubert* principles, and his testimony will be stricken and the jury instructed to disregard the same. *Daubert* motion hearing transcript at 124–26.

Interestingly, the Court in *Benfield* initially found the expert to be qualified to render opinions in the area of origin and cause of fires and allowed him to testify, based on his qualifications and credentials as a fire investigator. However, the judge struck the expert's testimony after it was presented based upon his methodologies in conducting the particular fire scene investigation in that case. Having stricken the expert's testimony, the judge then found that, as a matter of law, arson had not been proved by Michigan Miller's and directed a verdict against them on the arson issue.

The evidence was undisputed that the area of origin was on top of the dining room table. Therefore, the only issue was the cause of the fire. The expert testified that while he was conducting his investigation, he spoke with Ms. Benfield, who told him that when she was last in the house before the fire there was a hurricane lamp and a half-full bottle of lamp oil on the top of the table. He further testified that he examined photographs taken by the fire department before the scene was disturbed and observed an empty, undamaged bottle of lamp oil lying on the floor with the cap removed (also undamaged), indicating that it had been opened and moved from the table prior to the setting of the fire. He also explained his observations at the fire scene that enabled him to rule out all possible accidental causes. He concluded that the fire was incendiary, using the "elimination method" long recognized as a valid method of determining fire origin and cause. He could not, however, determine the source of ignition for the fire. More importantly, he did not "scientifically document" his findings on various points and relied primarily upon his 30 years' experience as a fire investigator, even as he held himself out as an expert in fire science adhering to the scientific method in conducting his investigation.

On cross-examination by Ms. Benfield's attorney, the expert was asked to define the scientific method and was asked the "scientific basis" for the taking of certain photographs apparently unrelated to the fire itself. The cross-examination continued by attacking each piece of evidence used by the expert that could not be said to be scientifically objective and scientifically verified. The investigator's determination of the smoldering nature of the fire and the time he estimated it burned before being discovered were discredited as not being based upon scientific calculations of heat release rate and fire spread, but merely the investigator's observations of the smoke damage and other physical evidence. The Court noted those points from the cross-examination in finding that the expert's methodology was not in conformity with the scientific method, relying instead almost exclusively on the expert's own training and experience, which it held to be inadmissible under *Daubert*.

On May 4, 1998 the Eleventh Circuit issued its ruling in the *Benfield* case. Contrary to the Tenth Circuit decision in *Markum* and the Federal District case in *Polizzi Meats,* the court found the investigator's fire scene analysis to be subject to the *Daubert* test of reliability. In reaching this conclusion, the court noted that the investigator in *Benfield* held himself out as an expert in the area of "fire science" and claimed that he had complied with the "scientific method" under *NFPA 921*. Thus, by his own admission he was engaged in a "scientific process," which the court held to be subject to *Daubert*.

Under the *Daubert* test of reliability, the Appeals court upheld the decision of the trial judge to strike his testimony. Noting that it is a matter of the trial court's discretion to admit expert testimony, such a decision will be affirmed on appeal absent a showing of an "abuse of discretion" or that the decision was "manifestly erroneous." Under such a daunting standard, the trial judge is effectively given the "final word" on whether or not both the qualifications and findings of an expert witness will be considered reliable enough to be presented to the jury. It is not simply a matter of having the power to decide if a witness is qualified to testify as an expert; the substance of the expert's testimony and his professional conclusions will first have to meet the approval of the trial judge before they can ever be presented to the jury. The trial judge acting as "gatekeeper" can summarily reject the findings and conclusions of an expert witness, preempting the jury from making that decision. In the *Benfield* decision, various scientifically unsupported and scientifically undocumented conclusions of the investigator were cited as grounds for the determination that his observations and findings failed the *Daubert* reliability test. A chandelier hanging over the dining room table where the fire started showed no signs of having caused the fire, but the investigator had not conducted any tests or examinations to eliminate it scientifically as a potential cause of the fire. His observations alone were held inadequate. Similarly, his opinion that the fire had likely been accelerated with the lamp oil contained in the bottle was rejected under *Daubert,* since he could not scientifically prove that there had been oil in the bottle before the fire and he had not taken any samples from the fire debris to prove scientifically its presence or absence at the time the fire was ignited. These and other observations of the investigator were held to demonstrate that there was no *scientific* basis for his conclusions, only his personal opinion from experience in investigating other fires.

It was not that the investigator was found to be "wrong" in the *Benfield* case. Indeed, there was never any evidence of an accidental cause of the fire. Ironically, although the Appeals Court upheld the decision of the trial judge to strike the testimony of the insurance investigator, it granted a new trial for Michigan Miller's on the arson defense. The Appeals Court felt that a *prima facie* case of arson had been shown at trial through the fire department investigator, who initially classified the fire only as "suspicious" (with virtually no challenge to the scientific documentation of his opinion), and the many incriminating circumstances surrounding the fire itself. Those circumstances included the fact Ms. Benfield claimed that she had not locked the deadbolts when she left the house, yet the deadbolts were locked when she returned to discover the fire. Ms. Benfield and her daughter (who had been out of town) had the only keys to those locks. Her assertion that the fire was extinguished by her boyfriend with a garden hose was refuted by the observations of the responding firefighters. Her insurance claim appeared to be significantly inflated. She had tried to sell the house and could not do so. She was trying to convince her estranged husband to transfer the house to her but could not do so. Ms. Benfield had given conflicting and contradictory accounts of her activities immediately before the fire. In listing all of these reasons, the Appeals Court found that there was compelling evidence of arson even as it discredited the findings of the insurance investigator that the fire was incendiary. The critical point in the case appears to be the distinction between the insurance investigator testifying as an expert in fire *science* and the fire department representative testifying as an expert in fire *investigation*.

This case history was provided courtesy of Guy E. Burnette, Jr., Tallahassee, Florida (Burnette 2003).

◆ 1.10 SUMMARY AND CONCLUSIONS

This chapter introduced the application of the scientific method formed by validated principles and research (fire testing, dynamics, suppression, modeling, pattern analysis, and historical case data). This approach relies on combined principles of fire protection engineering, forensic science, and behavioral science. In using the scientific method, the investigator can more accurately estimate a fire's origin, intensity, growth, direction of travel, and duration, as well as understand and interpret the behavior of the occupants.

Fire scene reconstruction can also bring in additional fire testing and similar cases to support the final theory. In addition to *NFPA 921*, the fire investigator should become familiar with other authoritative references and sources of data and information.

The strengths of this approach make a thorough examination of many hypotheses more effective. The reliability of the result of a specific method of investigation is a function of the number of hypotheses that have been tested and eliminated.

Fire engineering analysis also offers many advantages. In particular, it allows the investigator to evaluate the potential of several scenarios without repeated full-scale fire testing. A sensitivity analysis can also extend the analysis to different ranges of conditions.

■ ■

Problems

1.1. Obtain an adjudicated case from your local fire marshal's office involving the conviction of an arsonist who pleaded guilty to setting a fire. Analyze the fire investigator's opinion as to the origin and cause of the fire. Demonstrate how it did or did not meet the guidelines for the scientific method.

1.2. Obtain a photograph of a fire plume against a wall. Trace the lines of demarcation onto a plastic overlay. What can you learn from this exercise?

1.3. Search the Internet or current published periodical literature and list three examples of peer-reviewed fire testing research.

1.4. Search the Internet and list three federal and state court cases mentioning *NFPA 921* and *Daubert*.

■ ■

Suggested Reading

Cooke, R. A., and R. H. Ide. 1985. *Principles of fire investigation,* chaps. 8 and 9. Leicester, UK: Institution of Fire Engineers.

DeHaan, J. D. 2002. *Kirk's fire investigation,* 5th ed., chap. 17. Upper Saddle River, NJ: Prentice Hall.

CHAPTER 2   **Basic Fire Dynamics**

There is nothing like first-hand evidence.

—*Sir Arthur Conan Doyle*
"A Study in Scarlet"

The body of fire dynamics knowledge as it applies to fire scene reconstruction and analysis is derived from the combined disciplines of thermodynamics, chemistry heat transfer, and fluid mechanics. Accurate determination of a fire's origin, intensity, growth, direction of travel, duration, and extinguishment requires that investigators rely upon and correctly apply the basic principles of fire dynamics. They must realize that variability in fire growth and development is influenced by a number of factors such as available fuel load, ventilation, and physical configuration of the room.

Fire investigators can benefit from a working knowledge of published research involving the application of the knowledge that exists in the area of fire dynamics. The published research includes several recent textbooks and expert treatises (DeHaan 2002; Drysdale 1999; Karlsson and Quintiere 1999; Quintiere 1997), traditional fire protection engineering references (SFPE 2002a), U.S. Government research, done primarily by NIST, and fire research journals.

Bridging this knowledge gap is the primary purpose of this chapter. Many basic fire dynamics concepts are discussed in *NFPA 921* (NFPA 2001). Existing real-world examples and training programs illustrate their application and describe how basic fire dynamics principles apply to fuels commonly found at fire scenes. Case examples are offered in this chapter along with discussions on the application of accepted fire protection engineering calculations that can assist investigators in answering and interpreting fire behavior and conducting forensic scene reconstruction.

◆ 2.1 BASIC UNITS OF MEASUREMENT

The common units of measurement and dimensions used in fire scene reconstruction come from two systems: United States (U.S.) and metric. The United States is in the process of converting to the international standard, which is known as the International System of Units (SI).

This textbook uses the SI system as its standard, and in most cases the equivalent U.S. units follow in parentheses for comparison. Table 2.1 lists many of these

TABLE 2.1 ◆ Fundamental Dimensions Commonly Used in Fire Dynamics

Dimension	*Symbol*	*SI Unit*	*Conversion*
Length	L	m	1 m = 3.2808 ft.
Area	A	m^2	$1\,m^2 = 10.7639\,ft^2$
Volume	V	m^3	$1\,m^3 = 35.314\,ft^3$
Mass	M	kg	1 kg = 2.2046 lb
Mass density	ρ	kg/m^3	$1\,kg/m^3 = 0.06243\,lb/ft^3$
Mass flow rate	$\dot{m}$	kg/s	1 kg/s = 2.2046 lb/s
Mass flow rate per unit area	$\dot{m}''$	$kg/(s\,m^2)$	$1\,kg/(s\,m^2) = 0.205\,lb/(s\,ft^2)$
Temperature	T	°C	$T(°C) = [T(°F) - 32]/1.8$ $T(°F) = [T(°C) \times 1.8] + 32$
Temperature	K	K	$T(K) = T(°C) + 273.15$
Energy, heat	Q, q	J	1 kJ = 0.94738 Btu
Heat release rate	$\dot{Q}, \dot{q}$	W	1 W = 3.4121 Btu/hr = 0.95 Btu/s, 1 kW = 1 kJ/s
Heat flux	$\dot{q}''$	W/m^2	$1\,W/cm^2 = 10\,kW/m^2 = 0.317\,Btu/(hr\text{-}ft^2)$

Source: Derived from SFPE (2002), Quintiere (1998), Karlsson and Quintiere (1999), and Drysdale (1999).

commonly used fundamental dimensions from fire dynamics along with their typical symbol, unit, and conversion factor.

The most fundamental property of fire dynamics is heat. All objects above absolute zero ($-273°C, -460°F, 0\,K$) contain heat due to the motion of their molecules. Sufficient heat transfer (by conduction, convection, radiation) increases an object's ***temperature*** and may ignite it, which may spread fire to other neighboring objects.

◆ 2.2 THE SCIENCE OF FIRE

Fire is simply an oxidation reaction or combustion process involving the release of energy. Although a fire's energy release may be visible in the form of flames or glowing combustion, other nonvisible forms exist. Nonvisible examples include heat transfer in the form of radiation, rusting of iron, and yellowing of newsprint (Quintiere 1997).

FIRE TETRAHEDRON

In order for fires to exist, four conditions must coexist which may be represented by what is commonly known as the ***fire tetrahedron,*** shown in Figure 2.1. The basic components of the tetrahedron are the presence of ***fuel*** or combustible material, sufficient ***heat*** to raise the material to its ignition temperature and release fuel vapors,

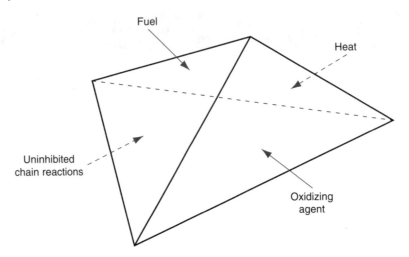

FIGURE 2.1 ◆ The fire tetrahedron.

enough of an ***oxidizing agent*** to sustain combustion, and conditions that allow ***uninhibited exothermic chemical chain reactions*** to occur. The science of fire prevention and extinguishment is based on the isolation or removal of one or more of these components.

The action of heat on a flammable liquid fuel causes simple evaporation (transforming it into a vapor). The action of heat on most solid fuels causes the chemical breakdown of the molecular structure, called ***pyrolysis,*** into vapors, gases, and residual solid (char). The pyrolysis also absorbs some heat from the reaction.

The flaming combustion of the gases from a liquid or solid takes place within an area above the fuel's surface. Heat and mass transfer processes create a situation where the fuel generates a region of vapors that can be ignited and sustained with the proper conditions (Figure 2.2). Smoldering combustion is a special case where oxygen from air combines directly with the solid surface of the fuel.

TYPES OF FIRES

There are four recognized categories of combustion generated by fire (Quintiere 1997). ***Diffusion flames*** make up the category of most natural fuel-controlled flaming fires such as candles, campfires, and fireplaces. These occur as fuel gases or vapors diffuse from the fuel surface into the surrounding air. In the zone where appropriate concentrations of fuel and oxygen are present, flaming combustion will occur.

The combination of fuel and oxygen prior to ignition results in a combustion process known as ***premixed flames***. These fuels consist of either vaporized liquid fuels or gases. Ignition of these fuels or gas mixtures is possible within regions bounded by upper and lower fuel/oxygen concentrations. In a jet, a special case of flames can occur when the fuel vapor is released under pressure into the air and mixes by mechanical action to produce an ignitable mixture.

Smoldering is typically a slow exothermic process where oxygen combines directly with the fuel on its surface (or within it, if it is porous). If enough heat is being produced, the surface may glow with incandescent reaction zones. Smoldering fires are

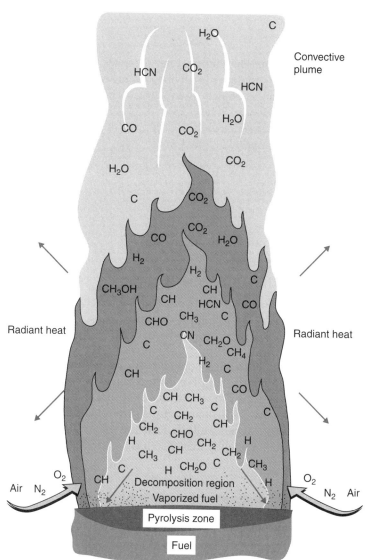

FIGURE 2.2 ◆ Schematic representation of the heat and mass transfer processes associated with a burning surface.
Kirk's Fire Investigation, 5th ed., by DeHaan, © 2002. Reprinted by permission of Pearson Education, Inc., Upper Saddle River, NJ.

characterized by visible charring and absence of flame. A discarded or dropped cigarette on the surface of a couch, mattress, or other upholstered furniture can cause a
smoldering fire.

Like smoldering, ***spontaneous combustion*** is another slow chemical reaction. Self-
heating can produce enough heat in a mass of fuel that results in flaming combustion
at a critical point known as thermal runaway. Spontaneous combustion usually involves naturally occurring fuels, such as those found in peanut and linseed oils.

◆ 2.3 HEAT TRANSFER

Through the process of **conduction**, thermal energy passes from a warmer to a cooler area of a solid material. In situations involving fire gases, heat conducted through walls, ceilings, and other adjacent objects often leaves signs of fire damage based upon the varying temperature gradients (Figure 2.3).

The general equations used in conductive heat transfer through a solid whose faces are at different temperatures, T_1 and T_2, are

$$\dot{q} = kA(T_2 - T_1)/l, \tag{2.1}$$

$$\dot{q}'' = \dot{q}/l, = kA(T_2 - T_1) \tag{2.2}$$

where

$\dot{q}$ = conduction heat transfer rate (kJ/s or kW),
$\dot{q}''$ = conductive heat flux (kW/m^2),
k = thermal conductivity of material (W/m-K),
A = area through which the heat is being conducted (m^2),
$\Delta T = (T_2 - T_1)$ wall face temperature difference, and
l = material thickness (m).

Conduction requires direct physical contact between the source and the target receiving the heat energy (Quintiere 1997). For example, heat from a fire in a living room may be conducted through a poorly insulated common wall to an adjoining room, producing surface demarcation patterns. In an investigation, examine and record damage to both sides of interior and exterior walls and surface materials. Interior wall finishing materials such as wooden paneling may also have to be removed to evaluate the

FIGURE 2.3 ◆ Example of internal conductive heat transfer of a car door from a fire inside the vehicle. *Photo by John DeHaan.*

TABLE 2.2 ◆ **Thermal Characteristics of Common Materials Found at Fire Scenes**

Material	Thermal Conductivity k (W/m–K)	Density ρ (kg/m³)	Specific Heat c_p (J/kg–K)	Thermal Inertia $k\rho c_p$ (W²s/m⁴K²)
Copper	387	8940	380	1.30×10^9
Steel	45.8	7850	460	1.65×10^8
Brick	0.69	1600	840	9.27×10^5
Concrete	0.8	1900	880	1.34×10^6
Concrete	1.4	2300	880	2.38×10^6
Glass	0.76	2700	840	1.72×10^6
Gypsum plaster	0.48	1440	840	5.81×10^5
PPMA	0.19	1190	1420	3.21×10^5
Oak	0.17	800	2380	3.24×10^5
Yellow pine	0.14	640	2850	2.55×10^5
Asbestos	0.15	577	1050	9.09×10^4
Fiber board	0.041	229	2090	1.96×10^4
Polyurethane foam	0.034	20	1400	9.52×10^2
Air	0.026	1.1	1040	2.97×10^1

Source: Data derived from Drysdale (1999, 33).

effects of heat transfer. See Table 2.2 for the characteristics of typical materials found at fire scenes.

Shown in Figure 2.4 are graphs of short, medium, and steady-state conduction through a solid material. During short-duration conduction, heat flux events produce high surface temperatures, with the inner core staying much cooler. Increasing the time duration of heat transfer increases the internal temperature of the solid, eventually producing a nearly linear temperature gradient.

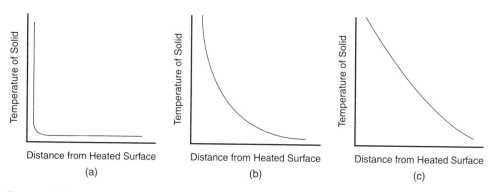

FIGURE 2.4 ◆ Temperature distribution through conduction for (a) short duration, (b) medium duration, and (c) steady state.

The second process, **convection**, is the transfer of heat energy via movement of liquids or gases from a warmer to a cooler area. Typically, heat from a moving gas transfers to a cooler surface. An indicator of convective heat transfer damage may be found in the region where a fire plume contacts the ceiling directly above the plume as shown in Figure 1.7.

The general equations used in convective heat transfer are

$$\dot{q} = h(T_2 - T_1)/A,\tag{2.3}$$

$$\dot{q}'' = \dot{q}/A = h(T_2 - T_1),\tag{2.4}$$

where

$\dot{q}$ = convective heat transfer rate (kJ/s or kW),

$\dot{q}''$ = convective heat flux (kW/m^2),

h = convective heat transfer coefficient (W/m^{2}K),

A = area through which the heat is being convected (m^2), and

$\Delta T = (T_2 - T_1)$ material temperature differences.

The term h is the convective heat transfer coefficient, expressed as W/m^{2}K. It is dependent on the nature of the surface and the velocity of the gas. Typical values for air are 5–25 W/m^{2}K for free convection and 10–500 W/m^{2}K for forced convection (Drysdale 1999, 33).

Closer examination of surface features will usually indicate the direction and intensity of the fire plume and ceiling jet, which contain heated products of combustion such as hot gases, soot, ash, and burning embers. The most extensive damage normally exists at the plume's impingement on the ceiling directly above the fuel source. Generally, materials farther from the fire plume will exhibit less damage. Figure 2.5 shows the external convective heating of a car door from a gasoline fire of short duration.

Radiation is the transmission of heat energy via electromagnetic waves from a warmer to a cooler surface. Heat transfer by radiation often damages surfaces facing the fire plume, including immobile objects such as furniture. The total thermal effect on a material's target surface may be a combination of conductive, convective, and radiative heat transfer rates. Convection and radiation are responsible for most victim burn injuries.

The equation for radiative heat flux is represented in general terms by

$$\dot{q}'' = \varepsilon \sigma T_2^4 F_{12},\tag{2.5}$$

where

$\dot{q}''$ = radiative heat transfer flux (kW/m^2),

ε = emissivity of the hot surface

σ = Stefan–Boltzman constant = 5.67×10^{-11} kW/m^2/K^4, and

F_{12} = configuration factor.

On heated solid and liquid surfaces the emissivity is 0.8 ± 0.2. The emissivities for gases and flames depend on their fuel and thickness, but most are approximately 0.5–0.7 (Quintiere 1998). The incident heat flux falling on an object or other surface is discussed further in a later section.

Guidelines for interpreting the damage due to radiation are simple—the investigator should generally start with the surfaces farthest from the apparent fire plume,

FIGURE 2.5 ◆ Example of external heating of a car door from a very short-duration gasoline fire that only scorched the paint on the door. The gasoline was poured outside of the door with its window closed. The glass shattered by convected and radiated heat. *Photo by John DeHaan.*

systematically advancing toward the apparent area and point of fire origin. Areas of more intense damage relative to these surfaces often were closer to the heat source as shown in Figure 1.5. By comparing these damaged surfaces, an investigator can often determine the fire's movement and intensity as it burned. After this evaluation, the investigator reverses his/her steps, tracing the damage back away from the point of fire origin.

Superposition is the combined effects of two or more fires or heat transfer effects, a phenomenon that may generate confusing fire damage indicators. Since the goal of an arsonist is often the rapid destruction of a building, multiple-set fires are often encountered. These fires, which initially consist of separate plumes, can confuse even the most experienced investigator, particularly when their sequence of ignition is unknown, and the effects of separate fires overlap.

Direct flame impingement is a case of superposition where the high radiant heat flux and high convective transfer of the flame quickly affect the exposed fuels. Such combined effects can quickly ignite fuels and raise others to very high surface temperatures.

Note also that separate multiple fires can occur when there is a collapse or "fall-down" of burning materials, such as hanging draperies or window shades. If multiple points of fire origin are present, the investigator should evaluate the combined heat transfer effects, or superposition of these plumes.

EXAMPLE 2.1   Heat Transfer Calculations

Problem. A fire in a storage room in an automobile repair garage raises the temperature of the brick walls and ceiling over a period of time to 500°C (932°F or 773 K). The ambient

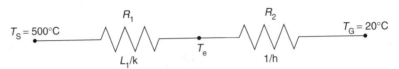

FIGURE 2.6 ◆ Solution of the storage-room fire transfer problem using an electrical analog.

temperature of the exterior air on the side of the brick wall to the garage is 20°C (68°F or 293 K). The thickness of the brick wall is 200 mm (0.2 m), and its convective heat transfer coefficient is 12 W/m²K. Calculate the temperature to which the exterior of the wall will be raised.

Suggested Solution. Assuming that the brick maintains its integrity during the fire, determine the steady-state temperature of the wall's surface in the garage (example modified from Drysdale 1999, 72). A one-dimension electrical circuit analog, shown in Figure 2.6, is used in this solution, because we can look at the way heat transfers through a material in the same way that current flows through resistors.

Solutions by electrical analogs are often used. The steady-state heat transfer solution is best illustrated by a direct-current electrical circuit analogy, where the temperatures are voltages ($V = T$), the thermal resistances of heat convection and conduction are represented as electrical resistances ($R = 1/h_h$ and $R = L_n/k_n$, respectively). The heat flux, $\dot{q}''$, of which the units are W/m², is analogous to the current, i.e., $I = \dot{q}''$.

The problem can be represented as shown in Figure 2.6, as a steady-state electrical circuit analog, where R_1 and R_2 are the thermal resistances of the brick wall and inner air of the garage.

Temperature in storage room	$T_S = 500°C$ (773 K)
Temperature in garage	$T_G = 20°C$ (293 K)
Thermal conductivity	$k = 0.69$ W/mK
Thickness of the brick	$l = 200$ mm (0.2 m)
Conductive thermal resistance	$R_1 = \Delta L/k = (0.2)/(0.69) = 0.290$ m²/WK
Convective heat transfer coefficient	$h = 12$ W/m²K
Convective resistance	$R_2 = 1/h = (1)/(12) = 0.083$ m²/WK
Total heat flux	$\dot{q}'' = (T_S - T_G)/(R_1 + R_2) = 1287$ W/m²
Exterior wall temperature	$T_e = 773\,K - (\dot{q}'')(R_1) = 773\,K - 374\,K = 399\,K$
	$399\,K - 273 = 126°C$ (258°F)

◆ 2.4 HEAT RELEASE RATE

The heat release rate (HRR) of a fire is the amount of heat released per unit time by a heat source. The term is often expressed in watts (W), kilowatts (kW), megawatts (MW), kilojoules per second (kJ/s), or BTUs per second (BTU/s) and is denoted in formulas as $\dot{Q}$ (the dot over the Q meaning per unit time). For general conversion purposes, 3412 Btu/hr = 0.95 BTU/s = 1 kJ/s = 1000 W = 1 kW.

The HRR is essentially the size or power of the fire. Three important effects of HRR make it the single most important variable describing fires and the hazards associated with its interaction with buildings and their occupants (Babrauskas 1996):

- ◆ creation of more heat
- ◆ correlation with other variables
- ◆ survivability of occupants

First, and most importantly, heat released by a fire is the driving force for that fire subsequently to produce more heat by creating more fuel by evaporation or pyrolysis. As long as adequate quantities of fuel and oxygen are available, the transfer of heat from the flames of the burning fuel causes the formation of more heat to be released from it and surrounding fuels. It is not guaranteed that all of the fuel burns. As with burning characteristics of fuels, there are physical limits to theoretical and actual heat release rates.

Second, another important role of the heat release rate in forensic fire scene reconstruction is that it directly correlates with many other variables. Examples include the production of smoke and toxic by-products of combustion, room temperatures, heat flux, mass loss rates, and flame height impingement.

Third, the direct correlation of high heat release rates and lethality of a fire is significant. High heat release rates produce high mass loss rates of burning materials, some of which may be very toxic. High heat fluxes, large volumes of high-temperature smoke, and toxic gases may overwhelm occupants, preventing their safe escape during fires (Babrauskas and Peacock 1992).

The first major introduction of fire investigators to the concept of heat release rates was in 1985, when Drysdale set forth formulas for the calculation of estimated flame heights and onset time for flashover. These equations attempted to answer the following time-honored questions.

- How hot was the fire, and could it ignite nearby combustibles or produce thermal injuries to humans?
- Was an ignitable liquid of sufficient quantity used in the fire, and how high did the flames reach?
- Were conditions right for flashover to occur?
- When did the smoke detectors and/or sprinklers activate?

An investigator's most basic question centers on the peak heat release rate, which can range from several watts to thousands of megawatts. Table 2.3 lists peak heat release rates for common burning items of potential interest to fire investigators.

The peak heat release rate is dependent on the fuel present, its surface area, and its physical configuration. The heat release rates of many typical fuels are available through NIST at *fire.nist.gov*. Other concepts in fire dynamics relate to the heat release rate. These include mass loss, mass flux, heat flux, and combustion efficiency.

TABLE 2.3 ◆ Peak Heat Release Rates for Common Objects Found at Fire Scenes

Material	*Total Mass (kg)*	*Peak Heat Release Rate (kW)*	*Time (s)*	*Total Heat Released (MJ)*
Pillow, latex foam	1.24	117	350	27.5
Wastepaper basket	0.94	50	350	5.8
Television set (T1)	39.8	290	670	150
Christmas tree (T17)	7.0	650	350	41

Source: Derived from Babrauskas (SFPE 2002a, sect. 3-1).

MASS LOSS

The heat release rate can be calculated from the ***mass loss rate*** and heat of combustion of an object. This "burning rate" is expressed as kg/s or g/s and is denoted $\dot{m}$ in most equations. The heat release rate can be calculated using the general equation (2.6). Mass loss rates can be experimentally determined by simply weighing a fuel package while it burns.

The heat release rate, $\dot{Q}$, can be calculated from

$$\dot{Q} = \dot{m}\, \Delta h_c, \tag{2.6}$$

where

$\dot{Q}$ = heat release rate (kJ/s or kW),

$\dot{m}$ = burning rate or mass loss rate (g/s), and

Δh_c = heat of combustion (kJ/g).

The mass loss rate $\dot{m}$ of a fuel typically depends on three factors: type of fuel, configuration, and area involved. For example, a campfire constructed with small sticks of dry wood arranged in a tent configuration will burn more quickly and yields a higher heat release rate over a shorter period of time than one using wet, large-dimensioned wood placed in a haphazard arrangement.

A material's heat of combustion is how much heat is released per unit of mass during combustion, expressed as MJ/kg or kJ/g. When denoted Δh_c, the term stands for the complete heat of combustion and is used when all of the material combusts leaving no fuel behind. The effective heat of combustion, denoted Δh_{eff}, is reserved for more realistic cases where combustion is not totally complete.

MASS FLUX

Another concept related to the heat release rate is a material's ***mass flux*** or mass burning rate per unit area, expressed as kg/(m^2 s) and denoted $\dot{m}''$ in equations. When the horizontal burning area of a material is known along with the burning rate per unit area, the heat release rate equation is written

$$\dot{Q} = \dot{m}'' \Delta h_c A, \tag{2.7}$$

where

$\dot{Q}$ = total heat release rate (kW),

$\dot{m}''$ = mass burning rate per unit area (g/m^2-s),

Δh_c = heat of combustion (kJ/g), and

A = burning area (m^2).

The heat of vaporization is the amount of heat generated when converting a solid or liquid fuel to a combustible vapor.

The heat needed to convert a solid or liquid fuel to a combustible form is called the latent ***heat of vaporization*** and represents heat that has to be created to initiate and sustain combustion. The energy needed to evaporate a mass unit of fuel, expressed as kJ/kg, is denoted H_L in equations. Solids, unlike liquids, have H_L values that are grossly time dependent. It is very hard to obtain accurate H_L values for solids. Values for Δh_c, $\dot{m}''$, and H_L are listed in Table 2.4 and are available in the reference literature for various fuels.

TABLE 2.4 ◆ Typical Heats of Mass Burning Rates, Vaporization, and Heats of Combustion for Common Materials

Common Material	Mass Flux $\dot{m}''$ (g/m² s)	Heat of Vaporization H_L (kJ/g)	Heat of Combustion Δh_c (kJ/g)
Gasoline	44–53	0.33	43.7
Kerosene	49	0.67	43.2
Paper	6.7	2.21–3.55	16.3–19.7
Wood	70–80	0.95–1.82	13–15

Source: Derived from SFPE (2002a) and Quintiere (1997).

HEAT FLUX

Another fire dynamics concept that is vital to interpreting ignition, flame spread, and burn injuries is **heat flux.** Heat flux is the rate at which heat is falling upon a surface or passing through an area, measured as kilowatts per square meter (kW/m²) and represented by the symbol $\dot{q}''$.

The radiant heat output of a surface temperature is predicted by equation (2.5), so the higher the temperature of a source, the greater the heat flux from that source. Note that it increases as a function of temperature, T (Kelvin), to the fourth power.

Two major pieces of information can be obtained if an investigator can estimate the heat flux and its duration. First, and most important, is whether or not thermal injury to a victim exposed to a fire is possible. The minimum heat flux needed to produce thermal injuries and to ignite some common fuels is presented in Table 2.5. The approximate rates of radiant flux for various human burn injuries are also listed in Table 2.5 (NFPA 2001, Table 3.5.3.2).

In the simplest of cases, radiant heat flux from a fire plume can be approximated as emerging from a single point source known as the **virtual origin,** a concept discussed in more detail in Chapter 3. A schematic illustration of the concept of radiant heat transfer from a point source of a flame to a target is shown in Figure 2.7 (NFPA 2000a, chap. 10–11).

Referring to Figure 2.7, the equation used to predict the incident heat flux falling upon an object or victim in a fire is

$$\dot{q}_0'' = \frac{P}{4\pi R_0^2} = \frac{X_r \dot{Q}}{4\pi R_0^2}, \tag{2.8}$$

where

$\dot{q}_0''$ = incident radiation flux on target (kW/m²),

P = total radiative power of the flame (kW),

R_0 = distance to the target surface (m), and

X_r = radiative fraction (typically from 0.20 to 0.60), representing the percentage of heat released from the source as radiant heat,

$\dot{Q}$ = total heat release rate of source (kW),

Radiant Heat Flux (kW/m^2)	Observed Effect on Humans and Wooden Surfaces
170	Maximum heat flux measured in postflashover fires
80	Thermal protective performance (TPP) test
52	Fiberboard ignites after 5 s
29	Wood ignites after prolonged exposure
20	Floor of residential family room at flashover
16	Pain, blisters, 2nd-degree burns to skin at 5 s
12.5	Piloted ignition of wood volatiles
10.4	Pain, blisters, 2nd-degree burns to skin at 9 s
6.4	Pain, blisters, 2nd-degree burns to skin at 18 s
4.5	Blisters, 2nd-degree burns to skin at 30 s
2.5	Exposure while firefighting, burns after prolonged exposure
<1.4	Exposure to sun

TABLE 2.5 ◆ Impact of a Range of Radiant Heat Flux Values

Source: Derived from *NFPA 921* (NFPA 2001, Table 3.5.3.2).

or, by rearranging,

$$R_0 = \left(\frac{X_r \dot{Q}}{4\pi \dot{q}_0''} \right)^{1/2}. \tag{2.9}$$

This equation is valid when the ratio of the distance to the targeted object to the radius(R) of the fuel package is

$$\frac{R_0}{R} > 4 \tag{2.10}$$

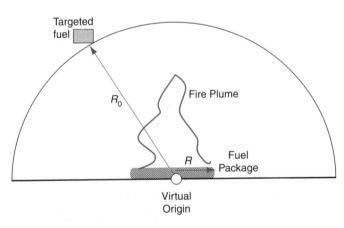

FIGURE 2.7 ◆ Schematic illustration of radiant heat transfer from a point source (virtual origin) of a fire plume to a target.

or when the radius of the burning surface of the fire, R, is small compared to the distance between the target and the fire, R_0, as shown in Figure 2.4.

For cases where

$$\frac{R_0}{R} \leq 4 \qquad\qquad (2.11)$$

the configuration factor F_{12} comes into play, which is usually determined by examining an appropriate chart, such as in Quintiere (1997, 57).

EXAMPLE 2.2   Heat Flux Calculation

Problem. A 500-kW fire represented as a point source as shown in Figure 2.7 has a surface radius of 0.2 m. Assume the radiative fraction to be 0.60. What is the heat flux falling on the target fuel at distances of R_0 equal to 1, 2, and 4 m?

Suggested Solution. The expected heat release rate can be calculated from equation (2.6).

Heat release rate $\qquad\qquad\qquad\qquad \dot{Q} = 500\,\text{kW}$
Radiative fraction $\qquad\qquad\qquad\qquad X_r = 0.60$
Radius of the burning fuel surface $\qquad R = 0.20\,\text{m}$
Equation ratio test $\qquad\qquad\qquad\quad R_0/R > 4$
Incident radiation flux on target $\qquad \dot{q}_0'' = X_r \dot{Q}/4\pi R_0^2$
At R_0 equal to 1, 2, and 4 m $\qquad\quad \dot{q}_0'' = 23.87, 5.96,\text{ and }1.49\,\text{kW/m}^2$

Note that as the distance is doubled, the radiant heat flux is decreased by a factor of one-fourth. This is known as the "inverse square" rule.

STEADY BURNING

Once a material is exposed to sufficient incident heat on the fuel and an ample air supply, the fire may spread evenly, become fully involved, and burn at ***steady state.*** The formulas for steady-state burning are

$$\dot{m}'' = \dot{q}''/H_L \qquad\qquad (2.12)$$

and

$$\dot{Q}'' = \frac{\dot{q}''}{H_L}\,\Delta h_c, \qquad\qquad (2.13)$$

where

$\dot{Q}''$ = heat release rate per unit surface area of burning fuel (kW/m^2),
$\dot{m}''$ = mass burning rate per unit area (g/m^2-s),
$\dot{q}''$ = net incident heat flux from the flame (kW/m^2), and
H_L = latent heat of vaporization (kJ/g).

These equations are useful for investigators because they allow estimates to be made of the steady-state burning rate. It should be noted that for most solid fuels, H_L is not constant, as a direct result of char formation and heat and mass transfer processes. It varies greatly with time and is not easily measured.

COMBUSTION EFFICIENCY

A substance's **combustion efficiency** is the ratio of the effective heat of combustion to the complete heat of combustion and is expressed by the term X, where

$$X = \Delta h_{eff}/\Delta h_c. \tag{2.14}$$

Thus fuels with more complete combustion produce values of X approaching 1. Fuels with lower combustion efficiencies, characteristic of flames containing incomplete burning producing sootier and more luminous fires, will result in values of X between 0.6 and 0.8 (Karlsson and Quintiere 1999, 64). For example, gasoline typically has a Δh_c of 46 but its Δh_{eff} is 43.8 due to the heat lost by evaporation ($X = 0.95$). Examples of soot producing substances include flammable liquids and thermoplastics.

When dealing with materials with larger burning surface areas or with incomplete combustion such as plastics and ignitable liquids, the combustion efficiency can be placed into the equation as the convective heat release rate:

$$\dot{Q}_c = X\dot{m}'' \Delta h_c A. \tag{2.15}$$

When calculating mean flame heights and virtual origins, which are explored in Chapter 3, the total heat release rate, $\dot{Q}$, is used. When calculating other properties of fire plumes such as plume radius, centerline temperatures, and velocity, use the convective heat release rate $\dot{Q}_c$ (Karlsson and Quintiere 1999, 64).

EXAMPLE 2.3 Ignitable Liquid Fire

Problem. An arsonist pours a quantity of gasoline onto a concrete floor, creating a 0.46-m (1.5-ft)-diameter pool. The arsonist ignites the gasoline, creating a fire very similar to a classic burning pool. Estimate the heat release rate from the burning pool using historical fire testing results.

Suggested Solution. Assuming an unconstrained pool fire, the expected heat release rate can be calculated from equation (2.7).

Heat release rate	$\dot{Q}$	$= \dot{m}'' \Delta h_{eff} A$
Area of the burning pool	A	$= (\pi/4)(D^2) = (3.1415/4)(0.46^2) = 0.164\,\mathrm{m^2}$
Mass flux rate for gasoline	$\dot{m}''$	$= 0.036\,\mathrm{kg/m^2}$-s
Heat of combustion for gasoline	Δh_{eff}	$= 43.7\,\mathrm{MJ/kg}$
Heat release rate	$\dot{Q}$	$= (0.036)(43,700)(0.164) = 258\,\mathrm{kW}$

The heat release rate of a real fire under these conditions would be expected to be in the range of 200–300 kW.

EXAMPLE 2.4 Estimating the Heat Release Rate per Unit Surface Area

Problem. Given an office environment where the fuel load consists of a mixture of wood and plastics, determine the estimated mass burning rate per unit surface area (mass flux) and the associated heat release rates per unit surface area.

Suggested Solution. Assume that the effective heat of combustion for a mixture of wood and plastics is approximately 16 kJ/g, the radiant heat flux ranges from 150 to 200 kW/m², and the latent heat of vaporization is 5 to 8 kJ/g.

HRR per unit surface area	$\dot{Q}''$	$= (\dot{q}''/L)\Delta h_c$
Incident heat flux	$\dot{q}''$	$= 150 \text{ to } 200\,\text{kW/m}^2$
Latent heat of vaporization	L	$= 5 \text{ to } 8\,\text{kJ/g}$
Heat of combustion	Δh_c	$= 16\,\text{kJ/g}$
Minimum $\dot{Q}''$	$\dot{Q}''$	$= (150/8)(16) = 300\,\text{kW/m}^2$
Maximum $\dot{Q}''$	$\dot{Q}''$	$= (200/5)(16) = 640\,\text{kW/m}^2$

◆ 2.5 FIRE DEVELOPMENT

Fires, particularly those within enclosed structures, burn in a predictable manner over time. For the purpose of fire scene reconstruction, development can be divided into four separate *fire phases,* each containing unique characteristics and its own time frame. These fire development curves are often referred to as the fire signature. The phases are as follows.

- Phase 1—Incipient ignition
- Phase 2—Growth
- Phase 3—Fully developed fire
- Phase 4—Decay

As shown in Figure 2.8, these phases form a characteristic "fire signature," which is often the subject of studies for items found as the source of the first fuel burned in typical structural fires. These phases are important when predicting a fire's growth, calculating the activation of smoke, heat, and automated sprinkler systems, estimating evacuation times, and determining the risk of heat exposure to its occupants.

There are also several physical features associated with each fire development phase including (1) low heat, some smoke, no detectable flame; (2) more heat, lots of

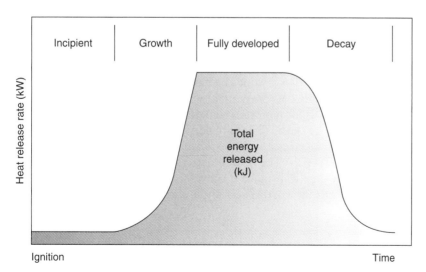

FIGURE 2.8 ◆ Fire development phases that often form a fire signature, consisting of (1) ignition, (2) growth, (3) fully developed fire, and (4) decay. *From The SFPE Handbook of Fire Protection Engineering, 3rd ed., 2002, by permission of Society of Fire Protection Engineers.*

smoke, flame spread; (3) massive heat and smoke output, extensive flame; and (4) decreased flame and heat, smoke, with increased smoldering.

At the transition from Phase 2 to Phase 3, the phenomenon of ***flashover*** sometimes occurs, depending upon the amount of heat produced during the initial stages of the fire and the size and configuration of the room of fire origin. In the absence of flashover, the fire may reach a maximum size because of limited available fuels, flame spread, or a ventilation limit. Flashover is discussed in detail later.

PHASE 1—INCIPIENT IGNITION

The fundamental properties that influence an object's ***ignitability*** are its density, ρ, heat capacity, c_p, and thermal conductivity, k. Table 2.2 lists these characteristic values for some common materials found at fire scenes (Drysdale 1999, 33).

An object's temperature is a measure of the relative amount of heat (energy) contained within that object. This heat is determined by the mass of that object and its heat capacity. The heat capacity of an object is a means of expressing how much heat must be added to the object to increase its temperature.

The product of these three characteristics ($k\rho c_p$) is often referred to as an object's ***thermal inertia*** and the calculated value $k\rho/c_p$ is called the ***thermal diffusivity.*** Like mechanical inertia, where the higher the inertia, the harder it is to move, the higher the thermal inertia, $k\rho c_p$, the harder a material is to ignite—that is, the more heat has to be applied or the longer the duration of exposure required before ignition.

The ignition of an object can be calculated when it is exposed to radiant or convective heat flux, $\dot{q}''$, with no initial contact with flames (autoignition). The calculation assumes that a first burning object (ignition source) provides sufficient radiant heat flux to ignite a second, nonburning fuel object nearby. Experimental fire testing has established three levels of ignitability: ***easy, normally resistant,*** and ***difficult*** (Babrauskas 1982). This fire testing included paper, wooden, polyurethane, and polyethylene fuels to establish these relationships.

Thin materials such as paper and curtains will be easily ignited when receiving a threshold radiant flux of $10 \, kW/m^2$ or greater. Thicker or more massive items with low thermal inertia values such as upholstered furniture will ignite when receiving a radiant flux of $20 \, kW/m^2$ or greater. Solid materials with thicknesses greater than 13 mm (0.5 in.) such as plastics or dense woods having large thermal inertia values will ignite when receiving a radiant flux of $40 \, kW/m^2$ or greater (Babrauskas 1982). Table 2.6 lists typical ignition properties for materials sometimes found at fire scenes.

Equations (2.16), (2.17), and (2.18) are derived from data plots (Babrauskas 1982) published in *SFPE Engineering Guide to Piloted Ignition of Solid Materials Under Radiant Exposure* (SFPE 2002b). The heat release rates, $\dot{Q}$, needed for a fire source to ignite another fuel package a distance, D, away can be classified into *easy, normally resistant,* and *difficult* as shown, respectively, in the following equations:

$$\dot{Q} = 30 \cdot 10^{\left(\frac{D + 0.8}{0.89}\right)}, \qquad \dot{q}'' \geq 10 \, kW/m^2, \tag{2.16}$$

$$\dot{Q} = 30 \left(\frac{D + 0.05}{0.019}\right), \qquad \dot{q}'' \geq 20 \, kW/m^2, \tag{2.17}$$

$$\dot{Q} = 30 \left(\frac{D + 0.02}{0.0092}\right), \qquad \dot{q}'' \geq 40 \, kW/m^2, \tag{2.18}$$

TABLE 2.6 ◆ Ignition Properties

Material	$k\rho c$ $(kW/m^2K)s^2$	T_{ig} (°C)	$q''_{critical}$ (kW/m^2)
Plywood, plain (0.635 cm)	0.46	390	16
Plywood, plain (1.27 cm)	0.54	390	16
Plywood, FR (1.27 cm)	0.76	620	44
Hardboard (6.35 mm)	1.87	298	10
Hardboard (3.175 mm)	0.88	365	14
Hardboard, gloss paint (3.4 mm)	1.22	400	17
Hardboard, nitrocellulose paint	0.79	400	17
Particleboard (1.27-cm stock)	0.93	412	18
Douglas Fir particleboard (1.27 cm)	0.94	382	16
Fiber insulation board	0.46	355	14
Polyisocyanurate (5.08 cm)	0.020	445	21
Foam, rigid (2.54 cm)	0.030	435	20
Foam, flexible (2.54 cm)	0.32	390	16
Polystyrene (5.08 cm)	0.38	630	46
Polycarbonate (1.52 mm)	1.16	528	30
PMMA, type G (1.27 cm)	1.02	378	15
PMMA, polycast (1.59 mm)	0.73	278	9
Carpet #1 (wool, stock)	0.11	465	23
Carpet #2 (wool, untreated)	0.25	435	20
Carpet #2 (wool, treated)	0.24	455	22
Carpet (nylon/wool blend)	0.68	412	18
Carpet (acrylic)	0.42	300	10
Gypsum board, common (1.27 mm)	0.45	565	35
Gypsum board, FR (1.27 cm)	0.40	510	28
Gypsum board, wall paper	0.57	412	18
Asphalt shingle	0.70	378	15
Fiberglass shingle	0.50	445	21
Glass-reinforced polyester (2.24 mm)	0.32	390	16
Glass-reinforced polyester (1.14 mm)	0.72	400	17
Aircraft panel, epoxy fiberite	0.24	505	28

Source: From Quintiere and Harkleroad (1984).

where

$\dot{Q}$ = heat release rate of the source fire (kW) and

D = distance from the radiation source to the second object (m).

The time to ignition for ***thermally thin*** and ***thermally thick*** materials can be estimated by several equations contained in engineering practice guides developed by the Society of Fire Protection Engineers (SFPE 2002a, 2002b).

The generally accepted equations for ***ignition time*** of thermally thin and thermally thick are (2.18) and (2.19), respectively. Under thermally thin conditions, $l \leq 1$ mm,

$$t_{ig} = \rho c l_p \left(\frac{T_{ig} - T_\infty}{\dot{q}''} \right). \qquad (2.19)$$

In thermally thick conditions, $l > 1$ mm,

$$t_{ig} = \frac{\pi}{4} k \rho c_p \left(\frac{T_{ig} - T_\infty}{\dot{q}''} \right)^2, \qquad (2.20)$$

where

t_{ig} = time to ignition (s),

k = thermal conductivity (W/m-K),

ρ = density (kg/m^3),

c_p = specific heat capacity (kJ/kg-K),

l = thickness of material,

T_{ig} = piloted fuel ignition temperature,

T_∞ = initial temperature, and

$\dot{q}''$ = radiant heat flux (kW/m^2).

The ignition modes of liquid versus solid fuels differ in their approach. When an ignitable liquid evaporates to the point where its vapor concentration in air is within its upper and lower flammable limits, the fuel's vapors are capable of undergoing piloted ignition. This pilot can be an electric arc or an open flame.

PHASE 2—FIRE GROWTH

Fire growth rates, as shown in Figure 2.5, can sometimes be modeled using mathematical relationships that show the flame spread and fire growth estimations using heat release rates. The growth of the flame front across a horizontal fuel surface is known as the ***lateral flame spread*** and can be expressed by the equation (Quintiere and Harkelroad 1984)

$$V = \frac{\dot{q}''}{\rho c_p A (T_{ig} - T_s)^2} \qquad (2.21)$$

or as

$$V = \frac{\phi}{k \rho c_p (T_{ig} - T_s)^2}, \qquad (2.22)$$

where

V = lateral velocity of flame spread (m/s),

A = cross-sectional area affected by the heating of $\dot{q}''$,

ϕ = ignition factor from flame spread data (kW^2/m^3) (experimentally derived),

k = thermal conductivity (W/m-K),

ρ = density (kg/m^3),

c_p = specific heat capacity (kJ/kg-K),

T_{ig} = piloted fuel ignition temperature (°C), and

T_s = unignited, ambient surface temperature (°C).

The values for ϕ and other variables needed for calculating lateral flame spread can be obtained from *ASTM E1321-97a* (ASTM 2002). Table 2.7 lists the horizontal flame spread of several common materials. Obviously, an external draft pushes the flame down against the "downwind" side and increases its radiant heat effect on the fuel in front of its path, so the flame front will grow faster if wind-aided.

Lateral or downward flame spread on thick solids such as wood is typically of the order of 1×10^{-3} m/s (1 mm/s). In contrast, upward spread on solid fuels is typically 1×10^{-2} to 1 m/s (10–1000 mm/s). Flame spread rates across liquid pools are typically 0.01–1 m/s, depending upon the ambient temperature and the flash point of the liquid and 0.1–10 m/s for premixed vapor/air mixtures.

An example of lateral flame spread over a fuel's surface is demonstrated in forest fires and can be affected by wind over a surface. For example, lateral flame spread in forests depends on several key variables: the type of fuel and its configuration, wind, humidity, and the slope of the terrain. The brush or duff along the forest floor is usually involved in the initial phases of the fire. The terrain and radiant and wind-induced convective heat transfer contribute significantly to the flame spread along the forest floor.

The Thomas (1971) formula, which can be used to approximate flame spread in wildland fires on flat ground, is

$$V = \frac{k(1 + V_\infty)}{\rho_b},$$ (2.23)

TABLE 2.7 ◆ **Example Lateral Flame Spread Data**

Material	Piloted Ignition Temperature T_{ig} (°C)	Ambient Surface Temperature T_s (°C)	Ignition Factor ϕ (kW^2/m^3)
Wool carpet			
Treated	435	335	7.3
Untreated	455	365	0.89
Plywood	390	120	13.0
Foam plastic	390	120	11.7
PMMA	378	<90	14.4
Asphalt shingle	356	140	5.4
Acrylic carpet	300	165	9.9

Source: Derived from ASTM (2002).

where

V = velocity of flame spread (m/s),

V_∞ = wind speed (concurrent) (m/s),

ρ_b = bulk density of the fuel (kg/m^3), and

k = 0.07 wildland fire, 0.05 wood crib (kg/m^3).

EXAMPLE 2.5 Estimating Lateral Flame Travel Velocity

Problem. An investigator determines that a fire ignited at the edge of a plywood floor. What was the estimated velocity of the lateral flame spread?

Suggested Solution. Use the Quintiere and Harkelroad equation and *ASTM 1321-97a* data, from Table 2.7.

Ignition factor	ϕ = 13.0 kW^2/m^3
Fuel thermal inertia	$k\rho c_p$ = 0.54 [kW^2-s/(m^4-K^2)]
Piloted fuel ignition temperature	T_{ig} = 390°C
Unignited, ambient surface temperature	T_s = 120°C
Velocity of flame spread	V = $\phi/k\rho c_p(T_{ig} - T_s)^2$ = 13.0/0.54(390 − 120)^2
	= 0.33 × 10^{-3} m/s = 0.0330 mm/s
	= 2 mm/min

EXAMPLE 2.6 Wildland Arson

Problem. Children playing with matches set a fire during a dry season of the year in a remote wildland area. The wind is blowing at a velocity of 2 m/s and the bulk density of the brush along the forest floor is 0.04 g/cm^3. Determine the velocity of the flame spread from this fire with and without considering the effect of the wind.

Suggested Solution. Use the Thomas equation.

Velocity of flame spread	V = $k(1 + V_\infty)/\rho_b$
Wind speed	V_∞ = 2 m/s
Bulk density	ρ_b = 0.04 g/cm^3 = 40 kg/m^3
Equation constant	k = 0.07 kg/m^3
Flame spread velocity (0 m/s)	= (0.07)(1 + 0)/(40)
	= 0.00175 m/s = 1.75 mm/s
	= 0.1 m/min
Flame spread velocity (2 m/s)	= (0.07)(1 + 2)/(40)
	= 0.00525 m/s = 5.25 mm/s
	= 0.315 m/min

Fire growth rates during the initial phases of development can sometimes be modeled using mathematical relationships. One common relationship assumes that the initial growth rate geometrically approximates the time that the fire has burned squared (also called t^2 *fires*). A perfect fire, unlimited by fuel or ventilation factors, would have an exponential growth rate. Note that this behavior applies only to the growth phase. This accepted growth rate formula is

$$\dot{Q} = (1055/t_g^2) \, t^2 = \alpha t^2, \tag{2.24}$$

TABLE 2.8 ◆ Four Growth Rates in t^2 Fires

Type of Fire	Example Objects	Time t_g (s)	Fire Growth Factor α (kW/s²)
Slow	Thick wooden objects (tables, cabinets, dressers)	600	0.003
Medium	Lower density objects (furniture)	300	0.012
Fast	Combustible objects (paper, cardboard boxes, drapes)	150	0.047
Ultrafast	Volatile fuels (flammable liquids, synthetic mattresses)	75	0.190

Source: Derived from *NFPA 92B* and *NFPA 72* (NFPA 2000b, 2002).

where

t = time (s),

t_g = time for fire to grow from ignition to 1.055 MW (1000 BTU/s),

$\dot{Q}$ = heat release rate (MW) at time t, and

α = fire growth factor (kW/s²) for the material being burned.

In an accepted principle used by *NFPA 72* (detectors) and *NFPA 92B* (smoke control systems), Table 2.8 and Figure 2.9 illustrate four types of growth times used to define fires by the time required to reach 1.055 MW (1000 BTU/s). The time, t_g, is obtained from repeated calorimetry tests of specific fuel packages (NFPA 2000b, 2002).

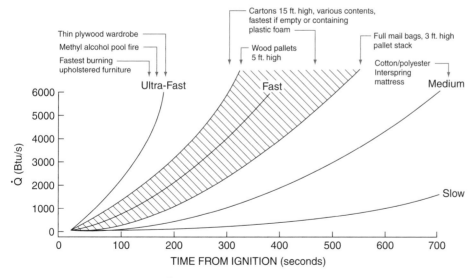

FIGURE 2.9 ◆ Relation of t^2 fires to several individual fire tests. Note that the heat release rate is in BTU/s and that 1000 BTU/s = 1.055 MW. *Courtesy of NIST, from Nelson and Tu [1991].*

	Heat Release Rate per Floor Area Covered (MW/m^2)	Characteristic Time to Reach 1 MW for t^2 Fires (s)
TABLE 2.9 ◆ Typical Ranges for Fire Behavior of Materials Stored in Warehouses		
Material		
Wood pallets,		
Stacked 1.5 ft high	1.3	155–310
Stacked 5 ft high	3.7	92–187
Stacked 10 ft high	6.6	77–115
Stacked 15 ft high	9.9	72–115
Polyethylene bottles, stacked 15 ft high	1.9	72
Polyethylene letter trays, stacked 5 ft high	8.2	189
Mail bags, filled, stored 15 ft high	0.39	187
Polystyrene jars in cartons, 15 ft high	14	53

Source: Derived from Quintiere (1998, Tables 6–5 and 6–6).

Table 2.9 shows typical ranges for the heat release rates per unit area and characteristic times for a fire to reach 1.055 MW (1000 Btu/s) in size. These data were obtained from actual fire testing of common materials stored in warehouses.

The slow growth curve ($t_g = 600$ s) usually involves thick solid wooden objects such as tables, cabinets, and dressers. The medium growth curve ($t_g = 300$ s) applies to lower-density solid fuels such as upholstered and light weight furniture. The fast upholstered & light weight growth curve ($t_g = 150$ s) involves combustible items such as paper, cardboard boxes, and drapes. The ultrafast growth curve ($t_g = 75$ s) is for flammable liquids, synthetic mattresses, and volatile fuels. Of course, the geometry of the objects also has an impact on growth rates.

Note that the t^2 growth rate applies after the initial incipient fire phase where the growth rate is nearly zero. In some cases involving polyurethane upholstered furniture, the steady-state phase does not exist, and the approximation for the heat release rate resembles a curve triangular in shape. The approach fitting triangular curves to model the heat release rate signatures of furniture can represent up to 91 percent of the total heat released (Babrauskas and Walton 1986).

EXAMPLE 2.7   Loading Dock Fire

Problem. An investigator responds to the scene of a fire on the back loading dock of a supply distribution center. The fire was extinguished by an automatic sprinkler system. A nearby security camera captured the silhouette of a young male running from the loading dock prior to the activation of the sprinkler system.

Upon examination of the fire scene, the investigator determines that the fire was intentionally set to a cart of polyethylene letter trays in two pallets stacked 1.56 m (5 ft) high on the loading dock. Determine for the investigator the heat release rate 30 s after the arsonist set the fire.

Suggested Solution. Use equation (2.24) for t^2 fire growth and data from Table 2.9 on typical ranges for fire behavior of polyethylene letter trays.

Time to reach 1 MW	t_g	$= 189\,\mathrm{s}$
Growth time of fire	t	$= 30\,\mathrm{s}$
Heat release rate	$\dot{Q}$	$= (1055/t_g^2)\,t^2$
		$= (1055/189^2)(30^2) = 26.55\,\mathrm{kW}$
and for t		$= 120\,\mathrm{s}$
	Q	$= (1055/189^2)(120^2) = 425\,\mathrm{kW}$

PHASE 3—FULLY DEVELOPED FIRE

The fully developed phase is also referred to as the "steady-state phase," either when the fire reaches its maximum burning rate or when there is insufficient oxygen to continue the combustion process (Karlsson and Quintiere 2000, 43). Fuel-controlled fires are associated with steady-state fires, particularly when sufficient oxygen is supplied.

When insufficient oxygen is available, the fire becomes ventilation controlled. In this case, the fire is usually in an enclosed room, and temperature may become more enhanced than in fuel-controlled fires. Fully developed fires in compartments are often postflashover fires where all the fuel is involved and the size of the fire becomes ventilation controlled.

PHASE 4—DECAY

The fire decay during Phase 4 usually takes place when approximately 20 percent of the original fuel is remaining (Bukowski 1995b). Although most fire service and fire protection specialists are interested in the first three phases of a fire, there are significant problems associated with the decay phase.

In high-rise buildings, for example, even after the fire is extinguished, trapped occupants may need to be rescued. Furthermore, fire investigators entering the building may also be exposed to residual amounts of toxic by-products of combustion, since there is a high concentration of CO and other toxic gases produced during the decay phase (often dominated by smoldering).

EXAMPLE 2.8   Reliable Test Data

Problem. A fire investigator is trying to develop reliable test data on the length of time a fire may have burned and an estimate of the corresponding heat release rate. The fire was reportedly started by a child playing with a cigarette lighter while sitting at the center of a mattress. The mattress was the only piece of furniture involved in the fire in the room, and a smoke detector in the hallway reportedly activated shortly after the fire.

Suggested Solution. The information needed by the investigator is found at the Fire on the Web Internet site of the Building and Fire Research Laboratory, National Institute of Standards and Technology (NIST), Gaithersburg, Maryland (*http://fire.nist.gov*). In the fire tests/data section, NIST places collections of actual test data for commonly found items in residential and commercial settings. The data files include still photos, video, graphs, miscellaneous data, and setup data for NIST fire modeling software programs.

Figure 2.10 shows the information needed by the investigator to compare this case to a similar fire under laboratory testing conditions. The fire test data showed that the peak heat release rate occurred at approximately 150 s (2.5 min) after ignition, resulting in a 750-kW fire. When forming the hypothesis for the fire, the investigator could evaluate whether the fire's time line is reasonably consistent with witness accounts, damage indicators, and heat release data. The investigator needs to remember that the test mattress may not be the same construction as the one involved in the fire and that even the same mattress will behave differently if ignited at

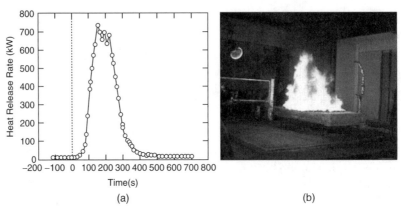

FIGURE 2.10 ◆ Reliable fire test data on (a) the heat release rate and (b) the fire after 120 s to support an investigation of a mattress fire. *Courtesy of NIST, Fire on the Web [http://fire.nist.gov].*

a corner or from underneath. Note that the test was conducted in a large enclosure so as to minimize heat radiation effects (which will increase the rate of burning) and ventilation effects (which could limit the maximum rate).

◆ 2.6 ENCLOSURE FIRES

If a fire is confined to a room, the flow of air, smoke, hot gases, and fire growth is usually constrained. Confining variables may include the ceiling height, ventilation openings formed by windows and doors, room volume, and location of the fire in the room or compartment. While the fire gases are constrained, a notable event called flashover may occur, as a consequence of the heat release rate of the combustible materials located within the room.

MINIMUM HEAT RELEASE RATE FOR FLASHOVER

In the growth of a compartment (room) fire, *flashover* is defined as the transition from the fire growing by igniting one item and then another (by direct flame contact, radiant heat, convective heating, or immersion in the hot gas layer) to the point where all fuel packages in the compartment are on fire and are burning as fast as oxygen can be made available through window and door openings. This transition may occur in just a few seconds in small, heavily fueled rooms, or over a period of minutes in large rooms, or may not take place at all due to insufficient heat release rate of the fire.

An authoritative source on this subject is found in the 2000 edition of *NFPA 555–Guide on Methods for Evaluating Potential for Room Flashover* (NFPA 2000). Prepared by the Technical Committee on Hazard and Risk of Contents and Furnishings, *NFPA 555* was approved as an American National Standard on July 26, 1996.

Fire researchers have documented other characteristics that occur during this transition phase. Definitions for flashover often include one or more of the following criteria (Milke and Mowrer 2001; NFPA 2000).

- Flames are emitted from openings in the compartment.
- Upper layer gas temperature rises to $\geq 600°C$.
- Heat flux at floor level reaches $\geq 20 \, kW/m^2$.

Because flashover represents a transition phase, defining the exact moment of occurrence is often a problem. Prior to flashover, the highest average temperatures are found in the upper or "hot gas" layer. If the average temperature of the hot gas layer exceeds 600°C (with or without the flaming ignition of the hot smoke called flameover or rollover), the radiant heat from the layer exceeds the minimum ignition radiant heat flux for exposed fuels and those fuels char and then ignite. If flashover occurs, temperatures throughout the room reach their maximum (1000°C is not uncommon) as the two-layer environment of the room breaks down and the entire room becomes a turbulently mixed combustion zone—from floor to ceiling.

The active mixing promotes very effective combustion, with oxygen concentrations dropping below 3 percent and producing very high temperatures. This environment in turn produces radiant heat fluxes of 120 kW/m^2 or higher, ensuring rapid ignition and burning of all exposed fuel surfaces, including walls, carpets, flooring, and low-lying fuels like baseboards. Also, combustion near ventilation points is enhanced.

Ignition of carpets produces floor-level flames that sweep under chairs, tables, and other surfaces that were initially protected from the downward radiant heat from the hot gas layer alone. Burning will then proceed throughout the room until the fuel supply is exhausted or some attempt is made at extinguishment.

An approximation for the minimum heat release rate required for flashover $\dot{Q}_{fo}$, can be found using the following equation, called the ***Thomas correlation*** (Thomas 1981):

$$\dot{Q}_{fo} = (378\,A_o)\,\sqrt{h_o} + 7.8A_w, \qquad (2.25)$$

where

$\dot{Q}_{fo}$ = heat release rate for flashover (kW),

A_o = area of the opening to the compartment (m^2),

h_o = height of the compartment opening (m), and

A_w = area of the walls, ceiling, and floor minus the opening (m^2).

EXAMPLE 2.9 Calculating Flashover

Problem. Given a 10 × 10-m room, with a 3-m-high ceiling and a 2.5 × 1-m opening (see Figure 2.11), determine the minimum heat release rate needed to cause flashover.

FIGURE 2.11 ◆ Example room with doorway opening for calculation of the minimum heat release rate needed for flashover.

Suggested Solution. Using equation (2.25), known as the Thomas correlation formula, the minimum heat release rate for a fire to progress to flashover in this room can be calculated as follows.

Area of compartment opening	A_o	$= (2.5 \text{ m})(1 \text{ m}) = 2.5 \text{ m}^2$
Height of compartment opening	h_o	$= 2.5 \text{ m}$
Area calculations	A_w	$= \text{floor} + \text{walls} + \text{ceiling} - \text{compartment}$ opening
	1 floor	$= (10 \text{ m})(10 \text{ m}) = 100 \text{ m}^2$
	4 walls	$= (4)(10 \text{ m})(3 \text{ m}) = 120 \text{ m}^2$
	1 ceiling	$= (10 \text{ m})(10 \text{ m}) = 100 \text{ m}^2$
		$= 320 \text{ m}^2 - 2.5 \text{ m}^2 = 317.5 \text{ m}^2$
Flashover heat release rate	$\dot{Q}_{fo}$	$= (378 A_o)(\sqrt{h_o}) + 7.8 A_w$
		$= (378)(2.5)\sqrt{(2.5)} + (7.8)(317.5)$
		$= 1494 + 2477 = 3971 \text{ kW} = 3.97 \text{ MW}$

ALTERNATIVE METHODS OF ESTIMATING FLASHOVER

Several alternative methods exist for estimating the minimum heat release rate necessary for flashover. The use of several estimates is necessary when applying the scientific method. Consider alternative approaches, evaluating the results of fire models, and comparing them with eyewitness accounts of the fire.

The first alternative approach assumes a flashover criterion temperature of $\Delta T = 575°C$ (Babrauskas 1980), noting that $\Delta T = T_{fo} - T_{\text{ambient}} = 600 - 25 = 575°C$.

$$\dot{Q}_{fo} = 0.6 A_v \sqrt{H_v}, \qquad (2.26)$$

where

$\dot{Q}_{fo}$ = heat release rate for flashover (MW),
A_v = area of the door (m²), and
H_v = door height (m).

The term $A_v \sqrt{H_v}$ is often referred to as the ***ventilation factor*** and frequently appears in similar equations. In actual experiments, two-thirds of the cases agree and are bounded between the following equations (Babrauskas 1988):

$$\dot{Q}_{fo} = 0.45 A_v \sqrt{H_v}, \qquad (2.27)$$

$$\dot{Q}_{fo} = 1.05 A_v \sqrt{H_v}. \qquad (2.28)$$

EXAMPLE 2.10   Calculating Flashover—Alternative Method 1

Problem. Repeat the calculation for minimum heat release rate needed to cause flashover as previously presented in Example 2.9 using an alternative method.

Suggested Solution. Use equations (2.26), (2.27), and (2.28), also known as the Babrauskas correlation formulas.

Area of the door opening	$A_v = (2.5 \text{ m})(1 \text{ m}) = 2.5 \text{ m}^2$
Height of door opening	$H_v = 2.5 \text{ m}$
Flashover heat release rate	$\dot{Q}_{fo} = (0.6)(A_v)(\sqrt{H_v})$
	$= (0.6)(2.5)\sqrt{(2.5)} = 2.37 \text{ MW}$

| Minimum heat release rate | $\dot{Q}_{fo} = (0.45)(A_v)(\sqrt{H_v}) = 1.78\,\text{MW}$ |
| Maximum heat release rate | $\dot{Q}_{fo} = (1.05)(A_v)(\sqrt{H_v}) = 4.15\,\text{MW}$ |

Note that the calculations of 3.97 MW (Example 2.9) and 2.23 MW (Example 2.11) fall within the minimum and maximum calculated range of 1.78 to 4.15 MW, from equations (2.27) and (2.28).

The second alternative approach is based upon test data and a flashover criterion temperature of $\Delta T = 500°C$ (Lawson and Quintiere 1985; McCaffrey et al. 1981):

$$\dot{Q}_{fo} = 610\sqrt{h_k A_s A_v \sqrt{h_v}}, \tag{2.29}$$

where

h_k = enclosure conductance to ceiling and walls [(kW/m²) K],
A_s = total enclosure surface area, excluding vent or door area (m²),
A_v = total area of vent or door opening (m²), and
H_v = height of vent or door opening (m).

The **McCaffrey and Quintiere approach** requires an estimate for the effective heat transfer enclosure **conductance coefficient,** h_k, of the enclosure. Assuming uniform heating of the enclosure with adequate thermal penetration, the conductance coefficient h_k can be estimated as

$$h_k = \frac{k}{\delta}, \tag{2.30}$$

where

h_k = enclosure conductance to ceiling and walls [(kW/m²)/K],
k = thermal conductivity of enclosure material [(kW/m)/K], and
δ = enclosure material thickness (m).

EXAMPLE 2.11   Calculating Flashover—Alternative Method 2

Problem. Repeat the calculation for the minimum heat release rate needed to cause flashover given an enclosure lined with 16-mm ($\frac{5}{8}$-in.)-thick gypsum board. Also calculate the estimated enclosure conductance.

Suggested Solution. Assuming uniform heating of the enclosure with adequate thermal penetration of the enclosure material, the typical values are as follows.

Heat release rate	$\dot{Q}_{fo} = 610\sqrt{h_k A_s A_v \sqrt{h_v}}$
Thermal conductivity	k = 0.00017 (kW/m)/K
Enclosure thickness	δ = 0.016 m
Enclosure conductance	h_k = k/δ
	= $0.00017/0.016 = (1.062 \times 10^{-2})[(\text{kW/m}^2)/\text{K}]$
Total surface area	A_s = 317.5 m²
Total vent area	A_v = 2.5 m²
Height of vent	h_v = 2.5 m
Flashover heat release rate	$\dot{Q}_{fo} = 610\sqrt{(0.01062)(317.5)(2.5)(\sqrt{2.5})}$
	= 2226.9 kW = 2.23 MW

IMPORTANCE OF RECOGNIZING FLASHOVER IN ROOM FIRES

Those who examine a fire scene after extinguishment or burnout should be aware of the importance of recognizing whether flashover did actually occur and the impact it may have on the burning of a room's contents. Chapter 3 addresses the interpretation of fire burn patterns, some of which may be created during the ***postflashover*** burning period.

Postflashover burning in a room produces numerous effects that were once thought to be produced only by accelerated (arson) fires involving flammable liquids such as gasoline. Walls scorched or charred from floor to ceiling can result from postflashover burning no matter how the fire actually began, as well as from a fireball of burning gasoline vapors (DeHaan 2002).

The combustion of carpets and underlying pads, once considered by fire investigators to be fairly unusual in accidental fires, is quite common in rooms filled with today's high-heat release materials such as polyurethane foam, and synthetic (thermoplastic) fabrics in upholstery, draperies, and carpets. This problem is exacerbated by the increasing use of highly combustible fibers like polypropylene for the face yarns of economical carpets as well as in the backings of nearly all wall-to-wall carpets. These carpets now melt, shrink, and ignite under radiant heat flux exposing the combustible urethane foam pad underneath to the flames. The involvement of common carpet in fires was the reason that flooring radiant panel tests were developed.

Fire damage under tables and chairs (once thought to be the result of flammable liquids burning at floor level) can be produced by the flaming fire of the carpet as it ignites during flashover. High temperatures and total room involvement, at one time thought to be linked to flammable liquid accelerants, are produced during postflashover burning without accelerants. The extremely high heat fluxes produced in postflashover fires can char wood or burn away other material at rates up to 10 times the rate at lower heat fluxes (Butler 1971).

Localized patterns of smoke and fire damage that can help locate fuel packages, characterize their flame heights, and reveal the direction of flame spread can be obliterated with prolonged postflashover burning, making the reconstruction of the fire very difficult. The fire investigator, then, should be aware of the possibility of flashover in modern room fires and the effects that postflashover burning can have.

In preflashover rooms, the most intense thermal damage will be in areas immediately around (or above) fuel packages. In postflashover fires, all fuels are involved and the most efficient (highest temperature) combustion will be occurring in turbulent mixing around the ventilation openings.

Interviews of first-in fire fighters may reveal if a room was fully involved, floor to ceiling, when entry was first made. Investigators must be aware of the conditions that can lead to flashover and how to recognize when it has or has not occurred. A thorough understanding of heat release rates and fire spread characteristics of modern furniture and the dynamics of flashover will be of great assistance.

◆ 2.7 OTHER ENCLOSURE FIRE EVENTS

As part of forensic fire scene reconstruction and analysis, other features of fire behavior in enclosures also become important to investigators. The commonly cited features are:

- smoke detector activation and
- heat and sprinkler system operation.

These enclosure fire features are can be used to solve several questions that arise during an investigation, such as How long did the fire effectively burn? and When during this time period did the smoke detector or sprinkler head activate? Answers to these questions can be placed on a time line of information needed to fill the voids, particularly when fire deaths occur.

SMOKE DETECTOR ACTIVATION

A common problem faced by fire investigators is determining when on a time line of events the activation of a thermal fire detector, smoke detector, or sprinkler system occurred. These milestone events are usually noticed by a witness or electronically reported by alarm systems and often serve as critical milestones in an investigation.

Figure 2.12 shows the measurements needed to calculate these activations. The dimension H is the distance to the ceiling above the fuel surface and r is the radial distance from the plume centerline to the heat detector or sprinkler head. The centerline is the imaginary vertical line emerging from the center of the fuel source to the ceiling. R is the equivalent radius of the burning fuel package. Virtual origins are discussed in detail in Chapter 3.

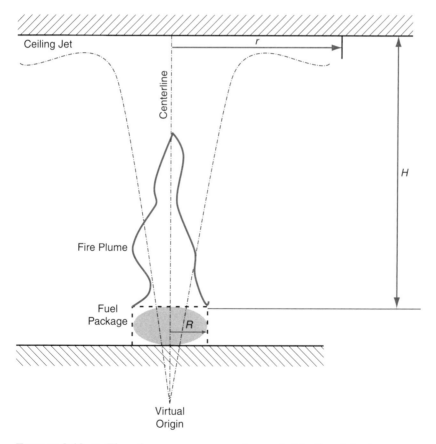

FIGURE 2.12 ◆ Fire plume measurements needed in fire detector, smoke detector, or sprinkler system calculations consisting of fuel surface to ceiling along centerline (H), radius of burning fuel surface (R) and radial distance of ceiling jet from centerline (r).

In the estimation of smoke detector activations for steady-state fires, two calculations are necessary. The first calculation is for the time for the fire gases to reach the ceiling at the plume centerline, or the **plume lag time.** The second calculation is the time to reach the detector from the plume centerline, or the **ceiling jet lag.** These equations are expressed as

$$t_{pl} = C_{pl} \frac{H^{4/3}}{\dot{Q}^{1/3}},$$ (2.31)

$$t_{cj} = C_{cj} \frac{r^{11/6}}{\dot{Q}^{1/3}H^{1/2}},$$ (2.32)

where

t_{pl} = transport lag time of plume (s),
t_{cj} = transport lag time of ceiling jet (s),
C_{pl} = plume lag time constant (experimentally determined),
C_{cj} = ceiling jet lag time constant (experimentally determined),
r = radius of distance from plume centerline (m),
H = height of ceiling above top of fuel (m), and
$\dot{Q}$ = heat release rate (kW).

EXAMPLE 2.12 Smoke Detector Activation by Wastepaper Basket Fire

Problem. A business owner closes up his shop and later claims that he may have caused a fire by carelessly discarding a lit match into a wastepaper basket before leaving. The fire ignited a small 0.3-m (0.98-ft)-diameter basket filled with ordinary papers. The fire developed and reached a steady heat release rate of 100 kW.

The initial temperature of the room was 20°C (293 K). The distance from the top of the wastepaper basket to the ceiling is 5 m. A smoke detector attached to the ceiling is located radially 6 m (19.6 ft) from the fire plume's centerline. Assuming that there are negligent radiation losses and a smooth ceiling, determine the time of the smoke detector's activation by heated ceiling jet gases.

Suggested Solution. Assume that steady-state conditions exist, along with immediate activation.

Height of ceiling above fuel	H = 5 m
Radius from plume centerline	r = 6 m
Plume lag time constant	C_{pl} = 0.60
Ceiling jet lag time constant	C_{cj} = 0.833
Heat release rate	$\dot{Q}$ = 100 kW
Plume transport time lag	t_{pl} = $C_{pl}(H^{4/3}/\dot{Q}^{1/3}$ = (0.60)(5^{4/3}/100^{1/3}) = 1.1 s
Ceiling jet transport lag time	t_{cj} = $C_{cj}(r^{11/6}/\dot{Q}^{1/3}H^{1/2})$
	= (0.833){6^{11/6}/[(100^{1/3})(5^{1/4})]} = 3.2 s
Detector activation time	t = $t_{pl} + t_{cj}$ = 1.1 + 3.2 = 4.3 s

---

SPRINKLER HEAD AND HEAT DETECTOR ACTIVATION

In estimating the time to activate a sprinkler head or heat detector, both the ceiling jet temperature and its velocity must be calculated along with the ratio r/H. The equations for these estimates are based upon data from a series of actual tests for 670 kW to 100 MW fires (Alpert 1972; NFPA 2000a, chap. 11-10).

The first calculation necessary is the centerline temperature directly above the plume produced by the burning fuel source. This equation is good when the ratio r/H is ≤ 0.18.

$$T_m = 16.9 \frac{\dot{Q}^{2/3}}{H^{5/3}} + T_\infty \qquad \text{for} \qquad r/H \leq 0.18. \qquad (2.33)$$

For r/H ratios higher than 0.18, the detector or sprinkler falls within the ceiling jet portion of the plume.

$$T_{mjet} = 5.38 \frac{(\dot{Q}/r)^{2/3}}{H} + T_\infty \qquad \text{for} \qquad r/H > 0.18, \qquad (2.34)$$

where

T_m = plume gas temperature above fire (K),
T_{mjet} = temperature of ceiling jet (K),
T_∞ = ambient room temperature (K),
$\dot{Q}$ = heat release rate from fire (kW),
r = radial distance from plume centerline to device (m), and
H = distance above fuel surface (m).

In order to study detector and sprinkler activation problems further, the maximum velocity of the ceiling jet, U_m, needs to be calculated. The following correlations depend upon the r/H ratios being lower or higher than 0.15.
For maximum velocity of the jet close to the centerline

$$U_m = 0.96 \left(\frac{\dot{Q}}{H}\right)^{1/3} \qquad \text{for} \qquad r/H \leq 0.15. \qquad (2.35)$$

For velocities of the jet farther away from the centerline

$$U_m = 0.195 \left(\frac{\dot{Q}^{1/3} H^{1/2}}{r^{5/6}}\right)^{1/3} \qquad \text{for} \qquad r/H > 0.15, \qquad (2.36)$$

where

$\dot{Q}$ = heat release rate (kW),
H = distance above fuel surface (m),
r = radial distance from plume centerline to device (m), and
U_m = gas velocity (m/s).

Determining the time to activation of the heat detector or sprinkler during steady-state fires relies upon a term called the ***response time index (RTI)***. This index assesses the time and ability for a heat detector to activate from an initial condition of being at ambient room temperature. Because the "detector" of a sprinkler has a finite mass, the *RTI* takes into account the time lag before its temperature rises.

$$t_{operation} = \frac{RTI}{\sqrt{U_m}} \log_e \left(\frac{T_m - T_\infty}{T_m - T_{operation}}\right), \qquad (2.37)$$

where

RTI = response time index ($m^{1/2} s^{1/2}$),
U_m = gas velocity (m/s),

T_m = plume gas temperature above fire (K),

T_∞ = ambient room temperature (K), and

$T_{operation}$ = operation temperature (K).

RTI values are specified by the manufacturer for each style or model of sprinkler.

EXAMPLE 2.13   Sprinkler Head Activation to Wastepaper Basket Fire

Problem. A closer examination of the debris in Example 2.12 shows that the wastebasket actually contains both paper and plastic, producing a steady-state fire of 500 kW. A sprinkler head is located directly above the basket, whose operation temperature is 74°C (347 K) and *RTI* is 200 m$^{1/2}$ s$^{1/2}$. The ambient temperature of the room is 20°C (293 K). Estimate the time to activate the sprinkler head.

Suggested Solution. Since the area of interest is directly above the fire plume, the radius from the centerline is zero, producing a ratio $r/H < 0.18$. Therefore, use the appropriate equations to calculate the temperature and velocity.

Heat release rate	$\dot{Q}$	= 500 kW
Distance above fuel surface	H	= 5 m
Ambient room temperature	T_∞	= 20°C + 273 = 293 K
Response time index	RTI	= 200 m$^{1/2}$ s$^{1/2}$
Temperature of operation	$T_{operation}$	= 74°C + 273 = 347 K
Temperature of ceiling jet	T_m	= $16.9(\dot{Q}^{2/3}/H^{5/3}) + T_\infty$
		= $(16.9)(500^{2/3}/5^{5/3}) + 293 = 365.8$ K
Ceiling jet velocity	U_m	= $0.96\left(\dfrac{\dot{Q}}{H}\right)^{1/3}$
		= $(0.96)(500/5)^{1/3} = 4.45$ m/s
Time to operation	$t_{operation}$	= $(RTI/\sqrt{U_m})\log_e[(T_m - T_\infty)/(T_m - T_{operation})]$
		= $(200/\sqrt{4.45})\log_e[(365.8 - 293)/(365.8 - 347)]$
		= 128 s = 2.1 min

Note that the calculations for ceiling jet behaviors are valid only for smooth, level ceilings. Open joists, ceiling ducts, or pitched surfaces will dramatically affect the movement of gases.

◆ 2.8 SUMMARY AND CONCLUSIONS

The body of fire dynamics knowledge as it applies to fire scene reconstruction and analysis is based upon the combined disciplines of thermodynamics, chemistry, heat transfer, and fluid mechanics. We have seen that the growth and development of fires are influenced by a number of variables such as available fuel load, ventilation, and physical configurations of the room. To estimate accurately a fire's origin, intensity, growth, direction of travel, and duration, investigators must rely upon and understand the principles of fire dynamics.

Fire investigators can also benefit from the application of the wealth of knowledge of fire dynamics contained in textbooks, expert treatises, and authoritatively conducted fire research. Case examples along with discussions on sound fire protection engineering calculations can assist investigators in interpreting fire behavior.

Future chapters will further tie together the application of fire dynamics to aid in identifying fire patterns caused by heat transfer. Also, fire dynamics will play a large part in constructing fire models, evaluating thermal injuries to humans, and evaluating fire testing methods. Case studies will include the use of fire dynamics concepts and calculations when conducting a fire scene reconstruction.

■ ■

Problems

2.1. Recalculate the heat release rate needed for flashover using several methods for a room measuring 5×5 m, with a 3-m ceiling and two 2.5-m-high $\times$ 1-m-wide openings.

2.2. Photograph or review a recent fire scene and estimate the heat release rate of the first article ignited.

2.3. Search for references to fires in warehouses in which the fire protection system detected and extinguished the incipient fire with only one sprinkler head activating. Obtain the floor plan to a warehouse and estimate the time to sprinkler head activation.

■ ■

Suggested Reading

Drysdale, D. D. 1999. *An introduction to fire dynamics,* 2nd ed. Chichester, UK: John Wiley & Sons.

Karlsson, B., and J. G. Quintiere. 1999. *Enclosure fire dynamics.* Boca Raton, Fla.: CRC Press.

Quintiere, J. G. 1998. *Principles of fire behavior,* chaps. 3–9. Albany, N.Y.: Delmar.

CHAPTER 3

Fire Pattern Analysis

There is no branch of detective science which is so important and so much neglected as the art of tracing footsteps.

—Sir Arthur Conan Doyle,
"Study in Scarlet"

The ability to document and interpret fire patterns accurately is essential to investigators reconstructing fire scenes, and they are often the only remaining visible evidence left after a fire is extinguished. This chapter describes how these fire patterns are analyzed and used by investigators in assessing fire damage and determining a fire's origin.

Heat transfer and flame spread are the major causes of change to the exposed surface and appearance of materials during a fire. Fire patterns are formed by the thermal intersection of fire plumes on exposed solid surfaces such as floors, ceilings, and walls. These burn patterns are influenced by a number of variables including the available fuel load, ventilation, and the physical configurations of the room. Many common combustible materials and ignitable liquids can produce these fire plumes and their resulting damage.

Case examples are offered here along with discussions on validated fire science and engineering calculations to assist investigators in interpreting fire patterns. Also addressed in this and subsequent chapters are various documentation techniques that can be helpful in fire scene reconstruction using the concepts of fire pattern analysis.

◆ 3.1 FIRE PLUMES

The single most important factor in fire scene reconstruction is the *fire plume,* which is usually a simple flaming fuel source emitting a vertical column of flames and hot products of combustion (Drysdale 1999). Fire plumes can originate from any fuel combination that has sufficient heat release rates to generate an upward column of flames and smoke. Fire plumes result from any significant fire. They can be produced by floor-level pools of ignitable liquids; however, many combustible solids, including certain types of foam mattresses and plastics, melt and collapse while burning and behave like liquid fuels.

The shape of burning pools producing fire plumes depends upon several variables including the geometry of the containment of the pool, the type of substrate on which the pool is resting, and, in some cases, the external winds. Since fire plumes are three-dimensional, their location can often be determined and documented by evaluating the patterns of heat transfer, flame spread, and smoke damage they cause to adjacent flooring, wall, and ceiling surfaces as illustrated in Figure 1.5. A fire investigator can gain additional insight through a more fundamental understanding of the nature, physics, and heat transfer characteristics of fire plumes.

"V" PATTERNS

The impingement, intensity, and direction of the fire plume's travel form lines or areas of demarcation on walls, ceilings, floors, and other materials. As the hot gases and smoke rise from a fire, they mix with surrounding air, with the mixing zone becoming wider as hot gases rise above the fuel. Entrainment mixes and spreads the rising column so it forms a "V" approximately 30° in width if unconfined. Therefore the width of the total unconfined plume is approximately one-half its height above the fuel surface. As the fire gases mix, they are diluted and cooled. The diameter of the hottest part of the plume along the centerline becomes smaller with increasing height.

Analyzing the shape of fire patterns caused by plumes can provide valuable information. For example, the most ceiling damage is often in the plume impingement area directly above the fuel source (as in Figure 1.5), a point from which a gas-movement vector points directly back to the fire's origin.

In simple two-dimensional views where the fire plume comes into contact with and damage wall surfaces, damage patterns forming lines of demarcation frequently appear. These lines are often referred to as ***V-shaped fire patterns***, based on their characteristic upward-sweeping curves.

NFPA 921 (NFPA 2001) and *Kirk's Fire Investigation* (DeHaan 2002) have dispelled past misconceptions regarding the shape and geometry of V patterns, once thought to relate to the rapidity of fire growth. The shape of a V pattern is actually related to the heat release rate, geometry of the fuel, ventilation effects, ignitability and combustibility of nearby surfaces, and intersection with horizontal surfaces (NFPA 2001, sect. 4.17.1).

If the V-shaped fire pattern's angle of demarcation is followed downward to its base, the area closest to the virtual origin of the plume can be located. Several mathematical relationships exist that provide assistance in documenting and analyzing the features of fire plume height, temperature, velocity, vortex shedding frequency, and virtual origin (Heskestad 1988; Quintiere 1998).

FIRE PLUME DAMAGE CORRELATIONS

Testing by the Factory Mutual Research Corporation (FMRC) reveals greater scientific insight into the formation of V patterns. In FMRC's testing, a closer examination of a fire plume's lines of demarcation reveals distinct areas of damage to the wall surfaces, also known as the ***fire propagation boundary***. This boundary is a visually distinguishable line where the heavy pyrolysis of the surface ends.

Actual fire testing by FMRC documents the close correlation of the fire propagation boundary with the ***critical heat flux boundary***. This is the location of the boundary where the minimum heat flux is at or below the point where a flammable vapor–air mixture is produced by pyrolysis at the surface of the solid (Tewarson 1995). Critical heat fluxes have also been experimentally determined by successive exposures of material samples to progressively decreasing incident heat fluxes until ignition no longer takes place (Spearpoint and Quintiere 2001).

Figure 3.1 illustrates this close correlation in FMRC's 25-ft corner tests of a growing fire peaking at a 3-MW heat release rate (Newman 1993). This test is useful for evaluating fires involving low-density, high-char forming wall and ceiling insulation

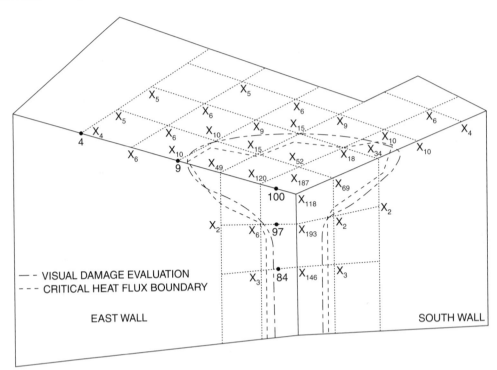

FIGURE 3.1 ◆ The correlation between critical heat flux boundary (dashed line) and visual lines of demarcation (dashed–dotted line) in an FMRC 25-ft corner fire test provides a scientific explanation for the formation of "V" patterns on walls. *From* The SFPE Handbook of Fire Protecting Engineering, *3rd ed., 2002, by permission of SFPE.*

materials. In Figure 3.1 the critical heat flux boundary as measured by radiometers attached to the surface is denoted as the dashed line and the visual damage evaluation as the dashed–dotted line. Note how closely these observations correlate. When evaluating the various types of fire patterns, the position and impact of the fire plume on the walls, ceilings, and surface areas are important to interpret these correlations correctly.

At first, the soot and pyrolysis products of the smoke condense on the cooler surfaces, with no chemical or thermal effect. As hot gases from a plume come into contact with the surface, heat is transferred by convective and radiative processes. As the heat is transferred, the temperature of the surface increases. The heat may be conducted into the surface. At some point, the temperature increases to the point where the surface coating begins to scorch, melt, or char (pyrolyze). At higher temperatures it may actually ignite. These temperatures are reached when the critical heat flux is reached for that material and sustained for a long enough time.

◆ 3.2 TYPES OF FIRE PLUMES

When a fire scene is examined, the only evidence typically remaining is the thermal impact of the fire burn pattern indicators. In cases where a fire was extinguished quickly by fire suppression or has self-extinguished, clearly defined burn patterns from the plume are left on the walls, floors, ceilings, and exterior finishes of the structure. Many of these patterns are formed by the intersection of fire plumes with the structure and targets that the plume comes into contact.

There are four major types of plumes of interest in fire investigation and reconstruction (Mowrer and Milke 2001):

- axisymmetric
- window
- balcony
- line

Understanding these plumes provides a keener insight into the dynamics of more complex fires. Examining the damage of these plumes must assume that the investigator has a working knowledge of their origin, type of fuel package, and ventilation. Furthermore, empirical testing has established mathematical relationships describing the behavior, shape, and impact of these classes of fire plumes.

AXISYMMETRIC PLUMES

Axisymmetric plumes have uniform radial distribution from a common vertical plume centerline. Generally, they occur in open areas or near centers of rooms where there is no nearby wall. Much research has been conducted on the correlations of the plume's height, temperatures generated, gas velocities, and entrainment. Figure 3.2 shows the

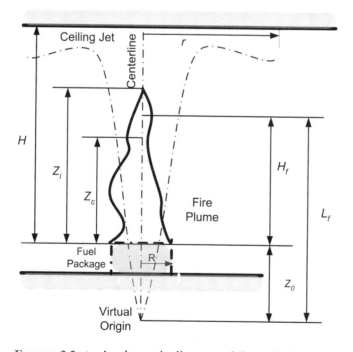

FIGURE 3.2 ◆ A schematic diagram of the typical axisymmetric fire plume and its interaction with a ceiling, where Z_0 is the distance from the fuel surface to the virtual origin, H_f is the mean flame height above the fuel surface, Z_i is the continuous flame height above the fuel surface Z_c is the intermittant flame height above the fuel surface, H is the distance to the ceiling from the fuel surface, R is the radius of the fuel package, and r is the distance of the ceiling jet from the plume centerline. Note that the statistical mean flame height is the portion of luminous flame that is visible during 50 percent of its total exposure time.

FIGURE 3.3 ◆ An axisymmetric fire plume under the NIST furniture calorimetry hood during research on flammable and combustible liquid spill and burn patterns. *Courtesy of Putorti [2001, 15].*

typical descriptions and measurements used for flame and plume characteristics describing the plume centerline, mean flame height, and virtual origin (NFPA 1997, chap. 11-10). These measurements are used in calculations that show relationships of the fire plume with heat release rates, flame height, temperatures, and smoke production.

Figure 3.3 is an example of an axisymmetric fire plume under the NIST calorimetry hood during research on flammable and combustible liquid spill and burn patterns (Putorti 2001, 15). These tests provide valuable data on heat release rates as well as the phenomenon of burn patterns on various flooring surfaces.

WINDOW PLUMES

The class of plumes that emerge from doors and windows into large open spaces is known as *window plumes* (NFPA 2000b, sect. 3.8.3). These plumes emerge from openings during the fire and are commonly ventilation controlled. In Figure 3.4, the plume emerges from the opening and measures Z_w above the soffit.

Estimating the heat release rate may be worthwhile when considering the observations of on-scene eyewitnesses and responding firefighters. For example, the description of the window plume may be sufficient to estimate the heat release rate based upon the observed area and height of the ventilation opening.

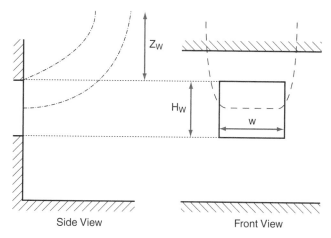

FIGURE 3.4 ◆ A schematic diagram of a window plume showing side and front orientations.

BALCONY PLUMES

The next class of plumes, which emerge under overhangs and from doors, is known as ***balcony spill plumes*** (NFPA 2000b, sect. 3.8.2). These plumes (Figure 3.5) are characteristic of fires occurring in enclosed rooms and emanating through patio doors or windows.

The width of the plume is estimated by retracing the horizontal and vertical measurements such that

$$W = w + b, \tag{3.1}$$

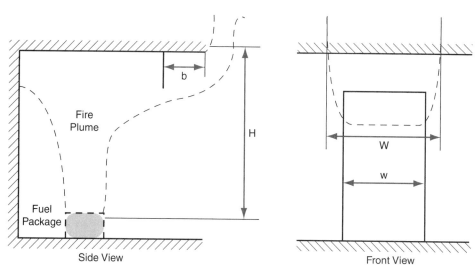

FIGURE 3.5 ◆ A schematic diagram of a balcony spill plume showing side and front orientations.

where

W = maximum width of the plume (m),

w = width of the opening from the area of origin to the balcony (m), and

b = distance from the opening to the balcony edge (m).

This calculation has been used in estimating the width and height of the fire plume to evaluate the properties of the fire plume and evaluating smoke production.

Figure 3.5 diagrams this class of plumes. Examples of balcony plumes include exterior doors leading to garden style apartments, internal hotel rooms in atrium style buildings, and multiple-level prison cells that face catwalks.

The characteristic of a balcony plume is that flames are deflected from the plume away from the upper floor as a result of the construction of the external overhang, as would an open porch or deck in an apartment or condominium. This is an example of the use of passive fire engineering approaches in building construction that reduces the chances of external plumes spreading fires to upper floors above the floor of fire origin.

Figure 3.6 shows an example of a window plume that comes into exterior contact with a porch overhang, creating a situation similar to a balcony spill. As the plume exits the window, it travels along the ceiling of the porch overhang and flames are projected out from the side of the overhang. This is common in residential fires, particularly when a porch or deck exists or has been added to the dwelling.

FIGURE 3.6 ◆ Window plume that comes into exterior contact with a porch overhang, creating a situation similar to a balcony spill. *Courtesy of D. J. Icove.*

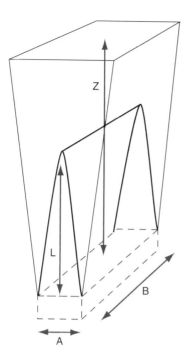

FIGURE 3.7 ◆ Illustration of a line source plume.

LINE PLUMES

Line plumes (Figure 3.7) have elongated geometric shapes that produce narrow, thin, and shallow plumes. The relationship of the heat release rate to the flame height is described in a later section. For plume heights (L) greater than five times the B measurement, the line plume more closely resembles an axisymmetric plume (Hasemi and Nishihata 1989).

Possible scenarios for line plumes include exterior fires that occur in ditches where spilled ignitable liquids collect and ignite. Line plumes can also be used to approximate elongated fires in open areas of atriums or warehouses.

◆ 3.3 FIRE PLUME CALCULATIONS

As indicated in Chapter 2, investigators can benefit from information on fire plumes, which are merely the physical manifestations of the combustion process. Sound scientific and engineering principles, based upon a mixture of both theoretical and actual fire testing data, describe fire plume behavior.

Information on fire plume behavior is expressed in terms of five basic calculated measurements often used in hazard and risk analysis (SFPE 2002, 387). They model the following fire plume characteristics.

- equivalent fire diameter
- virtual origin
- flame height
- plume centerline temperatures and velocities
- plume air entrainment

These basic characteristics constitute the bare minimum necessary information for fire reconstruction. Several common equations for performing a fire engineering analysis of fire plumes rely upon the energy or heat release rate and provide answers to the questions: How tall was the fire plume? What was the placement of the virtual source in relation to the floor?

The ability of a fire investigator to use these calculations is important in the application of basic fire science and engineering concepts. These calculations are essential in evaluating the various working hypotheses, such as how much fuel was burned, the fire's time duration, and the impact of ventilation of the room of origin.

Many of these calculations are found in the various fire modeling software tools, such as FPETool and CFAST. The calculations are briefly explained here to demonstrate their underlying concepts and potential applications.

EQUIVALENT FIRE DIAMETER

When calculating the flame height for fire plumes, the base of the fire is assumed to be circular. This is normally not the case when ignitable liquids are spilled onto a floor or when the fuel package is rectangular in shape. In these cases, calculation of the **equivalent diameter** is needed.

$$D = \sqrt{\frac{4A}{\pi}}, \tag{3.2}$$

where

D = equivalent diameter (m),

A = total area of burning fuel package (m^2), and

π = 3.1416.

Areas of irregular pools can be calculated by re-creating the shape and size of pools from fire scene photos and dimensions from fire scene notes using wrapping-paper cutouts. The cutouts can then be weighed using a postal scale and the area vs. weight determined from a reference paper sample of known area.

Rectangular-contained spills include those conditions where an oil-filled electric transformer is compromised and spills its contents into a rectangular diked area and ignites. The calculation would be needed to equate the rectangular spill with the diameter of a circle containing the same area (Figure 3.8).

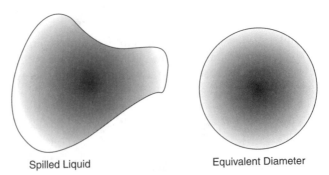

Spilled Liquid   Equivalent Diameter

FIGURE 3.8 ◆ The equivalent diameter concept used in estimating the diameter of a circle that represents the same area as a noncircular spill pattern.

The aspect ratio is formed by comparing the length-to-width ratio of the spill area. This equivalent diameter relationship generally holds true for pools having a width-by-length aspect ratio ≤2.5. Aspect ratios >2.5, such as would be seen with trench or line fires, should use other models (Mudan and Croce 1995).

In cases of short-duration fires, the area of a liquid fuel spill can be measured directly from the burned areas of the floor or covering. In postflashover or prolonged fires where the floor covering is readily combustible, the fire-damaged areas may extend well beyond the margins of the original pool and are therefore not a reliable guideline. DeHaan (2002) has shown a simple relationship among surface area, volume of liquid, and type of surface, which controls the equivalent depth of the pool. Putorti (2001) has also confirmed this general relationship.

For spills on nonporous surfaces such as sealed concrete, the depth of a free-standing pool of gasoline is about 1 mm (1×10^{-3} m). The area (m^2) is then calculated by dividing the volume of the spilled liquid (m^3) by the depth (m), where

$$\text{Area (m}^2\text{)} = \frac{\text{Volume (m}^3\text{)}}{\text{Depth (m)}}. \tag{3.3}$$

For the purposes of conversion, 1 liter $= 10^{-3}$ m^3.

On semiporous surfaces such as wood, the equivalent depth is 2–3 mm (2–3×10^{-3} m). On porous surfaces such as carpet, the maximum depth will be the thickness of the carpet, assuming full saturation of the carpet.

EXAMPLE 3.1   Equivalent Fire Diameter

Problem. A container of gasoline is accidentally tipped over onto an enclosed patio. The pool of spilled flammable liquid measures 3 m in length and covers a total area of 4 m^2. What is the equivalent diameter of a circle of the same area?

Suggested Solution. The equivalent diameter D is as follows.

Equivalent diameter	$D = \sqrt{4A/\pi}$
Total surface area	$A = 4\,\text{m}^2$
Constant	$\pi = 3.1416$
Equivalent diameter	$D = \sqrt{(4)(4)/3.1416} = 2.25\,\text{m}$

VIRTUAL ORIGIN

As shown in Figures 2.7, 2.12, and 3.2, the ***virtual origin*** is a location along the centerline of the fire plume that would accurately represent the actual fire as a point source. This assumption holds true only in cases where the ratio of the distance to the target, R_0, to the distance of the fire's radius from the fuel's surface at the point of origin, R, is greater than 4.

The virtual origin, Z_0, is calculated from the ***Heskestad equation*** (Heskestad 1982):

$$Z_0 = 0.083\,\dot{Q}^{2/5} - 1.02D, \tag{3.4}$$

where

D = equivalent diameter (m),

$\dot{Q}$ = total heat release rate (kW), and

Z_0 = virtual origin (m)

for the conditions

$$T_{amb} = 293 \text{ K},$$
$$P_{atm} = 101.325 \text{ kPa, and}$$
$$D \quad \leq 100 \text{ m}.$$

The virtual origin, depending upon the equivalent diameter and the total heat release rate, may fall above or below the fuel's surface. This placement depends primarily upon the heat release rate and effective diameter of the fuel. The resulting value can then be used in equations and models of other fire plume relationships.

Ignitable liquid spills of small diameters are more likely to exhibit virtual origins above the fuel surfaces. By inspection of equation (3.4), the smaller the value of the fire's equivalent diameter, D, and/or the larger the value of the heat release rate, $\dot{Q}$, the greater the chance for a positive virtual origin, Z_0, raising it above the fire surface. Since not all fires burn at floor level, the concept of Z_0 is useful in estimating where the effective origin of the fire plume is in relation to the fuel surface and the compartment itself.

EXAMPLE 3.2   Virtual Origin

Problem. A fire starts in a wastepaper basket filled with papers and quickly reaches a steady heat release rate of 100 kW. The basket measures 0.305 m (1 ft) in diameter. Determine the virtual origin of this plume using the Heskestad equation.

Suggested Solution. Use the equation for the virtual origin.

Virtual origin	$Z_0 = 0.083 \dot{Q}^{2/5} - 1.02D$
Equivalent diameter	$D = 0.305 \text{ m}$
Total heat release rate	$\dot{Q} = 100 \text{ kW}$
Virtual origin	$Z_0 = (0.083)(100)^{2/5} - (1.02)(0.305)$
	$= 0.524 - 0.311 = 0.213 \text{ m } (0.698 \text{ ft})$

Note that in this case the virtual origin is a positive number, indicating that it is above the surface of the fuel. If the same fuel is spread out over the floor such that its equivalent diameter is 1.0 m, while its $\dot{Q}$ remains 100 kW, its $Z_0 = -0.5$ m (below the floor). This means that the impact of the fire on the room will be as if the fire were burning farther away from the ceiling.

Discussion. If the trash can was filled with gasoline and $\dot{Q} = 500$ kW, recalculate the virtual origin. What can you assume about the virtual origin when the heat release rate is raised?

FLAME HEIGHT

Flames from plumes represent the visible portion of a fire's combustion process. The ability to see the visible portion of the combustion is based upon luminosity of heated soot particles and pyrolysis products. The buoyancy of these fire gases allows a fire plume to rise vertically. The height of the flames is described by several terms, including continuous, intermittent, and average.

Knowing the *flame height* allows the investigator to examine and confirm what potential damage could be expected based upon heat transfer to the ceiling, walls, flooring, and nearby objects. Several equations estimate the flame height, which are based upon fitting regression formulas to actual fire testing. As with all of these equations, users are cautioned that the equations are bounded by the limitations of the testing.

Within the flaming region, the two ***McCaffrey*** (1979) flame height measurements along the centerline are depicted as continuous, Z_c, and intermittent, Z_i:

$$Z_c = 0.08 \, \dot{Q}^{2/5}, \tag{3.5}$$

$$Z_i = 0.20 \, \dot{Q}^{2/5}, \tag{3.6}$$

where

Z_c = continuous flame height (m),

Z_i = intermittent flame height (m), and

$\dot{Q}$ = heat release rate (kW).

Crude estimates of heat release rates made by witnesses and responding firefighters can be drawn by algebraically rearranging the McCaffrey flame height equation for intermittent flame heights:

$$\dot{Q} = 56.0 \, Z_i^{5/2}. \tag{3.7}$$

EXAMPLE 3.3   Flame Height—McCaffrey Method

Problem. Using Example 3.2 involving the 100-kW wastepaper basket fire, determine the continuous and intermittent flame heights using the McCaffrey method.

Suggested Solution. Use the equation for the McCaffrey method.

Heat release rate	$\dot{Q}$ = 100 kW
Continuous flame height	$Z_c = 0.08 \, \dot{Q}^{2/5} = (0.08)(100)^{2/5}$
	= 0.505 m (1.66 ft)
Intermittent flame height	$Z_i = 0.20 \, \dot{Q}^{2/5} = (0.20)(100)^{2/5}$
	= 1.26 m (4.1 ft)

EXAMPLE 3.4   Heat Release Rate from Flame Height—McCaffrey Method

Problem. A witness to an accident of an overturned lawn mower sees the gasoline ignite and burn, forming an axisymmetric fire plume. He observes intermittent flames reaching 3 m (9.84 ft) into the air from the ground. Estimate the heat release rate, $\dot{Q}$, given off by the burning vehicle.

Suggested Solution. Use the equation for the McCaffrey method.

Heat release rate	$Z_i = 0.20 \, \dot{Q}^{5/2}$
Intermittent flame height	$Z_i = 3.0$ m
Estimated heat release rate	$\dot{Q} = 56.0 \, Z_i^{5/2}$
	$= (56.0)(3.0)^{5/2} = 873$ kW

EXAMPLE 3.5   Calculating the Area of a Pool Fire

Problem. From the data in Table 2.4, calculate the area of the pool fire at 873 kW from Example 3.4.

Suggested Solution. Use the equation for the heat release rate.

General equation	$\dot{Q} = \dot{m}'' \, \Delta h_c \, A$
Heat release rate	$\dot{Q} = 873$ kW

Mass burning rate per unit area	$\dot{m}'' = 53 \text{ g/m}^2\text{s}$
Heat of combustion	$\Delta h_{\text{eff}} = 43.7 \text{ kJ/g}$
Area of pool fire	$A = \dot{Q}/\dot{m}'' \Delta h_c = 873/(53)(43.7) = 0.38 \text{ m}^2$

One assessment of the flame height is its statistical ***mean flame height***, which is the portion of luminous flame that is visible during 50 percent of its total exposure time. When the human eye observes a pulsating flame, it tends to interpolate the variations into a rough approximation of the mean flame height.

A good estimate of the mean or average flame height, L_f, uses the Heskestad (1982) equation:

$$L_f = 0.235 \dot{Q}^{2/5} - 1.02D,$$ (3.8)

where

D = effective diameter (m),

$\dot{Q}$ = total heat release rate (kW), and

$L_f = Z + Z_0$ = mean flame height above the virtual origin (m).

EXAMPLE 3.6 Flame Height—Heskestad Method

Problem. From Example 3.2 involving the 100-kW wastepaper basket fire, determine the mean flame height, L_f, above the virtual origin, Z_0.

Suggested Solution. Use equation (3.8) for calculating the mean or average flame height.

Mean flame height	$L_f = 0.235 \dot{Q}^{0.4} - 1.02 D$
Diameter of basket	$D = 0.305 \text{ m}$
Total heat release rate	$\dot{Q} = 100 \text{ kW}$
Mean flame height	$L_f = 0.235 \dot{Q}^{0.4} - 1.02 D$
	$= (0.235)(100)^{0.4} - (1.0)(0.305)$
	$= 1.17 \text{ m} (3.84 \text{ ft})$ above the virtual origin

---

Because a nearby wall or corner reduces the cooling entrainment of room air into the rising plume, the height of the visible flame can be affected. *NFPA 921* lists relationships that are very useful to investigators and have been subjected to numerous studies and tests for determining the mean flame height near walls (NFPA 2001, sect. 3.5.5). These relationships are stated in the following formulas:

$$H_f = 0.174 (k\dot{Q})^{2/5},$$ (3.9)

$$\dot{Q} = \frac{79.18 H_f^{5/2}}{k},$$ (3.10)

where

H_f = flame height (m),

$\dot{Q}$ = heat release rate (kW), and

k = wall effect factor, with

k = 1: no nearby walls,

k = 2: fuel package at wall, and

k = 4: fuel package in corner.

EXAMPLE 3.7   Flame Height—*NFPA 921* Method

Problem. Using Example 3.2 involving the 100-kW wastepaper basket fire, estimate the height of the flame in three configurations—in the middle of the room, near a wall, and near a corner. Assume that this is an unconstrained fire burning in a large room with no ventilation constraints.

Suggested Solution. Using equation (3.9), the flame height H_f measured from the top of the fuel surface is as follows.

Total heat release rate	$\dot{Q} = 100\,\text{kW}$
Flame height	$H_f = 0.174\,(k\dot{Q})^{2/5}$
No nearby walls ($k = 1$)	$H_f = (0.174)(1 \times 100)^{2/5}$
	$= (0.174)(100)^{0.4} = 1.10\,\text{m (3.61 ft)}$
Near a wall ($k = 2$)	$H_f = (0.174)(2 \times 100)^{2/5}$
	$= 1.45\,\text{m (4.76 ft)}$
In a corner ($k = 4$)	$H_f = (0.174)(4 \times 100)^{2/5}$
	$= 1.91\,\text{m (6.27 ft)}$

PLUME CENTERLINE TEMPERATURES AND VELOCITIES

The temperature distribution along the centerline of a typical plume is shown in Figure 3.9a, and the temperature distribution along a horizontal cross section through the flames is shown in Figure 3.9b. Estimates of the centerline maximum temperature rise, velocity, and mass flow rate of a plume are useful in determining the impact of the fire plume in compartment fires. This impact includes the maximum effect of the plume as it is rising to the ceiling and may come into contact with other fuel packages. Several formulas exist to estimate these values.

The **Heskestad** method used for estimating the centerline maximum temperature rise, velocity, and mass flow rate of a plume is

$$T_0 - T_\infty = 25\left(\frac{\dot{Q}_c^{2/5}}{(Z - Z_0)}\right)^{5/3}, \tag{3.11}$$

$$U_0 = 1.0\left(\frac{\dot{Q}_c}{(Z - Z_0)}\right)^{1/3}, \tag{3.12}$$

$$\dot{m} = 0.0056\,\dot{Q}_c\frac{Z}{L_f} \quad \text{for} \quad Z < L_f, \tag{3.13}$$

$$\dot{Q}_c = 0.6\,\dot{Q} \quad \text{to} \quad \dot{Q}_c = 0.8\,\dot{Q}. \tag{3.14}$$

The **McCaffrey** (1979) method for estimating the centerline maximum temperature rise, velocity, and mass flow rate of a plume is

$$T_0 - T_\infty = 21.6\,\dot{Q}^{2/3}\,Z^{-5/2}, \tag{3.15}$$

$$U_0 = 1.17\,\dot{Q}^{1/3}\,Z^{-1/3}, \tag{3.16}$$

$$\dot{m}_p = 0.076\,\dot{Q}^{0.24}\,Z^{1.895}. \tag{3.17}$$

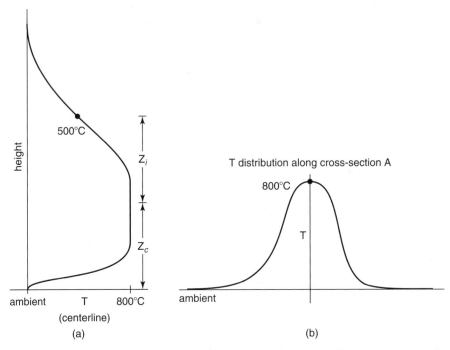

FIGURE 3.9 ◆ (a) Plot of average flame temperature vs. height shows the uniform maximum temperature in the center of the continuous flame zone and the decreasing average flame temperature in the intermittent region. (b) Plot of average flame temperature vs. horizontal position shows the maximum temperature at the axis of the flame plume.

An alternate method for estimating the maximum temperature rise and velocity of a plume is (Alpert 1972)

$$T_0 - T_\infty = \frac{16.9\,\dot{Q}^{2/3}}{H^{5/3}} \qquad \text{for} \qquad \frac{r}{H} \leq 0.18, \qquad (3.18)$$

$$T_0 - T_\infty = \frac{5.38\,(\dot{Q}/r)^{2/3}}{H^{5/3}} \qquad \text{for} \qquad \frac{r}{H} > 0.18, \qquad (3.19)$$

$$U = 0.96\frac{\dot{Q}^{1/3}}{H} \qquad \text{for} \qquad \frac{r}{H} \leq 0.15, \qquad (3.20)$$

$$U = \frac{0.195\,Q^{1/3}\,H^{1/2}}{r^{5/6}} \qquad \text{for} \qquad \frac{r}{H} > 0.15, \qquad (3.21)$$

where

T_0 = maximum ceiling temperature (°C),
T_∞ = ambient air temperature (°C),
$\dot{Q}$ = total heat release rate (kW),
$\dot{Q}_c$ = convective heat release rate (kW),
Z = flame height along the centerline (m),

Z_0 = measurement of the virtual origin (m),

L_f = $Z + Z_0$,

$\dot{m}$ = total plume mass flow rate (kg/s),

r = radial distance from the plume centerline (m),

H = ceiling height (m),

U = maximum ceiling jet velocity (m/s), and

U_0 = centerline plume velocity (m/s).

EXAMPLE 3.8   Plume Temperature, Velocity, and Mass Flow Rate at Ceiling Level—McCaffrey Method

Problem. A fire is discovered in a wastepaper basket in a room that has a ceiling height of 2.44 m (8 ft). The heat release rate estimate is 100 kW. Assume that the ambient air temperature is 20°C (68°F). Estimate the plume temperature, velocity, and mass flow rate at the ceiling using the McCaffrey method.

Suggested Solution. Use the McCaffrey equation.

Maximum plume temperature
$$T_0 = 21.6\ \dot{Q}^{2/3}\ Z^{-5/2} + T_\infty$$
$$= (21.6)(100)^{2/3}(2.44)^{-5/2} + 20 = 68.5°C$$

Centerline plume velocity
$$U_0 = 1.17\ \dot{Q}^{1/3}\ Z^{-1/3}$$
$$= (1.17)(100)^{1/3}(2.44)^{-1/3} = 4.033\ m/s$$

Plume mass flow rate
$$\dot{m}_p = 0.076\ \dot{Q}^{0.24}\ Z^{1.895}$$
$$= (0.076)(100)^{0.24}(2.44)^{1.895} = 1.244\ kg/s$$

ENTRAINMENT EFFECTS

When a fire is burning in the center of a room, the buoyant upward flow of the fire gases drags fresh room air into contact with it. This is the process of **entrainment**. Some of the heat is transferred to the air and is removed by convective processes, and there is usually turbulent mixing that dilutes the rising column of hot gases and cools it further. Looking down on the fire plume from above as in Figure 3.10a, we can see that the inward air currents are roughly equal around the entire perimeter so the plume remains symmetrical and upright.

If the fire is against a **wall**, it cannot draw room air into itself equally from all sides, so the air being drawn from the "free" side acts like a draft to force the plume sideways against the wall (shown in Figures 3.10b and c). The reduced entrainment means that less cooling air is brought in, slowing the cooling process. This means that the gases have more time to rise farther, stretching out the flame. This can occur on nearby wall surfaces even if they are not vertical. This is the mechanism for external flame spread between floors of modern high-rise buildings. When the windows of the room first involved fail, the plume exits the building and then entrains itself against the side of the building (as in Figure 3.10d). The windows and spandrel panels of the floors above are then exposed to the radiant and convective heat transfer ($>50\ kW/m^2$) of the plume and quickly fail. This allows the flames to enter the floor above and ignite the contents of that floor.

If there is not sufficient fire protection to suppress this newly ignited fire, then fire can leapfrog up the building. In older buildings where there is a ledge at each floor

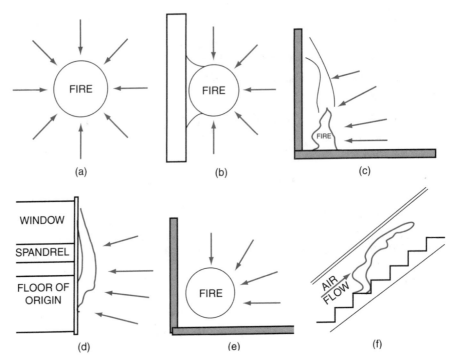

FIGURE 3.10 ◆ (a) Entrained air flow for an axisymmetric plume, from above. (b) Entrainment in wall configuration. (c) Side view. (d) Typical high-rise fire plume. (e) Entrainment in corner fire. (f) Trench effect, side view.

level, the plume is directed away from the side of the building so it never gets a chance to entrain. The chances of ignition by window failure of the floors above are much reduced by a ledge flame guard effect over sheer window-wall styles.

If the fire is built in a **corner**, the perimeter through which cooling air can be drawn is reduced to one-fourth its original size (note that there is still plenty of air reaching the combustion zone so the fire continues to burn at the same rate) (represented in Figure 3.10e). In a corner, then, the gases cool even more slowly and have even more time to rise before cooling to the point where they are no longer incandescent (500°C) and the flame becomes even taller. This effect is enhanced by the increased radiant heat reaching the fuel surface after being reflected off the nearby wall or corner surfaces, thus increasing the heat release rate.

A special case of this is called the **trench effect**, where a fire starts on the floor of an inclined surface with walls on both sides, such as a stairway or escalator (shown in Figure 3.10f). In this case, the flow of the air drawn into the fire from below is aided by the confinement of the buoyant flow. When the fire reaches a critical size (i.e., a critical airflow) the plume lies down in the trench and is stretched out by the directional entrainment. If the floor and/or walls of the trench are combustible, they ignite and the fire can spread the length of the stairway very quickly. This was the mechanism that drove a small fire on the floor of an all-wood escalator in the Kings Cross Underground Station in London in 1987 to develop quickly into a massive fireball that engulfed the large ticket hall at the top of the escalator (Moodie and Jagger 1991).

WINDOW PLUMES

The area and height of the ventilation openings of **window plumes** can be used to estimate the maximum heat release rate of the fire in a compartment of which ventilation is provided by that opening. A mathematical relationship based upon actual experimental data for wood and polyurethane can be used to predict the maximum heat release rate, assuming that the single opening is used for ventilation for the fire (Orloff et al. 1977; Tewarson 1988).

$$\dot{Q}_{max} = 1260\, A_w H_w^{1/2}, \tag{3.22}$$

where

$\dot{Q}_{max}$ = maximum heat release rate (kW),
A_w = area of ventilation opening (m²), and
H_w = height of ventilation opening (m).

This simple relationship for estimating the heat release rate may be worthwhile when considering the observations of on-scene eyewitnesses and responding firefighters. For example, the description of the window plume may be sufficient to estimate the heat release rate based upon the observed area and height of the ventilation opening.

LINE PLUMES

The relationship of the heat release rate to the flame height for **line plumes** in Figure 3.7, where $B > 3A$, is described below. For plume heights greater than five times the B measurement in Figure 3.7, the line plume more closely resembles an axisymmetric plume (Hasemi and Nishihata 1989).

$$L = 0.035 \left(\frac{\dot{Q}}{B} \right)^{2/3}, \quad \text{where} \quad B > 3A, \tag{3.23}$$

$$\dot{m} = 0.21z \left(\frac{\dot{Q}}{B} \right)^{2/3}, \quad \text{where} \quad B < 3A \quad \text{and} \quad L < z < 5B, \tag{3.24}$$

where

L = flame height (m),
$\dot{Q}$ = heat release rate (kW),
$\dot{m}$ = mass flow rate (kg/s),
A = shorter side of line plume base (m),
B = longer side of line plume base (m), and
z = ceiling height above fuel (m).

Possible scenarios for line plumes include exterior fires that occur with long trailers of gasoline or in ditches where spilled ignitable liquids collect and ignite. Line plumes can also be used to approximate elongated fires on sofas or in open areas of atriums or warehouses.

SMOKE PRODUCTION

The **smoke production rate** is estimated to be close to that of the mass flow rate of gas produced by a fire plume. This assumption is based upon the theory that an equal amount of air entrained into the rising plume equals the amount of smoke-filled gas (NFPA 2000a, chap. 11-10).

The best correlations to this theory come in cases involving circular equivalent pool fires and the enclosure is vented in the lower layer. The relationship for the descent of the upper layer is expressed as

$$U_t = \frac{\dot{m}_s}{\rho_l A_p},\tag{3.25}$$

where

U_t = rate of layer descent (m/s),
$\dot{m}_s$ = mass rate of smoke production (kg/s),
ρ_l = density of smoke layer (kg/m^3), and
A_p = enclosure floor area (m^3).

The enclosure smoke filling rate is important when estimating the smoke that accumulates below the ceiling. Factors contributing to how fast the smoke stratifies and descends from the ceiling primarily relate to the amount of smoke produced and the size and location of smoke vents. Fire investigators finding a room with smoke stratification deposits evenly distributed in an enclosed room may find this important when later determining the fire duration or the exposure to smoke a victim might have encountered at a particular time.

Using the Zukoski (1978) method for smoke production rates, the equation for estimating the mass flow rate of a plume above the flame at 20°C (68°F) is

$$\dot{m}_p = 0.065\, \dot{Q}^{1/3} Y^{5/3},\tag{3.26}$$

where

$\dot{m}_p$ = rate of smoke-filled gas production (kg/s),
$\dot{Q}$ = total heat release rate (kW), and
Y = distance from virtual point source to bottom of smoke layer (m).

EXAMPLE 3.9 Smoke Filling

Problem. The 100-kW fire discovered in a wastepaper basket in the previous example is in a closed room that has a ceiling height of 2.44 m (8 ft). The room measures 2.44 m (8 ft) square. Estimate the enclosure smoke filling rate for this fire.

Suggested Solution. Use the data previously calculated using the McCaffrey method. Assume the lower limit of the velocity of descent by using the ambient density of air at 1.22 kg/m^3 and 17°C.

Rate of layer descent	$U_t = \dfrac{\dot{m}_s}{\rho_l A_p}$
Mass rate of smoke production	$\dot{m}_s = 1.244$ kg/s
Density of smoke layer	$\rho_l = 1.22$ kg/m^3
Enclosure floor area	$A = (2.44)(2.44) = 5.95$ m^2
Rate of layer descent	$U_t = (1.244)/(1.22)(5.95) = 0.171$ m/s

When reconstructing fire scenes, skilled fire investigators use numerous indicators to estimate the areas and points of origin, distribution, and behavior of a fire such as direction of spread. These indicators, called *fire patterns*, are characteristically broken down into the following five major groupings summarized in Table 3.1. Some specific patterns are illustrated in Figure 3.11.

- ◆ demarcations
- ◆ surface effects
- ◆ penetrations
- ◆ loss of material
- ◆ victim injuries

DEMARCATIONS

Demarcations occur in locations where a combination of smoke, heat, and flames impinge upon materials, forming intersections of affected and unaffected areas. Examples include smoke layers deposited on walls and plume patterns on walls. See Item 1 in Figure 3.11.

Demarcations can vary depending on the type of exposed material, fire temperature, rate of heat release, and ventilation. Areas of material and nonmaterial loss are helpful in determining demarcations.

Heat and *smoke levels* (also referred to as *horizons*) refer to the height at which smoke or heat has stained or marked the walls and windows of a room. When using fire modeling programs, these levels can be predicted throughout the structure being modeled or correlated with the duration and size of the fire. Heat and smoke levels are also important when evaluating victim accounts of whether or not they could see an exit if standing in a room or whether they would be affected by smoke or heat as they navigated through the building.

In cases involving victims injured or killed in fires, the smoke and heat levels become important when evaluating whether or not victims stood up in a smoke-filled room and sustained their injuries from the cloud of heated toxic gases and smoky byproducts of combustion.

As the ceiling jet plume extends throughout the room, demarcations are formed by stratified heat and smoke damage. Shown as Item 1 in Figure 3.11, the ceiling plume extends to the right along the wall forming the start of a stratified line. Notice the sharp demarcation in the degree of damage between the affected and the unaffected areas on the wall surface.

Note also the acute angle the demarcation line forms with the ceiling. This angle is the result of buoyancy-driven flow along the ceiling and is a very reliable indicator of direction of spread.

Fire suppression efforts can also impact demarcations due to the application of hose spray to the walls, ceilings, and floors. Ventilation efforts during fire suppression, such as the venting of the roof and the opening of doors and windows, can also affect the demarcations formed by heat and smoke throughout the structure. Figure 3.12 illustrates the impact of a hose spray and its formation of an upward pattern on the wall surface by scouring the heat-affected paper and propelling it off the wall.

TABLE 3.1 ◆ Typology of Common Fire Patterns

Type of Pattern	Description	Variables	Examples
Demarcations	The intersection of affected and unaffected areas on materials where smoke and heat impinge	Type of material Fire temperature, duration, and rate of heat release Fire suppression Ventilation Heat flux reaching surface	"V" patterns on walls Discoloration patterns Smoke stains
Surface effects	Surface type determines the shape of boundary, areas of demarcation, and nature of heat application	Surface texture (rougher surfaces sustain more damage) Surface-to-mass ratio Surface coverings Combustible and noncombustible surfaces Thermal conductivity Temperature of surface	Spalling of floor surfaces Combustible surfaces scorched or charred by pyrolysis and oxidation Calcination Noncombustible surfaces changing color, oxidizing, melting, or burning cleanly
Penetrations	Penetration of horizontal and vertical surfaces	Preexisting openings in floors, ceilings, and walls Direct flame impingement Heat flux Duration of fire Suppression	Downward penetrations under furniture Saddle burns on exposed floor joists Failure of walls, ceilings, and floors Internal damage to walls and ceilings
Loss of material	Combustible surfaces with loss of material and mass	Type of materials Construction methods Room contents and fuel loads Duration of fire	Tops of wooden wall studs Fall-down of debris Heat shadowing Isolated areas of consumed carpet Beveling of corners and edges
Victim injuries	Areas, depths, and degrees of burns on victim's clothing and body	Location of victim in relation to other objects or victims Actions taken prior to, during, and after the fire	Absence of burn injuries protected by clothing or furniture Burns on lower torso Heat shadowing

Source: Expanded from Icove (1995).

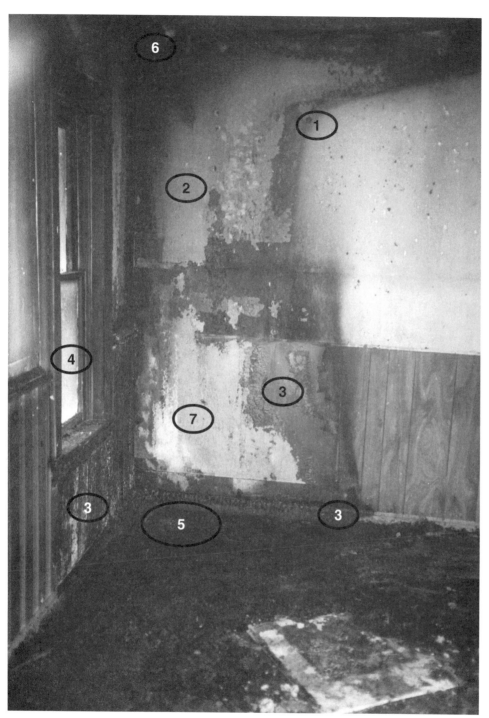

FIGURE 3.11 ◆ Fire pattern indicators found at fire scenes consisting of (1) demarcation, (2) calcination, (3) loss of material to wooden wainscoting and baseboard, (4) fractured glass, (5) ignitable liquid burn pattern on carpet, (6) penetration into the ceiling, and (7) area of clean burn where paper and pyrolysis products burned completely off wall and calcination indicating the most intense heat. *Courtesy of D. J. Icove.*

FIGURE 3.12 ◆ The impact of applying a hose spray to a wall during fire suppression efforts formed an upward pattern on the wall surface demarcations previously formed by heat and smoke. *Courtesy of D. J. Icove.*

The cooler the surface, the faster soot and pyrolysates will condense. Airflow will also control depositions. Figures 3.13(a) and (b) show the effects of airflow on soot deposition and how it can reveal hot gas spread.

SURFACE EFFECTS

Surface effects occur when heat transfer causes spalling of floor surfaces, scorching or charring of combustible surfaces, melting or changes in color, or oxidizing of noncombustible surfaces. Areas not damaged by the heat and flames are called protected areas and are significant in cases where objects or bodies have been found or have been removed prior to inspection. Surface effects can also include nonthermal deposits of soot or pyrolysis products that condense on the cooler surfaces. The cooler the surface, the faster these deposits occur. Variables influencing surface effects include the type and smoothness of the surface, the thickness and nature of the coverings, and even the object's surface-to-mass ratio.

An example of a surface effect is the discoloration of paint on the external metallic surfaces of a burned automobile. Heat transfer through conduction often produces discernible fire patterns on the exterior and interior surfaces of the vehicle (as in Figure 2.3).

Gypsum plasterboard or gypsum board is a material used in residential and commercial construction primarily for interior walls. It is sometimes referred to as drywall construction. The drywall board contains gypsum (calcium sulfate) bonded between two layers of paper. Fire-rated or X-type drywall is reinforced with fiberglass. The ASTM C 36 standard is the international standard for its manufacture.

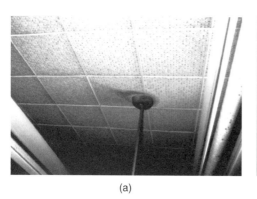 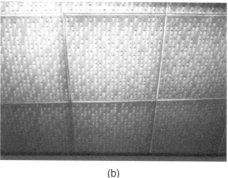

(a)   (b)

FIGURE 3.13 ◆ Movement of soot (a) around a light fixture and (b) through ceiling tiles indicates direction of air movement. *Courtesy of Jim Allen, Nipomo, California, by permission.*

Gypsum is a rock mineral whose qualities include the ability to be mixed with water and then dried into a hardened paste, resulting in a product that is 21 percent water, by weight. The ***calcination*** of gypsum wallboard is a process by which heat transfer causes the calcium sulfate to dehydrate in two discrete steps at characteristic temperatures.

Dehydration alone probably does not cause the gray-to-white change since this is a colorless transition. Accompanying the fire exposure is an accumulation of other carbonaceous pyrolysis products that turn the absorbent gypsum gray. Subsequent heating burns away (or evaporates) these deposits, causing whitening to progress through the thickness of the gypsum.

As it dehydrates, the gypsum shrinks and loses the water of hydration, and its mechanical strength degrades. It can fail, exposing combustible wall structure, when it loses so much strength from dehydration, shrinkage, or combustion of paper coverings that it can no longer support its own weight. Eventually, the fully dehydrated gypsum turns powdery and loses its mechanical strength, causing the wall to collapse. This process of calcination can also form lines of demarcation on its surface (DeHaan 2002). The effect of calcination on a gypsum board wall is shown as Items 2 and 7 in Figure 3.11.

Accompanying the calcination is the burning-away of the exposed paper and painted surfaces of the gypsum wallboard. The process of dehydrated gypsum falling or flaking off the surface during heat exposure is called ablation, which occurs at 600°C for ceilings and 800°C for walls (Konig and Walleij 2000). This process reduces the thickness of the board (Buchanan 2001, 341). Several documentation techniques are available for measuring the postfire loss of gypsum from the wall surface and estimating the radiant heat flux exposure time (Kennedy 2000; Schroeder and Williamson 2003).

Charring is the pyrolytic action of a fire, which converts the organic material (most often wood) into its volatile fractions, leaving behind carbonaceous char. The linear depth of the pyrolyzed material affected, altering the charcoal and lost material, is referred to as the ***char depth***. The rate at which wood chars is not linear. Wood chars quickly at first and then at a much slower rate due to various factors such as the insulation effect of the new char, reduced air–fuel interactions, ventilation, heat transfer, and physical arrangement.

The char rate is highly dependent on heat flux applied (Butler 1971). During a fire, heat fluxes on a surface may vary from near zero to 50 kW/m^2 in direct flame contact or 120–150 kW/m^2 under postflashover conditions. This results in a char rate that can vary from zero to several millimeters per minute.

Documentation of char depth is usually accomplished using a penetrating tool and survey grid diagrams (NFPA 2001, sect. 15-2.4). In these diagrams, the data from the measurements are converted to lines of equal depth forming ***isochar*** lines, each line connecting depths of the same value. Investigators are cautioned to realize that since variations in char rate can be caused by the differences in affected woods, coatings, and flame-retardant treatments, char depths should be measured on similar types of wood (baseboards, door frames, etc.).

Due to the nonlinear nature of charring, the measurement of char depth cannot establish precise times of fire exposure, but deeper relative depths of char may indicate areas of more intense or prolonged burning. Also, the appearance of the char surface often depends on the individual characteristics of the wood (DeHaan 2002). However, char depth relationships assuming a constant 50-kW/m^2 incident flux have been suggested (Silcock and Shields 2001).

Spalling is the chipping or crumbling of a concrete or masonry surface when it is exposed to varying heat, cold, or mechanical pressures. As a fire pattern damage indicator, spalling can indicate steep temperature gradients caused by radiated heat, a significant fuel load of ordinary combustibles, or other sources of localized heating.

Spalling occurs during rapidly rising temperatures, typically 20–30°C/min (Khoury 2000). Several factors influence spalling, but the moisture content and nature of the aggregate used (limestone vs. granite) are the leading variables. Spalling can also be caused by suddenly cooling a very hot concrete surface with a hose stream. High-strength concrete tends to spall more than normal concrete due to the higher density. The pores of the high-density concrete become filled more quickly with the high-pressure water vapor (Buchanan 2001, 228).

Lightweight concrete spalls more readily because of the vermiculite used as an aggregate. An example of extreme spalling of a concrete wall and ceiling surfaces is shown in Figure 3.14.

Figure 3.14 ◆ An example of typical spalling seen with high-strength concrete wall and ceiling surfaces, due primarily to its high density. *Courtesy of D. J. Icove.*

Localized combustion of flammable liquids on vinyl- or asphalt-tiled concrete surfaces produces a type of floor damage called ***ghost marks***. Ghost marks are subsurface staining of the concrete beneath the joints of tile floors caused by combustion of the solution of the cement/mastik by the applied flammable liquid. Radiant heat in postflashover fires and long-term heating by fall-down of burning materials can produce spalling but do not produce ghost marks (DeHaan 2002).

Flammable liquids themselves rarely produce significant spalling on bare concrete because the duration of a typical flammable liquid pool on bare concrete is only 1–2 min (DeHaan 1995). The radiant heat from the flame is, in part, absorbed by the liquid fuel during combustion and the maximum temperature of the surface cannot significantly exceed the boiling point of the liquid in contact with it. Since the boiling point range of gasoline is 40–150°C, the concrete will not usually be heated to the point where it will be affected.

If tile or carpet is present on the concrete, the flames of the gasoline can trigger localized charring, melting and combustion of the floor covering. An example of localized burning due to a flammable liquid pour pattern on floor tiles is shown in Figure 3.15. Since this combustion is more prolonged than the gasoline flame itself and there is no free-standing liquid to protect the concrete, there is better opportunity for heat to affect the concrete and cause it to spall if it is susceptible.

Fractured glass is another surface effect that usually occurs when glass is stressed by nonuniform heating. Common variables that affect glass breakage include the glass's thickness, the presence of any defects, the rate at which it was heated or cooled,

FIGURE 3.15 ◆ Surface effects due to a flammable liquid pour pattern (center of the photograph) charring the floor tiles. Two parallel pour patterns are documented, characteristic with a back-and-forth spilling of flammable liquids as the arsonist backs out of the building. *Courtesy of D. J. Icove.*

and the method used to secure the glass to the window frame (Shields, Silcock, and Flood 2001).

The potential for glass to fracture and break out has been the topic of several research initiatives. The prediction of the probability of breakage fits a Gaussian statistical correlation as shown in Figure 3.16 (Babrauskas 2000). This relationship shows that the mean temperature for the breakage of 3-mm-thick window glass is 340°C, with a standard deviation of 50°C.

The fire modeling program BREAK1, Berkeley Algorithm for Breaking Window Glass in a Compartment Fire, can calculate the temperature history of a glass window given several fire parameters combined with a physical composition of the glass (NIST 1991). The model also provides the temperature history of the glass normal to the glass surface and the window breakage time. Large-scale glass fracturing by heat is often observed as a room approaches flashover and the temperatures and heat fluxes dramatically increase.

One common mechanism for thermal fracturing of common window glass is the stress induced by nonuniform heating as a result of "thermal shadow" cast by framing or trim. When a window is exposed to radiant or convected heat, the exposed portions increase in temperature and begin to expand, but the areas shaded by the framing or window putty are not heated. Because glass is a relatively poor conductor of heat, those "shaded" areas remain cool and do not expand. This differentiation induces stress in the glass that can cause the glass to fracture often in patterns roughly parallel to the edges, as shown in Figure 3.17. The fractures radiating out from a single point on the right often are caused by a preexisting nick in the edge of the glass or by stress induced by a glazier point or frame nail (DeHaan 2002, 214).

The edges of ordinary glass fractured by mechanical stress are characterized by curved "conchoidal" ridge lines caused by the propagation of the fracture (as in

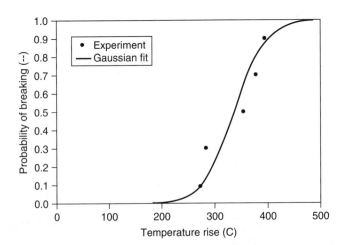

FIGURE 3.16 ◆ The prediction of the probability of the breakage of 3-mm-thick window glass fits a Gaussian statistical correlation with a mean temperature of 340°C and a standard deviation of 50°C. *Courtesy of Babrauskas [2000], by permission; www.doctorfire.com/glass.html.*

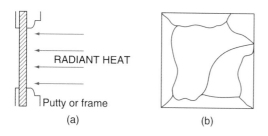

(a)

(b)

FIGURE 3.17 ◆ Heat fractures of a window pane caused by "shadowing" of radiant heat. Fractures radiating out from a point on the right often are caused by a preexisting nick in the edge of the glass or by stress induced by a glazier point or frame nail.

Figure 3.18a). The fracture pattern on the face of glass when broken by mechanical shock or pressure is typically a spiderweb pattern of mostly straight fractures (as in Figure 3.18b). The pattern usually consists of a number of radial fractures from the point of contact and concentric fractures connecting them.

Mechanical impacts of very brief duration cause minimal radial fractures because the glass does not have time to flex and may produce a domed plug of glass to be ejected from the side opposite the impact. Thermal stress usually results in more random patterns with wavy fractures. These fractures usually have mirror-smooth edges with minimal conchoidal ridges since they are usually produced at lower speeds than mechanical fractures.

If enough glass can be recovered to jigsaw fit pieces together for at least part of the pattern and the inside or outside face can be identified by soil, paint, decals or lettering, the direction of force can be established by the 3-R rule: The conchoidal lines on **r**adial fractures start at **r**ight angles on the **r**everse (side away from the force). The direction is reversed on concentric fractures. Toughened (safety) glass used in the side and rear windows of vehicles and in structural doors and shower doors fractures into small, rectangular blocks. As a general rule, the cause and direction of failure of safety glass cannot be established, but under some conditions fragments trapped in the frame may reveal the approximate location of a mechanical impact point.

The sequence of impacts can be established from intersecting and ending fractures. Sequence of failure versus time of exposure to fire can be established by the presence or absence of soot on the broken edges. It must be remembered that direct

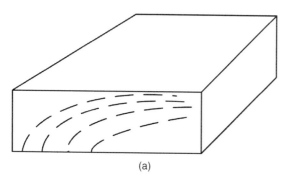

(a)

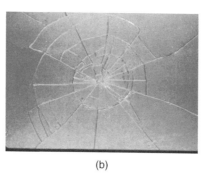

(b)

FIGURE 3.18 ◆ (a) Edge of glass fragment showing conchoidal fracture lines. (b) Face of glass sheet broken by mechanical impact near the center, with radial fractures and concentric fractures.

contact with flame can heat glass up to the point where all accumulated soot and smoke deposits are oxidized off, a so-called "clean burn."

Glass breakage patterns can also reveal a great deal of information about an explosion (DeHaan 2002). The distance to which the glass is blown is related to the thickness of the glass, the size of the window, and the pressures developed (Harris 1983).

A distinctive form of glass fracturing called ***crazing*** is usually formed by fire suppression water spray striking a heated glass surface. Crazing is characterized by a complex pattern (road map-like) of partial thickness fractures and concave pitting (Lentini 1992). Field experiments confirmed that when, during a fire, a water spray was applied to the hot glass panes on the right-hand side of the window, crazing occurred, as in Figure 3.19. Notice the melted glass in the lower left-hand pane. Common window glass softens when its temperature reaches 700–760°C (1300–1400°F).

Another descriptive form of surface burn patterns is called ***clean burn***. When a solid surface is exposed to a fire environment, products of combustion—water vapor, soot, and pyrolysis products—will condense on it. The cooler the surface, the faster they will condense (giving rise to the shadowy outlines of studs, nail heads, and other hidden features of a wall structure that cause a temperature differential). When a sooted surface is exposed to direct flame, it can reach high enough temperatures that the accumulated products can burn away, leaving a clean surface behind. This can be used to identify areas of direct contact between a surface and the flaming portion of the plume. Surfaces not in contact with the flame will rarely reach sufficiently high temperatures to allow the complete combustion of condensed soot or pyrolysates.

FIGURE 3.19 ◆ Crazing of glass forms by rapid cooling of glass, often during fire suppression, as seen in this field experiment where a water spray was applied to the window panes on the right. *Courtesy of D. J. Icove.*

Melted materials can also reveal much about the temperatures to which surfaces were exposed during the fire. Plastics, glass, copper, aluminum, and tin are the most common melted materials found at fire scenes that can document the wide range of temperatures produced by the fire (DeHaan 2002). Note that melting points of thermoplastics can vary depending upon the degree of their internal molecular structure. Listed in Table 3.2 are approximate melting temperatures of common materials.

Annealed furniture springs occur when the temperature exceeds the melting or annealing point and the springs lose their temper (Tobin and Monson 1989). This occurs when the temperature of the steel reaches over 500°C (1000°F). *Collapsed springs* are an indicator of this phenomenon, and an investigator should compare the relative degree of collapse to try to ascertain the direction and intensity of the fire (DeHaan 2002). Such collapse is not indicative of the presence of accelerants. The

TABLE 3.2 ◆ Approximate Melting-Point Temperatures of Common Materials

Material	*°C*	*°F*
Aluminum alloys (a)	570–660	1065–1215
Aluminum (pure) (c)	660	1220
Copper (c)	1083	1980
Cast iron (gray) (a)	1150–1200	2100–2200
50/50 solder (a)	183–216	361–421
Carbon steel (a)	1520	2760
Tin (c)	232	450
Zinc (c)	420	787
Magnesium alloy (HK31A) (a)	589–651	1092–1204
Stainless steel (a)	1400–1530	2550–2790
Iron (pure) (c)	1535	2795
Lead (pure) (c)	327	621
Gold (pure) (c)	1065	1945
Paraffin (wax) (d)	50–57	122–135
Human fat (e)	30–50	86–122
Polystyrene (b)	240	465
Polypropylene (d)	165	330
Polyester (Dacron) (d)	dec 250	480
Polyethylene (b,c)	130* (85–110)	185–230
PMMA (acrylic plastic) (b)	160*	320
PVC (b)	dec 250*	480
Nylon 66 (b)	250–260	487–500
Polyurethane foam (c)	dec–200	390

Sources: (a) Perry and Green 1984, Table 23-6, 23–40; (b) Drysdale 1999; (c) DeHaan 2002, 218–219; (d) *The Merck Index,* 1990; (e) DeHaan, unpublished.

investigator should also examine the framework (usually metal or wood) for other signs of thermal damage (NFPA 2001, sect. 4.14).

PENETRATIONS

Another common fire pattern is the occurrence of **penetrations** through horizontal and vertical surfaces. This occurs from direct flame impingement or intense radiant heat on walls, ceilings, and floors being sustained long enough to affect deeper layers of the material. Ceiling areas directly overhead of fire plumes may burn through or collapse, allowing flames to extend into confined spaces. In some cases, fall-down of combustible debris followed by extended smoldering can lead to localized penetration of wooden floors, sills, and toe plates in wood-framed structures.

Variables influencing penetrations include preexisting openings in floors, ceilings, and walls. Also, direct flame impingement coupled with high heat fluxes play an important role in creating the conditions necessary for penetrations. For example, the area directly above a fire plume intersecting with a ceiling, such as Item 6 in Figure 3.11, is an example of a penetration into the ceiling directly above the fire plume due to the high heat flux and temperatures that cause the failure of the ceiling materials.

Penetrations are not always directly above the fire plume. During flashover and rollover conditions, downward penetrations are sometimes found. Burning foam mattresses sometimes drop enough burning molten liquid onto the floor directly under the bed to cause burns through the floor. The same can occur with burning thermoplastics such as polyethylene trash containers or thermoplastic appliance liners. Traditional upholstery furniture such as sofas and chairs can support combustion long enough to burn through wooden floors beneath them after they collapse.

Failure of walls, floors, and ceilings during a fire may also contribute to forming penetrations through which the fire extends throughout a structure. These failures have been known to transport heat, flames, and smoke to other areas of the structure, leading inexperienced investigators to conclude improperly that two or more fires were set within the structure. Such penetrations can occur as gypsum wallboard fails, lath chars, plaster collapses, and metal ceilings or noncombustible ceiling coverings peel away.

LOSS OF MATERIAL

Loss of materials often occurs at the tops of wooden studs, flooring, carpets, and other objects located in the path of a fire's development and spread. This loss of material is helpful in determining the fire's intensity and direction of travel. **Heat shadowing** can affect the loss of materials by shielding the heat transfer and masking the potential lines of demarcation of other fire patterns. It can be driven by direct flame impingement or radiant heat alone. The rapid charring caused by postflashover radiant heat can be enhanced by ventilation from doors or windows to produce penetrations. These can occur through floors near doors or windows or even in the center of rooms where postflashover mixing is most efficient (DeHaan 2002).

Items 3 and 7 in Figure 3.11 are two examples of loss of materials due to direct impingement with the fire plume. The most significant is the relative loss of the wooden wainscoting material. The closer to the plume centerline, the deeper the charring. Localized damage to the wooden baseboard documents the radiant heat from above.

Loss of material may occur on the tops of wooden studs, corners and edges of furniture and moldings, and floors and carpeted surfaces. Other forms of loss of materials come from fall-down of debris from above. Typical examples include burning

FIGURE 3.20 ◆ The bodies of incapacitated victims during fires may block heat from reaching the surfaces on which they are lying, resulting in a "heat or smoke shadowing" fire pattern. In these instances, a silhouette of the victim is left on the bed, carpet, or chair. *Courtesy of D. J. Icove.*

drapery or curtains, which ignite other objects in the room and sometimes mislead novice investigators into thinking that multiple fires may have been set.

VICTIM INJURIES

Victim injuries may occur during discovery of, interaction with, or escape from a fire and may reveal actions taken during critical time periods. This is increasingly true in situations resulting in fire deaths where the only remaining fire pattern indicators are the areas and degrees of burns on the victim's body and clothing.

Radiant heat travels in straight lines and is absorbed by most materials. It is reflected by most materials to some degree but it can be reflected from metallic surfaces with sufficient energy to induce thermal damage to secondary target surfaces. The bodies of victims incapacitated during fires may block radiant heat from reaching the surfaces on which they are lying, resulting in a "heat or smoke shadowing" pattern. In these instances, a silhouette of the victim is left on the bed, carpet, or chair, as illustrated in Figure 3.20.

◆ 3.5 INTERPRETING FIRE PLUME BEHAVIOR

A vector is a mathematical pointer that has a unique direction and magnitude. An applied technique known as *fire vectoring* provides the investigator with a tool for understanding and interpreting the combined *movement* and *intensity* of plumes that create fire pattern damage. Direction indicators of various types have been in use by fire investigators since the earliest investigations.

FIRE VECTORING

The concept of heat and flame vector analysis was first seriously set forth in 1985 by Kennedy and Kennedy. However, few investigators fully understood the application of the concept and systematically applied it. Scientific research into fire scene investigations has addressed the topic only in the last decade.

Heat and flame vectoring, referred to in this textbook as simply *fire vectoring*, is becoming a standard technique for using fire patterns to determine a fire's area and point of origin (Kennedy 2000; NFPA 2001, sect. 15.2.3). Several examples throughout this textbook use this simple, yet effective principle for tracing a fire's development, which was fully explained and used to document the 1997 USFA Burn Pattern Study (FEMA 1997).

The underlying concept of fire vectors is that they consist of two characteristics: direction of movement and relative intensity. Each fire vector must use these two concepts.

Movement patterns are caused by flame, heat, and by-products of combustion produced when the fire spreads away from its initial source. Evaluating areas of relative damage from the least to the most damaged areas can reveal the fire's movement, a technique for tracking a fire back to its possible room, area, and point of origin.

Intensity patterns are the result of the fire's sustained impingement on exposed surfaces, such as walls, ceilings, furnishings, wall coverings, and floors. Various intensities of heat transfer result in thermal gradients along these surfaces, sometimes forming lines of demarcation dividing burned and unburned areas.

Using this technique, the vector's position and length quantitatively indicate the intensity of the particular burn pattern, while its direction indicates the apparent flow from hottest to coolest. Once the vectors are documented and reviewed, their cumulative effect will assist in the determining of the source of the fire. Each vector is assigned a number to allow notations in the report. A fire vector diagram used in USFA test 5 is shown in Figure 3.21 (Federal Emergency Management Agency [FEMA] 1997a, 47).

VIRTUAL ORIGIN

When a majority of the vectors point to the fire plume's base, it is often appropriate to consider this to be the location of the initial fuel package. Vectoring is helpful in estimating the location of the *virtual origin* or source of the fire.

The virtual origin is a point located along the plume's centerline that would give the same radiative output as the actual fire. This location is sometimes lower than the floor level. The virtual origin can be mathematically calculated and its location can be helpful in reconstructing and documenting the fire's virtual source, point of origin, area, and direction of travel.

As we saw in Chapter 2, the virtual origin of a fire is located below the floor level when the area of the fuel source is large compared to the energy released by the fuel, as shown in previous examples. Likewise, a fuel releasing a high energy over a small area will produce a virtual origin above the floor level (Karlsson and Quintiere 1999). An example of a high-energy release over a small area is a small gasoline pool fire.

TRACING THE FIRE

The fire investigator's task would be quite easy if one were present for the duration of the incident. Unfortunately, fire scenes are often cold and in disarray after the extinguishment and subsequent overhaul to assure that the fire will not rekindle.

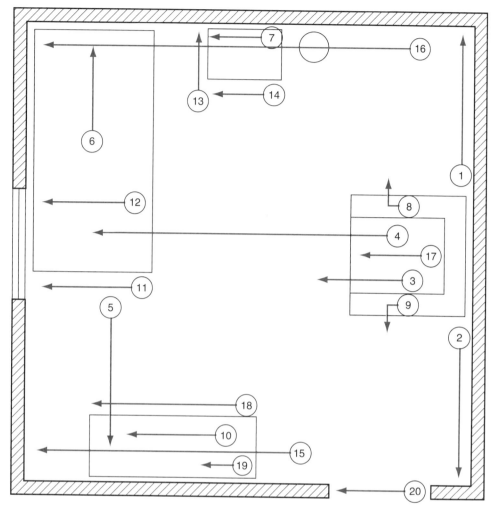

FIGURE 3.21 ◆ The fire vectoring technique uses the length of a drawn vector to indicate quantitatively the direction and intensity of the particular burn pattern. *Courtesy of FEMA [1997a, 47].*

The investigator's role then becomes to reconstruct the sequence of the fire backward to its area and point of origin. A firm understanding of fire behavior becomes imperative during this examination and documentation of the fire scene.

The ability to trace the fire's behavior and growth is important to the investigation. Like an experienced tracker, the investigator relies upon pattern damage as roadmarks to track the fire's path. Based on experience, the following rules of fire behavior are offered to understand and interpret this pattern damage (DeHaan 2002).

- ◆ A fire plume's hot gases (including flames) are much lighter than the surrounding air and therefore will rise. In the absence of strong winds or physical barriers (such as noncombustible ceilings) that divert flames, fire will tend to burn upward. Radiation from the plume will cause some downward and outward travel.
- ◆ Combustible materials in the path of the plume's flames will be ignited, thereby increasing the extent and intensity of the fire by increasing the heat release rate. The more intense the fire grows, the faster it will rise and spread.

- A flame plume that is large enough to reach the ceiling of a compartment is likely to trigger full involvement of a room and increase the chances for flashover to occur. If there is not more fuel above or beside the initial plume's flame to be ignited by convected or radiated heat, or if the initial fire is too small to create the necessary heat flux on those fuels, the fire will be selflimiting and often will burn itself out.
- In evaluating a fire's progress through a room, the investigator must establish what fuels were present and where they were located. This fuel load includes not only the structure itself, but its furnishings, contents, and wall, floor, and ceiling coverings (as well as combustible roofing materials), which feed a fire and offer it paths and directions of travel.
- Variations on the upward spread of the fire plume will occur when air currents deflect the flame, when horizontal surfaces block the vertical travel, or when radiation from established flames ignite nearby surfaces. If fuel is present in these new areas, it will ignite and spread the flames laterally.
- Upward, vertical spread is enhanced when the fire plume finds chimneylike configurations. Stairways, elevators, utility shafts, air ducts, and interiors of walls all offer openings for carrying flames generated elsewhere. Fires may burn more intensely because of the enhanced draft.
- Downward flame spread of the fire may occur whenever there is suitable fuel in the area. Combustible wall coverings, particularly paneling, encourage the travel of fire downward as well as outward. Fire plumes may ignite portions of ceilings, roof coverings, draperies, and lighting fixtures, which can fall onto ignitable fuels below and start new fires that quickly join the main fire overhead.
- Fire plumes that are large enough to intersect with ceilings form ceiling jets that extend radially along the ceiling surface. Radiation from overhead ceiling jets or hot gas layers can ignite floor coverings, furniture, and walls even at some distance, creating new points of fire origin. The investigator is cautioned to take into account what fuel packages were present in the room from the standpoint of their potential ignitability and heat release rate contributions.
- Suppression efforts can also greatly influence fire spread and the investigator must remember to check with the fire suppression personnel present as to their actions in extinguishing the fire. Positive pressure ventilation or an active attack on one face of a fire may force it back into other areas that may or may not already have been involved and push the fire down and even under obstructions such as doors and cabinets.
- Heat and smoke plumes tend to flow through a room or structure much like a liquid, that is, upward in relatively straight paths and outward around barriers.
- When using fire vectoring, the total fire damage to an object observed after a fire is the result of both the intensity of the heat applied to that object and the duration of that exposure. Both the intensity and the exposure of that heat may vary considerably during the fire.
- The highest-temperature area of a plume will produce the highest radiant heat flux and, therefore, affect a surface faster and more deeply than cooler areas. This can show the investigator where a flame plume contacted a surface or in which direction it was moving (since it will lose heat to the surface and cool as it moves across).
- The contribution a fire makes to the growth process in a room depends not only on its size (heat release rate) and direction of travel but also on its location in the room—in the center, against a wall, in a rear corner away from ventilation sources, or close to a ventilation opening.

Fire investigation and reconstruction of a fire's growth pattern back to its origin are based on the fact that fire plumes form patterns of damage that are, to a large extent, predictable. With an understanding of fire plume behavior, combustion properties, and general fire behavior, the investigator can prepare to examine a fire scene for these particular indicators. As with the application of the scientific method, each indicator is an independent test for direction of travel, intensity, duration of heat application, or point of origin. The investigator should remember that there is no one indicator that proves the origin or cause of a fire. They must be evaluated together and yet they may not all agree. Finally, the application of fire vectoring can also enrich and document the opinion rendered as to the original location of the fire plume.

◆ **3.6 FIRE BURN PATTERN TEST**

Fire testing can produce many types of fire pattern damage phenomena (Table 3.1) including ***demarcations***, ***surface effects***, ***penetrations***, and ***loss of materials***. Fire testing is one method to expand the range of adequate knowledge of how to recognize and interpret these fire patterns.

As an example of the testing that is now ongoing with the assistance of NIST and the Federal Emergency Management Agency's U.S. Fire Administration (FEMA/USFA), the U.S. Tennessee Valley Authority (TVA) Police coordinated the logistics of and participated in a series of fire pattern burn tests in Florence, Alabama. The tests used two forms of structures, a vacant single-family and a multiple-use assembly building (FEMA 1997; Icove 1995; NIST 1997). The U.S. TVA Police concentrated on extending the results of fire pattern burn testing on larger-scale structures, while NIST and FEMA worked on the single-family dwelling scenarios.

The primary goals of the project test included demonstrating the production of fire burn patterns, exploring fire scene evidence and documentation techniques, and validating basic fire modeling—all concepts that form the underlying theme for forensic fire scene reconstruction. Due to the immense amount of diverse information gleaned from the Florence test burns, various aspects of this fire burn pattern appear in examples and problems throughout this text.

FIRE TESTING GOALS

The overall objective for conducting the test burns on the multiple-use facility included the simulation of arson fires using plastic containers of ignitable liquids, since arsonists often use these containers when setting fires to structures (Icove et al. 1992). Often when a burned, self-extinguished firebomb is found, investigators ask the questions: How long did the device burn? What damage can be expected? and Where is the best place to collect evidence of the device once it is damaged?

To simulate the arson, a typical firebomb consisting of a 3.8-liter (1-gal) plastic container filled with unleaded gasoline was placed in the corner of the multiple-use facility's main assembly room, as shown in the floor plan in Figure 3.22. The main room was carpeted with an institutional-grade synthetic carpeting.

As thermocouple data for the test were collected, the firebomb was remotely ignited with an electric match, with the resulting fire burning and then self-extinguishing, leaving a 0.46-m (1.5-ft)-diameter circular burned area. Three thermocouple trees

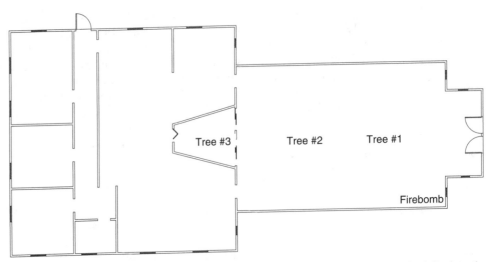

FIGURE 3.22 ◆ Floor plan of structure used to simulate the use of a typical firebomb placed in the corner of the room as indicated. *Courtesy of D. J. Icove.*

recorded the room temperature profiles, as shown in Figure 3.23. Note that the fire lasted approximately 300 s.

The postfire damage patterns are depicted schematically and photographically in Figures 3.24a and b. The exact location of the firebomb and its subsequent burning pool diameter are indicated. Figure 3.24a illustrates the fire patterns from demarcations, surface effects, penetrations, and loss of material. The schematic uses the relative

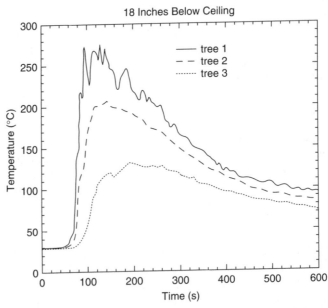

FIGURE 3.23 ◆ Three thermocouple trees recorded the room temperature profiles after the firebomb was remotely ignited with an electric match.

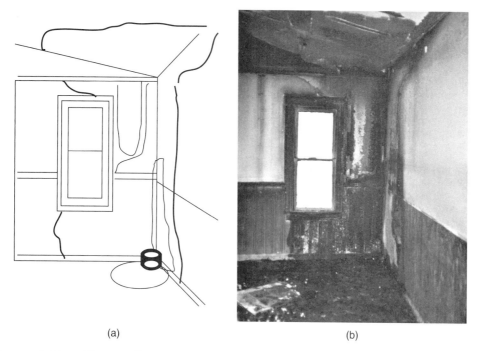

(a) (b)

FIGURE 3.24 ◆ The postfire pattern damage from the firebomb test depicted both (a) schematically and (b) photographically. (a) Dark lines indicate major demarcations of burned/unburned areas. Light lines indicate isochar demarcations. (b) Note that the wall finish above the wainscoting was painted drywall and played no significant role in vertical flame spread. Colored lines can be used to delineate areas of surface deposit, penetration, and consumption. *Courtesy of D. J. Icove.*

damages revealed on the exposed surfaces to form isochar lines corresponding to nearly equal depths.

These isochar lines are useful in assisting in the placement and interpretation of fire vectors. The lines also document the ignitable liquid burn patterns found on carpeted and other floor surfaces.

ESTIMATED HEAT RELEASE RATE

The peak estimated heat release rate, $\dot{Q}$, for the gasoline pool fire can be calculated using equation (2.2).

Heat release rate	$\dot{Q} = \dot{m}'' \Delta h_c A$
Area of the burning pool	$A = (3.1415/4)(0.457^2) = 0.164\,\mathrm{m}^2$
Mass flux for gasoline	$\dot{m}'' = 0.036\,\mathrm{kg/m}^2\text{-s}$
Heat of combustion for gasoline	$\Delta h_c = 43.7\,\mathrm{MJ/kg}$
Heat release rate	$\dot{Q} = (0.036)(43700)(0.164) = 258\,\mathrm{kW}$

The real-world estimate for the heat release rate would be 200–300 kW.

VIRTUAL ORIGIN

A fire vector analysis of this firebomb test case can be used by the investigator to understand and interpret the combined movement and intensity of plumes that created the fire pattern damage shown in Figure 3.25. Note that the vectors originate from an area above where the firebomb was located on the floor, which actually corresponds to the theoretical virtual origin of the fire plume.

The calculation for the location of the virtual origin confirms this observation. As previously calculated in Example 2.3, the heat release rate for a 0.46-m (1.5-ft)-diameter gasoline fire is approximately 258 kW. The calculation for virtual origin, as taken from the Heskestad equation, is as follows.

Virtual origin	$Z_0 = 0.083 \, \dot{Q}^{2/5} - 1.02 \, D$
Equivalent diameter	$D = 0.457$ m
Total heat release rate	$\dot{Q} = 258$ kW
Virtual origin	$Z_0 = (0.083)(258)^{2/5} - (1.02)(0.457)$
	$= 0.766 - 0.466 = 0.3$ m (1 ft)

For this case example, Z_0 is determined to be 0.3 m (1 ft), which is above the floor level as documented in the vector diagram in Figure 3.23. As stated previously, research has shown that a fuel releasing a high energy over a small area, such as in this case example, may produce a virtual origin above the floor level (Karlsson and Quintiere 1999).

FLAME HEIGHT

The *NFPA 921* flame height calculation takes into account the placement of the fire within the compartment, particularly for corner fires. By observation of the damage caused by the fire plume as well as its positive virtual origin above the floor, we can assume that the flame height at least touches the ceiling, which is 3.18 m (10.43 ft).

Total heat release rate	$\dot{Q}$	$= 258$ kW
Flame height	H_f	$= 0.174(k\dot{Q})^{2/5}$
In a corner ($k = 4$)	H_f	$= (0.174)(4 \times 258)^{2/5}$
		$= 2.79$ m (9.15 ft)
Including the virtual origin	H_{total}	$= Z_0 + H_f$
		$= 0.299 + 2.79 = 3.09$ m (10.14 ft)

This corresponds to a greater-than-ceiling height, thus creating the conditions for a ceiling jet across the ceiling. Therefore, the fire plume was of ample height to reach the ceiling and radially disperse ceiling jets extending out from the centerline.

FIRE DURATION

Determining the initial growth period is useful in fire scene reconstruction. In validating estimates of fire duration, it is always fortunate to have historical testing, actual loss history data, and realistic fire models.

When comparing the development of real-life fires to models, the investigator relies upon observations, documentation, analysis of historical fire test data, and similar cases that relate to the fire under study. A wealth of information is available from handbooks available in the fire protection engineering field (SFPE 2002a, sect. 3-1).

Fire duration was one of the key questions to be answered in the firebomb test case. After the firebomb ignited, the plastic jug melted, releasing its contents of gasoline into a circular pool of burning liquid. The question now arises as to how long the pool of gasoline can be expected to burn. Note that the question of whether the surface of the spill resides on a nonporous concrete surface or a carpeted material is addressed in a later section.

The burning duration of pool fires can usually be estimated based on the steady mass burning rate, assuming rapid growth and a large fuel supply. Gasoline has a

FIGURE 3.25 ◆ The fire vector analysis of the firebomb test case assists in the understanding and interpreting the fire plume and points to an area above the floor that corresponds to the plume's theoretical virtual origin. *Courtesy of D. J. Icove.*

density of 740 kg/m^3 or 0.74 kg/liter. The steady-state mass loss rate (mass flux) of gasoline is 0.036 kg/m^2-s (SFPE 1995, Figure 3-1.2[a]).

The following calculation is the working hypothesis for the estimated burning duration of the gasoline, assuming that it will self-extinguish after all the available fuel is consumed.

Mass of gasoline	$m = (1 \text{ gal})(3.785 \text{ liters/gal})(740 \text{ g/liter}) = 2.7 \text{ kg}$
Mass burning rate for gasoline	$\dot{m}'' = 0.036 \text{ kg/m}^2 \text{ s}$
Burning area	$A = \pi r^2 = (3.1416)(0.23 \text{ m})^2 = 0.166 \text{ m}^2$
Burn rate of gasoline	$\dot{m} = A \dot{m}'' = (0.166 \text{ m}^2)(0.036 \text{ kg/m}^2 \text{ s})$
	$= 0.00598 \text{ kg/s}$
Approximate burning duration	$= \dot{m}/m = (2.7 \text{ kg})/(0.00598 \text{ kg/s}) = 451 \text{ s}$

The real-world estimate for the average burning duration should be 7 to 8 min.

REGRESSION RATES

A flammable liquid pool will burn from the top down at a fairly predictable rate depending on its chemical composition and physical properties if all other factors (pool diameter and depth) are constant. A thin layer of liquid will burn away very quickly, often so quickly that only the most transient thermal effects will be observed. The depth of a pool is determined by the quantity of liquid, its physical properties (viscosity), and the nature of the surface.

On a level and smooth nonporous surface, a low-viscosity liquid like gasoline will form a pool approximately 1 mm deep if not otherwise limited. On a porous surface the liquid will tend to penetrate as far as it can and then spread horizontally by gravity flow and capillary action. The size of the pool will then be controlled by the amount of liquid and the depth of the porous material, the porosity of the substrate, and the rate at which it is poured. As a rough guideline, the thickness of carpet can be used as a maximum pool depth, since saturated carpet represents a pool whose depth equals the thickness of the carpet and of any porous pad beneath (DeHaan 1995).

A large quantity of gasoline dripping slowly from a leaking gas tank may present a visible pool on soil or sand not much larger than the diameter of the individual drops but many feet deep. The same quantity dumped quickly onto the same ground may produce a pool of great diameter but much shallower. A similar measure of liquid can be broadcast in an arc to generate a shallow layer (film) over a large area. The viscosity and surface tension of the liquid and the speed with which the material is ejected will determine the nature (thickness) of this film.

Once ignited, the rate of burning per unit area (mass flux) is controlled by the amount of heat that can reach the surface of the pool and the size of the perimeter through which air can be entrained. For gasoline pools of 0.05–0.2 m in diameter this produces a minimum regression rate of 1–2 mm/min. (Blinov, in Drysdale 1999, 153). For very small pools the rate is much higher (due to the laminar flame structure) so the heat effects (scorching) beneath the pool are more superficial because there is insufficient time to produce them. Since the fire size is driven by the evaporation rate per area (mass flux), the larger the pool in surface area, the larger the fire (the larger $\dot{Q}$). Gasoline pools over 1 m in diameter have a maximum regression rate of 3–4 mm/min due to the limitations of turbulent mixing in the plume.

The flame has a variety of effects on nearby surfaces depending on the geometry and nature of heat application. Because the fuel-soaked substrate is being cooled by evaporation, its temperature cannot be more than a few degrees hotter than the boiling point of the liquid and so will not do much more than scorch.

Low-melting point materials like synthetics can melt as well as scorch. The areas of carpet or floor immediately outside the pool will be exposed to the radiant heat of the plume. Without the protection of the evaporating liquid, materials will melt and scorch, particularly if small or thin (carpet fibers). Materials in direct contact with the flame front will be scorched and ignited.

Thermal effects of materials near the flame front often produce what is called a halo or ring effect around the outside of the pool as shown in tests (Figure 3.26) by NIST (Putorti 2001). If the substrate is ignitable, a ring of damage may extend some distance from the pool and even the center of the protected area may be burned as the protective layer of liquid fuel evaporates. In traditional carpets this area will be noticeable and sufficient residues of the liquid fuel may survive in the center of the burned area to be recovered and identified. If the fire burns itself out or is quickly extinguished by a sprinkler system or oxygen depletion, the pool area can be roughly estimated from the dimensions of the burned margins. If the fire is sustained in the carpet or the fire (externally) reaches flashover, the area of burned carpet can grow well beyond the area of the original liquid pool. This is particularly true for the new generation of synthetic fiber carpets with polypropylene backing and polypropylene face yarn pile.

Wool or nylon carpets with jute backing will tend to self-extinguish but polypropylene will not. Combustion tests have shown that flames on such carpets can propagate at a rate of approximately 0.5–1 m^2/hr, generating very small flames at the margins of the burning area. (Note that such carpets will pass the methenamine tablet test since its igniting flame is about 50 W and of brief duration [similar to a dropped match], but if there is any significant additional heat flux from a larger ignition source [wad of paper] or a continuously burning object like a piece of furniture, these can propagate a fire over a large area if given enough time [DeHaan 2002, 105].)

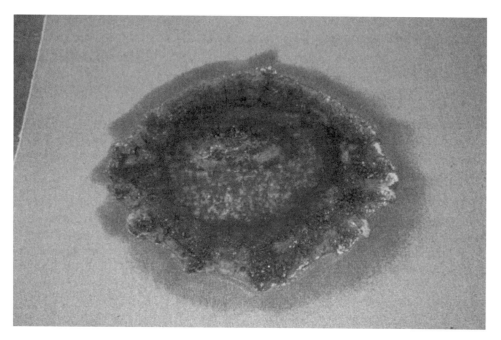

FIGURE 3.26 ◆ Materials in contact with the flame front often produce what is called a halo or ring effect in the carpet around the outside of the pool. *Courtesy of Putorti [2001].*

ADJUSTMENTS TO FIRE DURATION

This test examined the calculated burning duration of 451 s against the actual burning duration shown in Figure 3.23. The test showed that the actual duration is less than the estimated duration. A modification to this working hypothesis must be done to account for the carpet present rather than the concrete floor as originally assumed in the initial calculation.

Experiments on carpet saturated with pentane (a component of gasoline) show that its rate of evaporation (nonburning) is 1.5 times higher than that for a free-standing pool at the same temperature (DeHaan 1995). Using this information, the steady-state mass loss rate by evaporation of gasoline saturated into carpet is approximately 1.5 times the free-standing value of 0.036 or 0.054 kg/m²-s. Assuming that combustion processes will process in the same manner on this carpet as evaporation, recalculating the burning duration for a carpeted surface gives a corrected estimate. The new burning duration was estimated as 316 s and the heat release rate as 387 kW.

Knowledge of the burning characteristics of common flammable and combustible liquids on various surfaces, including those of pool fires such as the one in our test case, is crucial in fire testing and analysis. Synthetic carpets may melt and reduce the mass flux. Also, heat release rate contributions from the molten plastic container pool and carpet were not included. These would be expected to add to the heat release rate and to the duration of the fire.

Recent NIST research on flammable and combustible liquid spill and burn patterns showed that the peak spill fire heat release rates for burning liquids on nonporous surfaces were found to be only 12.5 to 25 percent of those from equivalent deep pool fires. The peak heat release rates for fires on carpeted surfaces were found to be approximately equal to those for equivalent pool fires (Putorti 2001). These results are in conflict with other similar tests and further testing may resolve the disagreement in predictions.

FLAME HEIGHT ADJUSTMENT FOR FIRE LOCATION

The flame height is a function of the fire location within the room. As shown in equation (3.9), the flame height, H_f, depends on the fire's placement either in the center ($k = 1$), against the wall ($k = 2$), or in the corner ($k = 4$) of the compartment.

In the initial approach, the fire was assumed to have the full impact of the corner on the flame height, predicting a total flame height of 3.1 m (10.1 ft). With the new heat release rate at 387 kW and the corner configuration, the new flame height estimate would be recalculated to be 3.28 m plus the virtual origin of 0.43 m, or 3.71 m (12.2 ft).

POOL FIRES AND DAMAGE TO SUBSTRATES

When a flammable liquid burns it does so by evaporation from the liquid, creating a cloud of vapor denser than air. Brownian motion at elevated temperatures causes this layer to have a finite thickness and to diffuse with overlying air to form a steep concentration gradient with distance from the liquid surface. Wherever this gradient is within the flammable range of the vapors, a flame can be supported. A diffusion flame is one supported by fuel vapors diffusing from the fuel surface into the surrounding air/oxygen.

The distance between the fuel surface and the flame front varies with the temperature (and thereby the vapor pressure) of the fuel. Radiant heat from the layer of flame travels in all directions. Of the heat radiated downward, some heat is absorbed by fuel, keeping the temperature high enough that there is a continual supply of vapors to support a flame. Some heat goes through the fuel into the substrate beneath, being absorbed to increase the temperature of the substrate. Due to intimate contact between the substrate and the liquid fuel, heat is transferred to the fuel and then distributed through the fuel (if it is a deep enough layer) by convective circulation.

The temperature of the surface under the pool will not be more than a few degrees above the boiling point of the liquid overlying it. If the fuel's boiling-point temperature is low enough ($<200°C$), the liquid can burn off without any visible effect if the surface is smooth, with no pores, joints, or seams, and has a relatively high decomposition temperature. The higher the boiling point of liquid, the better the chances of thermal damage to the floor. Damage can be pyrolysis (scorching), melting, or both. If a fuel contains a mixture of compounds with different boiling points, the low-boiling liquid will tend to burn off first, leaving the higher-boiling point compounds to continue to heat. A mixture like gasoline covers the range of boiling points $40–250°C$ ($100–450°F$). As the mixture burns, the boiling point increases and therefore the limiting temperature factor increases. At $250°C$ ($450°F$), there will be only scorching of wood surfaces but some synthetic floor coverings can be significantly damaged. Field fire tests demonstrating the absence of significant thermal effects on wood with gasoline pools have recently been published (DeHaan 2002).

FINAL HYPOTHESIS

The final hypothesis noted for this test, which Figure 3.23 reflects, is the consumption of the gasoline in the time–temperature curve, leveling out approximately 400 s after ignition, which is consistent with the above estimate. This demonstrates the effects of flammable liquids used in arson and other fire-related crimes—a very rapid increase in localized temperature followed by a rapid consumption to self-extinguishment.

Expert conclusions or opinions on the strength of these hypotheses rest partially on the use of accepted and historically proven fire testing techniques that validate this methodology and can be easily replicated. Additionally, this methodology has been peer reviewed and published. There are established error rates for applying this methodology specifically to known variables such as room size and fire development times. There also exist standardized methods maintained by independent and unbiased organizations for applying and interpreting these relationships. Finally, this methodology is generally accepted in the scientific community.

◆ **3.7 SUMMARY AND CONCLUSIONS**

Fire pattern damage analysis is a vital investigation technique for fire scene reconstruction. The visual interpretation of damage created by fire plumes can isolate and accurately identify the area of fire origin.

Locating and identifying the first fuel package ignited is a critical step in the accurate reconstruction of a fire incident. Careful analysis of fire patterns can significantly aid the scene investigator in this effort. Because effects like charring, melting, ignition, and protection are predictable, their location and distribution offer a sound basis for locating fuel packages, which can be confirmed by interviews or prefire photos.

Systematic steps can be used by fire investigators to support and document thermal damage patterns, identify the fire's direction and intensity, confirm significant witness observations, and verify the results of fire modeling. These systematic steps invoke the scientific method to test and evaluate various hypotheses of the fire's origin and spread.

From these observations, the authors conclude the following:

- As long as there is not too much damage, ample fire pattern heat indicators exist and can be documented.
- Evidence of fire pattern damage at scenes can be found for fire plumes that can indicate a fire's source, area and point of origin, and direction of travel.
- Fire engineering analysis using the scientific method is necessary to perform a comprehensive analysis.
- Investigators should not overlook signs of physical evidence documenting human activity (hand- or fingerprints, shoeprints, blood spatters, broken glass, discarded hoses/fire extinguishers, and burn injuries).
- Fire analysis and computer-assisted modeling can provide further insights into fire scene behavior.

Recommendations include the continued use of *Kirk's Fire Investigation* and *NFPA 921* as the universally accepted guides to documenting all fire and explosion investigations. Continued funding of research into fire pattern analysis and reconstruction technologies, as well as international conferences on the forensic aspects of fire investigation, is an important means by which the body of knowledge in the field may be increased.

Problems

3.1. Find a photograph of a fire plume in a recent news story or article. Estimate the dimensions of the plume and attempt to calculate its heat release rate. What information on the fire can you infer from these calculations that relate to the story or article?

3.2. Photograph a recent fire scene and illustrate as many fire patterns as you can find.

3.3. From the plume information in the photos and the identified first fuel package, estimate the heat release rate and virtual origin of the initial fire.

3.4. Use the equations provided in this chapter to determine the radial heat fluxes at various distances from the fire's virtual origin.

Suggested Reading

DeHaan, J. D. 2002. *Kirk's fire investigation,* 5th ed., chap. 7. Upper Saddle River, N.J.: Prentice Hall.

National Fire Protection Association. 2001. *NFPA 921—Guide for fire and explosion investigations,* chaps. 4 and 15. Quincy, MA: NFPA.

Fire Scene Documentation

4 CHAPTER

When you have eliminated the impossible, that which remains, however improbable, must be the truth.

—Sir Arthur Conan Doyle

Fire scenes often contain complex information that must be thoroughly documented, a task of paramount importance to the investigator. A single photograph and scene diagram are often not sufficient to capture vital information on fire dynamics, building construction, evidence collection, and avenues of escape for the building's occupants.

Thorough documentation is accomplished through a comprehensive effort of forensic photography, sketches, drawings, and analysis. The purposes of forensic fire scene documentation include documenting visual observations, emphasizing development characteristics, and authenticating physical evidence found at the scene.

This chapter provides the fundamental concepts of forensic fire scene documentation, with emphasis placed upon a systematic set of guidelines. Reviewed are concepts and techniques that produce viable and legally acceptable documentation for both investigative reports and courtroom presentations. Various computer-assisted photographic and sketching technologies that help to ensure accurate and representational diagramming are also explored.

Report writing guidelines are not included in this textbook; only the documentation needed to assemble authoritative reports. For a detailed discussion and examples, see *Kirk's Fire Investigation,* 5th ed. (DeHaan 2002) and *Combating Arson-for-Profit: Advanced Techniques for Investigators* (Icove, Wherry, and Schroeder 1998).

◆ 4.1 NATIONAL PROTOCOLS

The systematic documentation of a fire scene from its initial stages plays an essential role in the effort to record and preserve the events and evidence for all parties involved, particularly for forensic and other experts who are retained later that may be asked to render their opinion in the case. This systematic approach captures all available information needed for later use in criminal, civil, or administrative matters. The bottom line is that systematic documentation, when correctly conducted, will be well suited to pass the *Daubert* challenges for courtroom admissibility and ensures that an independent qualified investigator will arrive at the same opinions. Several national interrelated protocols cite the need for systematic documentation of fire scenes. Two of the recent protocols include those by the U.S. Department of Justice and the National Fire Protection Association (NFPA).

U.S. DEPARTMENT OF JUSTICE

One example of a peer-reviewed national protocol for fire investigation was developed and published in 2000 by the National Institute of Justice (NIJ), which serves as the applied research and technology agency of the U.S. Department of Justice. The NIJ's (2000) *Fire and Arson Scene Evidence: A Guide for Public Safety Personnel* is the product of their Technical Working Group on Fire/Arson Investigation, which consisted of 31 national experts from law enforcement, prosecution, defense, and fire investigation communities. This document has become the standard for documenting and recording the collection of fire and arson scene evidence. Both of the authors participated in the preparation and editing of the NIJ guide.

The expressed intent of the NIJ guide is to expose as many of the public-sector personnel (mainly fire, police, and prosecution) as possible to the process of identifying, documenting, collecting, and preserving critical physical evidence at fire scenes. This document has become the most widely distributed public guide on fire scene processing since the 1980 historic publication by the National Bureau of Standards (NBS; forerunner of the National Institute of Standards and Technology), *Fire Investigation Handbook*.

NATIONAL FIRE PROTECTION ASSOCIATION

The NFPA's *Guide for Fire Incident Field Notes*, which was republished in 1998, serves as a standardized protocol (*NFPA 906*). In existence since its first publication in 1988, *NFPA 906* is a standardized protocol for recording field notes of fire scenes. The NIJ guide cites *NFPA 906* and includes the data collection forms in its appendix. A synopsis of the forms contained in *NFPA 906* appears in Table 4.1, including an additional form from *NFPA 921* needed for data on compartment fire modeling. Similar forms are found in the appendixes to *Kirk's Fire Investigation* (DeHaan 2002). These forms are reprinted in Figures 4.1–4.16.

The intended users of *NFPA 906* include all persons having responsibility for investigating fires. This includes the fire company officer, the incident commander, the fire marshal, or a private investigator. These data collection forms consist of an organized investigative protocol. The expressed purpose of these forms is to serve as an input for collecting and recording preliminary information needed in the preparation of a formal incident or investigative report and in constructing a compartment fire model.

The reports cover structure, vehicle, and wildland fires; information on casualties, witnesses, evidence, photographs, and sketches; and documentary data on insurance and public records. A cover case management form is used to track the progress of the investigation. The maintenance of *NFPA 906* is now the responsibility of NFPA's Technical Committee on Fire Investigations, which also oversees *NFPA 921* (NFPA 2001).

TABLE 4.1 ◆ Forms Used in Assuring the Collection of Uniform and Complete Field Data for Constructing Written Reports

Form	Name	Description
906-0	Case Supervision	Top cover sheet to track the progress of the investigation
906-1	All Fires	Collects general identification and contact information
906-2	Structure Fires	Used for structure fires
906-3	Motor Vehicle Fires	Used for motor vehicle fires
906-4	Wildland Fires	Used for grass, brush, or wildland fires
906-5	Casualties	Records information on persons injured or killed in the fire
906-6	Witness Statements	Identifies and records expected testimonies of witnesses
906-7	Evidence	Documents and records recovered and seized evidence
906-8	Photographs	Logs the descriptions of all photographs taken by the investigators
906-9	Sketches	Contains a scene destruction sketch
906-10	Insurance Information	Records information on insurance coverage, adjustments, and loss
906-11	Records/Documents	Records information on incident, property, business, and personal records that are available
921	Fire Modeling	Data for compartment fire modeling

Source: Derived from NFPA (1998, 2001.)

◆ 4.2 SYSTEMATIC DOCUMENTATION

Each fire investigator should adopt a systematic procedure or protocol for documenting through sketches, photographs, witness statements, and written documentation of the examination of a fire scene along with recording the chain of custody of all evidence removed. A systematic four-phase process (Icove and Gohar 1980) was first introduced with the publication of the NBS (1980) *Fire Investigation Handbook*.

This process recommends a careful documentation of the fire scene in the following four phases.

- ◆ Phase 1—Exterior
- ◆ Phase 2—Interior
- ◆ Phase 3—Investigative
- ◆ Phase 4—Panoramic

text continued on p. 134

CASE SUPERVISION FIELD NOTES 906-0		AGENCY	FILE NUMBER

This cover sheet will assist in keeping track of the progress of the investigation. Indicate what has been done, what needs to be done, assignments, dates and so forth, in the Remarks sections. The lower portion should be used to record routine checks or rechecks and other information pertinent to the investigation.

FIELD NOTES FORMS

					REMARKS
ANY FIRE	906-1	☐ COMPLETE	_____ DATE	☐ N/A	
STRUCTURE	906-2	☐ COMPLETE	_____ DATE	☐ N/A	REMARKS
VEHICLE	906-3	☐ COMPLETE	_____ DATE	☐ N/A	REMARKS
WILDLAND	906-4	☐ COMPLETE	_____ DATE	☐ N/A	REMARKS
CASUALTY	906-5	☐ COMPLETE	_____ DATE	☐ N/A	REMARKS
WITNESS	906-6	☐ COMPLETE	_____ DATE	☐ N/A	REMARKS
EVIDENCE	906-7	☐ COMPLETE	_____ DATE	☐ N/A	REMARKS
PHOTOGRAPH	906-8	☐ COMPLETE	_____ DATE	☐ N/A	REMARKS
SKETCH	906-9	☐ COMPLETE	_____ DATE	☐ N/A	REMARKS
INSURANCE	906-10	☐ COMPLETE	_____ DATE	☐ N/A	REMARKS
RECORDS/DOCUMENT	906-11	☐ COMPLETE	_____ DATE	☐ N/A	REMARKS

INCIDENT AND CASUALTY REPORTS UPDATED ☐ YES _____ DATE ☐ NO ☐ NOT NECESSARY

DATE	ACTIVITY	BY

FIGURE 4.1 ◆ NFPA 906-0, case supervision form: the top cover sheet to track the progress of the investigation. *Reprinted with permission from NFPA 906—Guide for Fire Incident Field Notes, 1998 Edition. Copyright © 1998, National Fire Protection Association, Quincy, MA 02269. This reprinted material is not the complete and official position of the National Fire Protection Association on the referenced subject, which is represented only by the standard in its entirety.*

ANY FIRE FIELD NOTES 906-1		AGENCY	FILE NUMBER

INCIDENT

ADDRESS/LOCATION			DAY	DATE	TIME	FIRE DEPT. INCIDENT NO.		
WEATHER AT TIME OF FIRE	GENERAL CONDITIONS				TEMP.	WIND DIR.		WIND SPEED
PROPERTY DESCRIPTION	STRUCTURE (906-2) ☐	VEHICLE (906-3) ☐		WILDLAND (906-4) ☐		OTHER ☐		

OWNER/OCCUPANT

OWNER'S NAME	PHONE NO.
OWNER'S ADDRESS	
OCCUPANT'S NAME	PHONE NO.
OCCUPANT'S ADDRESS	
DOING BUSINESS AS	PHONE NO.

NOTIFICATION FOR INVESTIGATION

DAY	DATE	TIME	FROM WHOM	
RECEIVED BY			ASSIGNED TO	
ARRIVED AT SCENE	DAY	DATE	TIME	SCENE SECURED ☐ NO (COMMENT ON CONDITION) ☐ YES (BY WHOM):
AUTHORITY TO ENTER	EMERGENCY	CONSENT ☐ VERBAL ☐ WRITTEN	WARRANT ☐ ADMIN. ☐ CRIM.	OTHER (Describe)
DEPARTED SCENE	DAY	DATE	TIME	COMMENTS

OTHER AGENCIES INVOLVED

FIRE DEPT.	INCIDENT NO.	CONTACT PERSON	PHONE NO.
POLICE DEPT.	FILE NO.	CONTACT PERSON	PHONE NO.
OTHER	CASE NO.	CONTACT PERSON	PHONE NO.

ESTIMATED TOTAL LOSS

$	ESTIMATED BY

REMARKS

FIGURE 4.2 ◆ NFPA 906-1, "Any Fire Field Notes": collects general dispatch notification, owner, occupant, and investigative and contact information. *Reprinted with permission from NFPA 906—Guide for Fire Incident Field Notes, 1998 Edition. Copyright © 1998, National Fire Protection Association, Quincy, MA 02269. This reprinted material is not the complete and official position of the National Fire Protection Association on the referenced subject, which is represented only by the standard in its entirety.*

STRUCTURE FIRE
FIELD NOTES 906-2a

AGENCY	FILE NUMBER

TYPE AND STATUS

PROPERTY USE

STATUS (OCCUPIED, UNOCCUPIED, VACANT)　　　　　　　　　COMMENTS

AREA DESCRIPTION

☐ RURAL　☐ FARM　☐ URBAN　☐ SUBURBAN　☐ OTHER _____

☐ ZONED　☐ UNZONED　☐ IMPROVING　☐ DECLINING　☐ STABLE　☐ OTHER _____

CONSTRUCTION

FOUNDATION
☐ SLAB　☐ CRAWL SPACE　☐ BASEMENT(S)　☐ OTHER _____

DIMENSIONS
_____ FT LENGTH　_____ FT WIDTH　_____ FT HEIGHT　_____ STORIES　_____ NO. UNITS

TYPE OF CONSTRUCTION	EXTERIOR WALLS	INTERIOR WALLS	FLOORS	ROOF

SECURITY (Time of Fire)

DOORS
☐ SECURE　☐ NOT SECURE　　　PER:

WINDOWS
☐ SECURE　☐ NOT SECURE　　　PER:

OTHER
☐ SECURE　☐ NOT SECURE　　　PER:

COMMENTS ON SECURITY

ALARM/PROTECTION SYSTEMS

ALARMS
☐ YES　☐ NO　　TYPE ALARM

ALARM COMPANY	CONTACT PERSON	PHONE NO.

COMMENTS

PROTECTION SYSTEMS
☐ YES　☐ NO　☐ OPERATED　☐ DID NOT OPERATE　　COMMENTS

DESCRIPTION OF SYSTEM(S)

UTILITIES (Time of Fire)

			UTILITY COMPANY NAME	CONTACT	PHONE NO.
ELECTRIC	☐ ON	☐ OFF			
GAS	☐ ON	☐ OFF			
WATER	☐ ON	☐ OFF			
PHONE	☐ ON	☐ OFF			
OTHER	☐ ON	☐ OFF			

FIGURE 4.3 ◆ NFPA 906-2a, structure fire form: collects structure description, security, protection, and utility contact information. *Reprinted with permission from* NFPA 906—Guide for Fire Incident Field Notes, 1998 Edition. *Copyright © 1998. National Fire Protection Association, Quincy, MA 02269. This reprinted material is not the complete and official position of the National Fire Protection Association on the referenced subject, which is represented only by the standard in its entirety.*

STRUCTURE FIRE
FIELD NOTES 906-2b

AGENCY	FILE NUMBER

EXTERIOR OBSERVATIONS

INTERIOR OBSERVATIONS

HEATING SYSTEM

TYPE

LOCATION

COMMENTS

ELECTRICAL SERVICE

☐ FUSES   ☐ BREAKERS

ENTRY LOCATION

SERVICE PANEL LOCATION

COMMENTS

OTHER HEATING EQUIPMENT

TYPE(S)

LOCATION

COMMENTS

STRUCTURE CONTENTS

COMMENTS

AREA OF ORIGIN

COMMENTS

FIGURE 4.4 ◆ NFPA 906-2b, structure fire form: collects exterior, interior, heating, electrical service, contents, and area of fire origin information. *Reprinted with permission from NFPA 906—Guide for Fire Incident Field Notes, 1998 Edition. Copyright © 1998, National Fire Protection Association, Quincy, MA 02269. This reprinted material is not the complete and official position of the National Fire Protection Association on the referenced subject, which is represented only by the standard in its entirety.*

STRUCTURE FIRE FIELD NOTES 906-2c	AGENCY	FILE NUMBER

IGNITION SEQUENCE

HEAT SOURCE

MATERIAL IGNITED

IGNITION FACTOR

IF EQUIPMENT INVOLVED
MAKE   MODEL   SERIAL NO.

COMMENTS

FIRE SPREAD

MATERIALS

AVENUES

COMMENTS

SMOKE SPREAD

MATERIALS

AVENUES

COMMENTS

REMARKS

FIGURE 4.5 ◆ NFPA 906-2c, structure fire form: collects ignition sequence and fire and smoke spread information. *Reprinted with permission from* NFPA 906—Guide for Fire Incident Field Notes, 1998 Edition. *Copyright © 1998, National Fire Protection Association, Quincy, MA 02269. This reprinted material is not the complete and official position of the National Fire Protection Association on the referenced subject, which is represented only by the standard in its entirety.*

MOTOR VEHICLE FIELD NOTES 906-3		AGENCY	FILE NUMBER

VEHICLE DESCRIPTION

COLOR(S)	YEAR	MAKE	MODEL	LICENSE — NO., STATE, EXPIRES	VIN NO.

OWNER/OPERATOR

OWNER'S NAME	OWNER'S ADDRESS	OWNER'S PHONE NO.
OPERATOR'S NAME/LICENSE NO.	OPERATOR'S ADDRESS	OPERATOR'S PHONE NO.

EXTERIOR

PRIOR DAMAGE	FIRE DAMAGE
TIRES/WHEELS (Missing, Match, Condition)	
PARTS MISSING	

FUEL SYSTEM

PRIOR DAMAGE		FIRE DAMAGE	
TYPE FUEL	CONDITION OF TANK	FILLER CAP CONDITION	FUEL LINE CONDITION

ENGINE COMPARTMENT

PRIOR DAMAGE	FIRE DAMAGE

FLUID LEVELS OIL _____ TRANSMISSION _____ RADIATOR _____ OTHER _____

PARTS MISSING

INTERIOR

PRIOR DAMAGE	FIRE DAMAGE

IGNITION SYSTEM   KEY IN IGNITION ☐ YES ☐ NO

PERSONAL CONTENTS MISSING

ACCESSORIES MISSING

ODOMETER READING	SERVICE STICKER INFORMATION

VEHICLE SECURITY

ALARM	DOOR AND TRUNK LOCKS	WINDOW POSITIONS

ORIGIN/IGNITION SEQUENCE

AREA

HEAT SOURCE

MATERIAL IGNITED

IGNITION FACTOR

FIGURE 4.6 ◆ NFPA 906-3, motor vehicle form: collects vehicle description, owner/operator, exterior/interior, security, and area of fire origin information. *Reprinted with permission from NFPA 906—Guide for Fire Incident Field Notes, 1998 Edition. Copyright © 1998, National Fire Protection Association, Quincy, MA 02269. This reprinted material is not the complete and official position of the National Fire Protection Association on the referenced subject, which is represented only by the standard in its entirety.*

WILDLAND FIRE FIELD NOTES 906-4		AGENCY	FILE NUMBER

PROPERTY DESCRIPTION

FIRE DAMAGE ☐ LESS THAN ACRE _____ NO. ACRES	OTHER PROPERTIES INVOLVED
SECURITY ☐ OPEN ☐ FENCED LOCKED ☐ GATES	COMMENTS

FIRE TRAVEL FACTORS

TYPE FIRE ☐ GROUND ☐ CROWN	FACTORS ☐ WIND ☐ TERRAIN	COMMENTS

AREA OF ORIGIN

PEOPLE IN AREA

AT TIME OF FIRE ☐ YES ☐ NO ☐ UNDETERMINED	COMMENTS

IGNITION SEQUENCE

HEAT OF IGNITION

MATERIAL IGNITED

IGNITION FACTOR

IF EQUIPMENT INVOLVED

MAKE	MODEL	SERIAL NO.

COMMENTS

FIGURE 4.7 ◆ NFPA 906-4, wildland fire form: collects property, fire ignition sequence, area of origin, and travel information. *Reprinted with permission from NFPA 906— Guide for Fire Incident Field Notes, 1998 Edition. Copyright © 1998, National Fire Protection Association, Quincy, MA 02269. This reprinted material is not the complete and official position of the National Fire Protection Association on the referenced subject, which is represented only by the standard in its entirety.*

CASUALTY FIELD NOTES 906-5	AGENCY	FILE NUMBER

DESCRIPTION

NAME	ADDRESS					PHONE NO.		
RACE	SEX	AGE	DATE OF BIRTH	HEIGHT	WEIGHT	HAIR	EYES	OTHER
DESCRIBE CLOTHING								

TYPE OF INJURY

☐ MINOR   ☐ MODERATE   ☐ SEVERE   ☐ FATAL   DESCRIBE INJURY

CIRCUMSTANCES

WHO FOUND VICTIM? WHERE?

VICTIM'S ACTIVITY JUST PRIOR TO AND AT TIME OF IGNITION

VICTIM'S ACTIVITY AFTER TIME OF IGNITION

CASUALTY TREATMENT

☐ TREATED AT SCENE BY?

SENT TO	VIA	TREATED BY
REMARKS		

FATALITY

BODY POSITION		
BODY REMOVED TO	BODY REMOVED BY	AUTHORITY TO MOVE BODY GIVEN BY
MEDICAL EXAMINER/CORONER	ADDRESS	PHONE NO.
CAUSE OF DEATH		
AUTOPSY BY	ADDRESS	PHONE NO.

DATE OF AUTOPSY	CASE NO.	BLOOD TEST ☐ YES ☐ NO	X-RAYS ☐ YES ☐ NO	REPORTS IN POSSESSION ☐ YES ☐ NO

NEXT OF KIN

NAME	RELATIONSHIP	ADDRESS AND PHONE
NOTIFIED BY (How, Date, and Time)		

REMARKS

FIGURE 4.8 ◆ NFPA 906-5, casualty form: collects casualty victim, injury, circumstances, treatment, fatality, and family information. *Reprinted with permission from NFPA 906—Guide for Fire Incident Field Notes, 1998 Edition. Copyright © 1998. National Fire Protection Association, Quincy, MA 02269. This reprinted material is not the complete and official position of the National Fire Protection Association on the referenced subject, which is represented only by the standard in its entirety.*

WITNESS STATEMENT FIELD NOTES 906-6		AGENCY	FILE NUMBER

IDENTIFICATION

NAME				ADDRESS		PHONE NO.
RACE	SEX	AGE	DATE OF BIRTH	SOC. SECURITY NO.		DRIVER'S LIC. NO.
EMPLOYER				ADDRESS		PHONE NO.
RELATIONSHIP TO INCIDENT				CAN BE CONTACTED AT		
STATEMENT TAKEN BY				LOCATION, DATE, AND TIME OF STATEMENT		

STATEMENT

FIGURE 4.9 ◆ NFPA 906-6, witness statement form: collects descriptive identification information and expected witness testimony. *Reprinted with permission from NFPA 906—Guide for Fire Incident Field Notes, 1998 Edition. Copyright © 1998, National Fire Protection Association, Quincy, MA 02269. This reprinted material is not the complete and official position of the National Fire Protection Association on the referenced subject, which is represented only by the standard in its entirety.*

EVIDENCE FIELD NOTES 906-7		AGENCY	FILE NUMBER

DESCRIPTION	WHERE FOUND/WHEN	REMOVED TO/BY
1.		
2.		
3.		
4.		
5.		
6.		
7.		
8.		
9.		
10.		
11.		
12.		

REMARKS

FIGURE 4.10 ◆ NFPA 906-7, evidence form: collects and itemizes evidence, location found, and chain of custody information. *Reprinted with permission from NFPA 906— Guide for Fire Incident Field Notes, 1998 Edition. Copyright © 1998, National Fire Protection Association, Quincy, MA 02269. This reprinted material is not the complete and official position of the National Fire Protection Association on the referenced subject, which is represented only by the standard in its entirety.*

PHOTOGRAPH FIELD NOTES 906-8		ROLL NO.	AGENCY	FILE NUMBER

*ONLY ONE ROLL OF FILM PER FORM.

NEG. NO.	DESCRIPTION	NEG. NO.	DESCRIPTION
1		21	
2		22	
3		23	
4		24	
5		25	
6		26	
7		27	
8		28	
9		29	
10		30	
11		31	
12		32	
13		33	
14		34	
15		35	
16		36	
17		37	
18		38	
19		39	
20		40	

REMARKS

FIGURE 4.11 ◆ NFPA 906-8, photograph log form: collects, itemizes, and obtains descriptions of photographs. *Reprinted with permission from NFPA 906—Guide for Fire Incident Field Notes, 1998 Edition. Copyright © 1998, National Fire Protection Association, Quincy, MA 02269. This reprinted material is not the complete and official position of the National Fire Protection Association on the referenced subject, which is represented only by the standard in its entirety.*

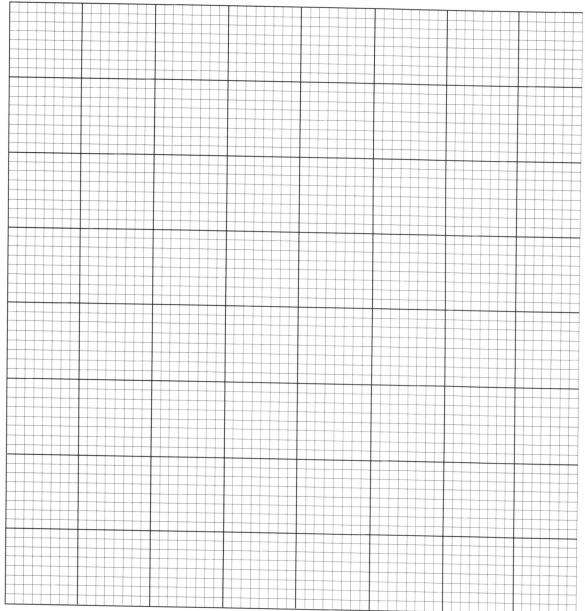

FIGURE 4.12 ◆ NFPA 906-9, fire scene sketch form: collects on-scene rough sketch of property for use in documenting evidence collected, photographs, and fire pattern damage information. *Reprinted with permission from* NFPA 906—Guide for Fire Incident Field Notes, 1998 Edition. *Copyright © 1998, National Fire Protection Association, Quincy, MA 02269. This reprinted material is not the complete and official position of the National Fire Protection Association on the referenced subject, which is represented only by the standard in its entirety.*

INSURANCE INFORMATION FIELD NOTES 906-10	AGENCY	FILE NUMBER

COMPANY

NAME	ADDRESS	PHONE NO.
1.		
POLICY NO.	EFFECTIVE DATE	EXPIRATION DATE
NAME	ADDRESS	PHONE NO.
2.		
POLICY NO.	EFFECTIVE DATE	EXPIRATION DATE

COVERAGE

STRUCTURE/VEHICLE	CONTENTS, PERSONAL PROPERTY	BUSINESS INTERRUPTION, LOSS EARNINGS, LIVING EXPENSES
1. ☐ NEW ☐ RENEWAL	NAME OF INSURED	ADDRESS OF INSURED
2. ☐ NEW ☐ RENEWAL	NAME OF INSURED	ADDRESS OF INSURED
PREVIOUS INSURANCE CARRIER NAME	ADDRESS	PHONE NO.

$ _____ STRUCTURE/VEHICLE $ _____ CONTENTS $ _____ OTHER ? _____

PREVIOUS LOSSES, CANCELLATIONS

INSURANCE AGENT

NAME	ADDRESS	PHONE NO.
1.		
NAME	ADDRESS	PHONE NO.
2.		

ADJUSTER/INVESTIGATOR

NAME OF COMPANY ADJUSTER/INVESTIGATOR	ADDRESS	PHONE NO.
1.		
NAME OF COMPANY ADJUSTER/INVESTIGATOR	ADDRESS	PHONE NO.
2.		
NAME OF PUBLIC ADJUSTER	ADDRESS	PHONE NO.

TOTAL PAID LOSS

STRUCTURE	CONTENTS/PERSONAL PROPERTY	OTHER (Explain)
1. $	1. $	1. $
STRUCTURE	CONTENTS/PERSONAL PROPERTY	OTHER (Explain)
2. $	2. $	2. $

REMARKS

FIGURE 4.13 ◆ NFPA 906-10, insurance information form: collects company, policy coverage, agent, adjuster, investigator, and loss payout information. *Reprinted with permission from NFPA 906—Guide for Fire Incident Field Notes, 1998 Edition. Copyright © 1998, National Fire Protection Association, Quincy, MA 02269. This reprinted material is not the complete and official position of the National Fire Protection Association on the referenced subject, which is represented only by the standard in its entirety.*

RECORDS/DOCUMENTS FIELD NOTES 906-11	AGENCY	FILE NUMBER

Use this form as a checklist to indicate which records have been considered in the investigation. The Remarks sections should be used to note availability, contacts, and so forth.

INCIDENT RELATED

FIRE DEPT. NAME	INCIDENT NO.	REMARKS
POLICE DEPT. NAME	FILE NO.	REMARKS
INSURANCE CO. NAME	CASE NO.	REMARKS
GAS CO. NAME	REMARKS	
ELECTRIC CO. NAME	REMARKS	
MEDIA COVERAGE	REMARKS	
MEDIA COVERAGE	REMARKS	
MEDIA COVERAGE	REMARKS	
OTHER — INCIDENT RELATED	REMARKS	
OTHER — INCIDENT RELATED	REMARKS	

PROPERTY RECORDS

MORTGAGE HOLDER	REMARKS
LIEN HOLDER	REMARKS
TAX RECORDS	REMARKS
CONTRACTS/LEASES	REMARKS
TITLES/REGISTRATIONS	REMARKS
ZONING/CODES	REMARKS
DEEDS	REMARKS
OTHER	REMARKS
OTHER	REMARKS

BUSINESS/PERSONAL

ACCOUNTING	REMARKS
INVENTORY	REMARKS
BANKS/CREDIT UNIONS, ETC.	REMARKS
BUSINESS AND PERSONAL TAX	REMARKS
CRIMINAL HISTORY	REMARKS
CIVIL LITIGATIONS	REMARKS

FIGURE 4.14 ◆ NFPA 906-11, records/documents form: collects incident-related, property, business, and personal information obtained during the investigation. *Reprinted with permission from NFPA 906—Guide for Fire Incident Field Notes, 1998 Edition. Copyright © 1998, National Fire Protection Association, Quincy, MA 02269. This reprinted material is not the complete and official position of the National Fire Protection Association on the referenced subject, which is represented only by the standard in its entirety.*

EXAMPLE FORM FOR DATA FOR COMPARTMENT FIRE MODELING

Room Number _____ Use _____

Size (use diagrams if possible)   Wall/floor/ceiling

Construction

  Length _____   _____

  Width _____   _____

  Height _____   _____

Lining Materials (that represent over 10% of room lining)
(Include thickness, density, and other material characteristics if known)

  Wall Material   Percentage of Walls or Area Involved

  _____   _____

  _____   _____

  _____   _____

  Ceiling Material

  _____   _____

  _____   _____

  Floor or Floor Covering Material

  _____   _____

  _____   _____

Doors, Windows, and Other Openings [Enter all heights as distance above floor. If door sill is at floor, enter zero (0).]

Openings	to Top	to Sill	Width	Changes During Fire (How?)[1]
_____	___	___	_____	_____
_____	___	___	_____	_____
_____	___	___	_____	_____
_____	___	___	_____	_____
_____	___	___	_____	_____
_____	___	___	_____	_____

[1] For example: "Window broke at 10:33" or "Door was closed until opened by escaping occupant, then left open — Exit Time 10:30."

(Page 1 of 2)

FIGURE 4.15 ◆ NFPA 921 fire modeling form, page 1: collects compartment size, lining material, and opening information. *Reprinted with permission from* NFPA 921—Guide for Fire and Explosion Investigations. *Copyright © 2001, National Fire Protection Association, Quincy, MA 02269. This reprinted material is not the complete and official position of the National Fire Protection Association on the referenced subject, which is represented only by the standard in its entirety.*

DATA FOR COMPARTMENT FIRE MODELING

Heating, Ventilation, and Air Conditioning (HVAC). Include air flows from HVAC systems. Give rates and positions of supply and return or exhaust in this room. Also sizes and types of ducts/diffusers.

_____

_____

_____

Tightness of Walls, Closed Windows, Door Fits, etc. (Unless fit is very loose, classify as tight, average, or loose. If fit is very loose, try to get size, number, and location of cracks, etc.)

Doors _____

Windows _____

Inside Walls _____

Exterior Walls _____

Fire History (List all significant events involving progress of the fire.)

Time (hard or soft)	Event
e.g. 1:10 am	sofa involved, flames 3 feet high
1:17 am	room flashover
1:19 am	large fire plume into hallway
1:23 am	smoke out of third floor window

_____

_____

_____

_____

_____

Initial Fuel Item(s) Description

Description	Size	Material
e.g. sofa	full	polyurethane, with cotton upholstery

_____

_____

_____

_____

_____

Suspected Ignitor (List ignitor if known with qualification on confidence.)

Ignitor: __e.g. cigarette_____

Confidence: _probable_____

(Page 2 of 2)

FIGURE 4.16 ◆ NFPA 921 fire modeling form, page 2: collects HVAC, leakage, fire history time line, fuel package, and ignition information. *Reprinted with permission from NFPA 921—Guide for Fire and Explosion Investigations. Copyright © 2001, National Fire Protection Association, Quincy, MA 02269. This reprinted material is not the complete and official position of the National Fire Protection Association on the referenced subject, which is represented only by the standard in its entirety.*

Table 4.2 shows the recommended updated guidance and purpose for this systematic documentation philosophy. Note that this approach covers fire investigations without any preconception as to whether the fire is accidental, natural, or incendiary in origin.

TABLE 4.2 ◆ Systematic Documentation Techniques in Fire Investigation

Step	Technique	Guidance	Purpose
1	Exterior	Photograph perimeter of property. Sketch exterior. Use GPS to obtain location.	Establishes venue and location of fire scene in relation to surrounding visual landmarks Documents exposure damage to adjacent properties Reveals structural conditions, failures, violations, or deficiencies Establishes extent of fire damage to exterior of scene Establishes egress and condition of doors and windows
2	Interior	Record and sketch extent of fire damages, potential ignition sources, and data needed for fire reconstruction and modeling.	Traces fire travel and development from exterior to suspected point(s) of fire origin Documents heat transfer damage, heat and smoke stratification levels, and breaches of structural element Documents condition of power utility and distribution, furnace, water heater, and heat producing appliances Documents position and damage to windows, doors, stairwells, and crawl space access points Documents fire protection equipment locations and operation (sprinklers, heat and smoke detectors, extinguishers) Documents readings on clocks and utility equipment Documents alarm and arc fault information
3	Investigative	Document clearing of debris and evidence prior to removal, fire burn and charring patterns, packaged evidence.	Assists in recording fire pattern and plume damage, isochar lines Establishes condition of physical evidence, utilities, distribution, and protection equipment (breakers, relief valves) Documents integrity of the chain of custody of evidence
4	Panoramic	Produce multidimensional sketches.	Clearer peripheral views of exterior and interiors Establishes photograph viewpoints of witnesses

Source: Updated from Icove and Gohar (1980).

In an effort to show the utility of the four-phase documentation approach, the use of the *NFPA 906* forms is introduced. This discussion also includes several areas where *NFPA 906* does not capture information needed for full fire scene reconstruction and engineering analysis of the incident.

Agencies that use the *NFPA 906* field notes often copy the blank forms directly from the standard and bind them into packets that are placed into case jackets. Several additional copies of the witness statement form are often included for multiple interviews.

The "Case Supervision Field Notes" (Figure 4.1) include a case progress sheet (Form 906-0) that is intended to be the cover sheet of the investigative notes. This cover sheet is to be used as a working document for both investigator and supervisor. Included on the form are check boxes indicating whether a form is contained in the case file, when it was completed, and remarks as to its status. The disposition of evidence, court times, and other major information may be noted in the activity section of the report.

Another high-level case document is Form 906-1, "Any Fire Field Notes" (Figure 4.2). This form captures the series of events as to how an agency is notified of the incident, the conditions upon arrival, the owner/occupant of the property, other agencies involved, and an estimated total financial loss. Also documented are the time of arrival, the legal authority to enter the scene, and the time the scene was released.

The weather conditions prior to the fire incident sometimes become important, particularly when high winds, temperature fluctuations, or lightning come into play. The National Weather Service, Office of Climate, Water, and Weather Services, has a forensic services program to support investigations, particularly those involved in litigation. Certified climatological records (including radar images, satellite photos, and surface analysis) can be obtained from the National Weather Service Headquarters in Silver Spring, Maryland.

Alternative free and low-cost weather notification and historical systems are available in the United States. These services provide radar and severe weather notifications. One popular Web site is the Weather Underground.

The location of lightning strikes is an important database when addressing whether to rule in or rule out the possibility that lightning caused the fire. The U.S. National Lightning Detection Network locates strikes across the United States. For a fee, a private firm, called Vaisala, will provide a report on all lightning strikes in a particular area for a given time frame. Vaisala (2003) can also supply data and custom software to its customers at www.vaisala.com.

◆ **4.3 EXTERIOR**

The first major phase is to document the exterior of the structure, vehicle, forest, wildland, boat, or object prior to probing into the cause of the fire. While encircling the exterior of the structure or vehicle, a cursory field search can also be made for additional evidence.

The purpose of this phase is to establish the location of the fire scene in relation to surrounding visual landmarks. The exterior views should reveal the extent of fire damage, collapse, structural conditions, failures, code violations, deficiencies, or potential safety concerns. This technique also documents exposure damage to adjacent properties from the fire. In large investigations, a useful technique is the use of an overhead crane, aerial ladder truck, or aircraft to obtain photographs of the scene.

If there has been any explosion, the distance to which fragments of glass or structure were thrown must be measured and documented via both diagramming and photography. These concepts are discussed in later sections.

The "Structure Fire Field Notes" (Form 906-2a; Fig. 4.3) documents structure fire cases by recording the type of property, geographic area, construction techniques, security, alarm protection, and utilities. Documenting the security at the time of the fire is an important consideration in arson cases where the issue of "exclusive opportunity" is raised along with the condition of doors, windows, and protection systems. Also important is documenting the condition of doors, windows, and evidence found as part of the external documentation.

The condition of the utilities at the time of the fire is important, particularly when considering all other sources of ignition. The status of the utilities may indicate whether the owner/occupant(s) was(were) living in the structure. It is important to document who actually may have given the instructions to disconnect the utilities if it was done prior to the fire. Determining when gas and electric services were cut off during fire suppression is also important.

◆ 4.4 INTERIOR

The second phase involves the documentation of interior damage by showing the extent and progress of the fire, through the room(s), area(s), and suspected point(s) of fire origin. These pictures and sketches are made prior to excavating the fire debris and serve to document the conditions of the scene as found upon the investigator's arrival.

The "Structure Fire Field Notes" (Forms 906-2b and -2c; Figures 4.4 and 4.5), "Motor Vehicle Field Notes" (Form 906-3; Figure 4.6), and "Wildland Fire Field Notes" (Form 906-4; Figure 4.7) are used to document further the property, utility service, contents, preliminary area of fire origin, estimated ignition sequences, and fire and smoke spread factors.

DOCUMENTATION OF DAMAGE

While conducting the internal review, the investigator documents damage to all rooms, including heat and smoke stratification levels, heat transfer effects, and breaches of structural elements (walls, floors, ceilings, and doors). The investigator may find it useful to map out and delineate the areas of damage corresponding to the pattern types described in Table 3.1. This can be accomplished with color-coded chalk or tape on the surfaces themselves, followed by photography. Areas of surface "demarcations," thermal effects, penetrations, and loss of material can also be outlined using colored markers on sketches or photographs. One system might be the use of yellow lines to outline areas of surface deposits, green lines for thermal effects, blue lines for penetrations, and red lines to outline areas where the material has been consumed completely.

At the same time, this allows for the examination and documentation of the condition of power utility and distribution, furnace, water heater, and heat producing appliances, as well as the contents and conditions of the rooms. The thickness of window glass, the sizes of windows, and whether they are single- or double-glazed must also be noted and documented. The sill height and soffit depth of each door and window in the rooms involved must be noted, as well as their opening dimensions.

FIRE PROTECTION SYSTEMS

It is also important to document for later evaluation the fire protection equipment (fire alarms, smoke detectors, automatic suppression systems) in place that would have helped contain the fire as well as alert the occupants. The time readings on electric and mechanical clocks can serve to document the approximate times that they were damaged and halted by heat or by interruption of power.

The data systems in modern fire alarm systems can be interrogated to capture the time, sequence, and zone of alarms or sprinkler activations. The zones and manner of activation should be established for alarm sensors.

ARC FAULT MAPPING

When a flame or heat attacks an insulated cable, the insulation begins to pyrolyze and degrade. In rubber, cloth, or some plastics, when the insulation chars, the carbonaceous residue becomes conductive (sometimes called "arcing through char"). In some plastics like PVC, when the insulation reaches temperatures in excess of 200°C, the inorganic fillers in the plastic allow current to begin to flow. The resistance decreases as the insulation gets hotter. In either case, current begins to flow between the hot or energized wire and the neutral or ground (or to a grounded conduit, appliance, or junction box).

Because of the resistance provided by the charred insulation, this is not the same as a direct short or fault (which has nearly zero resistance). The current flow can vary from very low to very high (as high as 150 A has been measured). Because the conductors are not bonded together, they are free to move, so these current flows are usually very short in duration, in many cases too short to cause overcurrent protective devices (OCPD) to function. This means that arcing can occur at many places along a single energized wire and multiple times until the OCPD trips or the wire is severed by melting (sometimes called fusing). Once the wire is separated, no current can flow beyond that point.

These facts provide a way of locating a possible area of origin of a fire by tracing the wiring from its source (such as the breaker or fuse box) and mapping the locations of all failures of the wire (since wires can be caused to fail by fire effects even when nonenergized). The failures are then characterized as arcing (energized) or melting (nonenergized).

The locations of the arcing failures farthest downstream along a circuit from the power source are indicators of the point at which the fire first attacked the wiring, as in Figure 4.17. These failures, then, may be very useful in indicating a possible area of origin. This approach was pioneered by Dr. Robert Svare and has been tested in numerous live-burn structure tests.

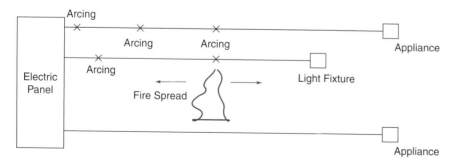

FIGURE 4.17 ◆ Mapping the locations of arcing failures farthest downstream along a circuit from the power source, which are indicators of the point at which the fire first attacked the wiring.

It has been shown to be effective even in multistory buildings and installations of "ring-mains" wiring (as in the United Kingdom) (Carey 2002) It is not effective in three-phase wiring systems, where there are two energized wires in every circuit. It requires diligent and careful tracing of wiring and circuits and accurate evaluation of the nature of the failure. In buildings protected by zone alarm systems, arc fault indicators can be used together with data from the alarm system (alarm, sensor, or sprinkler activation) to estimate areas of possible origin.

◆ 4.5 INVESTIGATIVE

The third phase is the systematic investigation concentrating on the debris clearing operations, char and burn patterns, and position of evidence prior to its removal from the fire scene. Should crimes associated with the fire such as burglary, theft, and homicide have taken place, evidence of these acts should also be documented.

Nothing should be moved, including bodies of victims, until photographically recorded and supported by fire scene notes. Investigative photographs ensure the integrity of the probe and custody of evidence.

CASUALTIES

Information on casualties is recorded on the "Casualty Field Notes" form (906-5; Figure 4.8). This report includes a description of the victim, type of injury, circumstances, treatment received, disposition of the body and its examination, next of kin, and other appropriate remarks.

This form does not presently include general information about the victim obtained at autopsies such as burn injuries, tests for blood alcohol, hydrogen cyanide, and carboxyhemoglobin levels, and other conditions often documented in fire death investigations. That information is discussed in length in Chapter 7.

WITNESSES

Documented information from witnesses is captured on the "Witness Statement" form (906-6; Figure 4.9). Complete information as to witness identification, home and work addresses, contact, and expected testimony is initially recorded on this form by investigators.

In cases involving on-scene witnesses to the fire, thorough interviews may reveal additional information relating to the initial stages of the fire and the environmental conditions at the time (rain, wind, extreme cold, etc.). Furthermore, it is important to document details as to the position of the witnesses in relation to the fire scene when recording their visual observations. This process can be enhanced by walking witnesses through the scene (when safe) or back to the location from which they witnessed the fire. This can help establish or confirm lines of sight and prompt more complete statements.

EVIDENCE COLLECTION AND PRESERVATION

The chain of custody is intended to trace the item of evidence from its discovery to court. Its purpose is to authenticate the evidence as it is found as well as to prevent its loss or destruction. The documentation includes photographing the evidence at its discovery and preparing a written list itemizing the transfer of the evidence once it leaves the scene, as with the "Evidence Field Notes" (Form 906-7; Figure 4.10).

PHOTOGRAPHY

The "Photograph Field Notes" form (906-8; Figure 4.11) is used to record the description, frame, and roll number of each photograph taken at the fire scene. The form is designed to be filled out as the photographs are taken. The frame and roll numbers are used later on the fire scene sketch to indicate the location and direction from which they were taken. Documentation of each roll of film should appear on a separate form. The "remarks" field is used to document the disposition of the film.

SKETCHING

Fire scene sketches, whether or not to scale, are important supplements to photographs. A sketch graphically portrays the fire scene and items of evidence as recorded by the investigator. The "Sketch Field Notes" form (906-9; Figure 4.12) is for use by investigators when drawing simple two-dimensional sketches of fire scenes. Square-grid paper is available in many sizes at stationery supply stores. These sketches may be rough exterior building outlines or detailed floor plans.

DOCUMENTARY RECORDS

The collection of insurance information (Form 906-10; Figure 4.13) and documentary records (Form 906-11; Figure 4.14) is a requirement to ensure that a thorough investigation is conducted. This is particularly true in the event of the tracking and management of insurance information of multiple policy holders as in the case of commercial buildings. Care should be taken accurately to catalog and secure incident, property, business, and personal documentary records.

COMPARTMENT FIRE MODELING DATA

The documentation of data needed for compartment fire modeling exceeds the data collected in the *NFPA 906* forms. This additional detail is covered in the two-page *NFPA 921* fire modeling form shown in Figures 4.15 and 4.16. Included is information on the compartment for the model such as room size, construction, surface materials, openings in doors and windows, HVAC, case time line, fuel packages, and suspected source of ignition.

PHYSICAL EVIDENCE

Physical evidence is sometimes called the "silent witness" because it can provide reliable answers to questions other investigative techniques cannot address, fill in details, and corroborate other information. Clearly, the investigative phase of systematic investigations brings together all of the essential elements of forensic evidence collected from the fire scene.

Generally accepted forensic guidelines are designed to prevent contamination, loss, or destruction of the evidence and provide a reliable chain of custody for that evidence. For instance, when dealing with dried blood, the object itself should be collected whenever possible and allowed to air-dry before packaging. Place bloody objects into individual sealed paper bags, boxes, or envelopes after drying. Keep dry and refrigerate if possible. Do not use plastic bags since they will not allow the sample to ventilate. All investigators should be aware of the safety issues involving blood-borne pathogens.

All firearms should be placed into individual manila envelopes and rifles should be tagged. Firearms should be hand-carried to the laboratory. Special handling instructions include the collecting technique, noting the cylinder position in revolvers and recording the serial number, make, and model of the weapon. Fire debris or containers containing volatile liquids must be sealed in appropriate packaging and kept cool to minimize evaporation. Further details are included in *Kirk's Fire Investigation*, 5th ed. (DeHaan 2002).

The science of criminalistics becomes important when the scene is complex or serious (deaths or injuries) and every avenue of information seeking must be explored. Evidence may include physical items such as debris or incendiary devices, witness interviews, casualty injuries or postmortems, and fire scene photographs and sketches. Criminalistics is the science of examining all these items of physical evidence to link them to a common origin, identify them, or aid in the reconstruction of events.

◆ 4.6 PANORAMIC PHOTOGRAPHY

A majority of the photographs taken at fire scenes focus on fire patterns and other evidence. In structure fires, an aerial or perspective view of the building can reveal the overall impact that the fire had on the structure, including how the building itself was breached by both the fire and efforts to extinguish the blaze. An example of a panoramic photograph of a fire scene created by stitching several photographs together is shown in Figure 4.18.

Panoramic cameras, popular in the late 1800s, were often used to capture the ravaging effects of fires. Many of the present-day cameras, which simply crop the top and bottom portions of the image, create a false appearance of a panoramic photograph. However, superwide-angle and 360° rotating cameras are now available but are costly (Curtin 1999).

FIGURE 4.18 ◆ Construction of a panoramic view using individual photographs and stitching software. *Courtesy of D. J. Icove.*

Panoramic photographs can also be used to establish viewpoint photographs. This type of photography can be used to document what a witness might have seen or not seen from a particular location. The peripheral vision or field of view of the normal human eye is not readily duplicated by a normal camera lens, so sometimes panoramic imaging is the only way to replicate the view a witness may have had.

Simply stated, using a steady tripod, a series of overlapping photographs can be taken and prints later overlaid to form a mosaic view of the fire scene. Overlaying several photographic shots in the form of a mosaic can compensate for the lack of a true panoramic photograph. For best results a 35- to 55-mm-focal length lens is used. Wide-angle and telephoto lenses introduce unwanted distortions. Once an investigator has become familiar with the results and limitations of the camera, an attempt should be made to develop an expertise with panoramic photography.

Techniques used in the past included the careful mounting of overlaid photos onto a cardboard backing, where the photographs are then rephotographed. Note that even though this technique can produce clearly defined fire scenes, it should not replace existing methods. Newer technologies allow for an electronic merging of scanned images, where an image processing algorithm stitches individual images together. These software programs are frequently found packaged with digital camera accessories and are also included in professional image editing programs, such as QuicktimeVR and Adobe Photoshop.

In some photographic stitching programs using fisheye camera lenses, the viewing software assumes a central observation point, and the user interactively browses around the scene and can even link to views in adjacent rooms. Full 360° immersive images allow viewing of floors, ceilings, and walls in any direction. Images can then be linked together or linked to traditional photographs and renderings, audio, or other file types. Technologies such as these could one day play an important role in jury presentations (iPIX 2003).

EXAMPLE 4.1   The School Fire

School fires with massive structural destruction are difficult to reconstruct. A fire scene reconstruction was requested of a large Southeastern school fire that occurred some years ago. A majority of the contents and walls had been removed and placed outside of the structure prior to examination of the scene.

Using a combination of eyewitness statements, raw news videos, scene analysis, and knowledge of plume geometry, a fire scene reconstruction was possible. Investigation revealed that a fire started on the school office floor adjacent to filing cabinets where student records were stored.

A methodical examination of the debris, the dynamic time sequence of the fire's movement, and the damage caused by the fire confirmed the office to be the room of fire origin. A careful fire scene analysis of pattern damage revealed a roughly elliptical penetration at the floor of the office where the fire originated. At the center of this elliptical pattern was the apparent point of fire origin.

Yellow chalk was used to sketch on the floor the location of the desk, cabinets, and walls. These chalk lines were photographed from the second floor and displayed as a mosaic overlay forming a panoramic view of the fire scene. After piecing the panoramic photos together, the walls were marked using graphic chart tape on a plastic overlay. Plastic overlays are important for courtroom presentation, since they can be removed if objections are raised to their use. A photograph of this exhibit with the overlay present is shown to the left of the schematic floor plans in Figure 4.19. Such "ghost" walls can now be inserted using the graphics packages described previously.

FIGURE 4.19 ◆ The panoramic photo overlay from Example 4.1 using graphic chart tape on a plastic overlay to show missing walls and area of fire origin. Floor plans on right show (bottom to top) the successive fire spread as seen through exterior windows. *Courtesy of D. J. Icove.*

Other visual exhibits offering a perspective view include the use of scale models for courtroom presentations, which is one of the most persuasive exhibits that can be introduced. Care must be made to define and capture accurately the overall dimensions of the building being recreated. Architectural firms often have technicians on staff that can prepare these models. Care should be taken to allow for a removable roof on the model's structure, giving access to the building's interior layout.

◆ 4.7 APPLICATION OF CRIMINALISTICS AT FIRE SCENES

As a simple example, an investigator arrives at a crime scene to find a broken window, an unlatched door, and blood smears, all visible evidence and easily documented (Figure 4.20). The reconstruction of this evidence would be that the door was originally locked and its window broken by mechanical force (determined by glass fracture patterns to have

FIGURE 4.20 ◆ The reconstruction shows that the intruder cut himself/herself on the glass and, while fumbling for the lock and latch, left blood smears behind on the door. Glass fracture patterns document that the entry was made from the outside. *Courtesy of Lamont "Monty" McGill, McGill Consulting, Gardnerville, NV, by permission.*

come from the outside), and the intruder cut himself/herself on the glass and, while fumbling for the lock and latch, left blood smears behind on the door, as shown in Figure 4.20.

Analysis of the blood (or any fingerprints left on the door, glass, or latch) could identify the individual present. Glass found on the clothing could be compared to window glass from the door to confirm a two-way transfer (suspect blood to scene, scene glass to suspect) and distribution of cuts or abrasions (and glass/paint fragments) would confirm the method of entry. The services offered by many public and private criminalistics laboratories can be of great help to the fire investigator in the reconstruction of scenes and events and the testing of hypotheses about what happened, where, when and in what sequence. They can also link a person with a scene or a victim, and a scene with a person or a vehicle.

CRIMINALISTICS

There is a set of generally accepted forensic guidelines for collecting and preserving physical evidence recovered from fires during a fire scene investigation (DeHaan 2002).

The application of these best-accepted practices is derived from procedures used by professional criminalists.

Criminalistics has been defined as the application of the methods and knowledge of the natural sciences (physics, chemistry, biology, botany, etc.) to legal inquiries. Although not commonly in use in the United States until the 1950s, the term derives from the German word *Kriminalistik* (from the 1880s) for forensic evidence analytical techniques.

Criminalistics involves the extensive use of physical science to analyze and identify materials but, more importantly, to compare, classify, and individualize items of physical evidence, often with the intent of establishing a common origin (or excluding one). This physical evidence can take any form—impressions from shoes, tools, or friction ridge skin (on fingers, palms or feet), tissue, blood or other physiological fluids, glass, paint, soil, grease, oil, dyes, inks, documents, physical matches, or residues of chemicals associated with fires or explosions.

Criminalistics often goes further than simple comparison or identification to include the analysis of human behavior and physical dynamics of force, impact, and transfer and the reconstruction of scenes in both physical and dynamic aspects.

GRIDDING

Methodical techniques to assist in forensic evidence collection and documentation have been widely used by archaeologists when processing field sites. Fire investigators can adopt similar sketching and photographic techniques (Bailey 1983).

Coordinate systems aligned perpendicular to major structural walls can aid in the placement of grid lines, especially when the zero point is a corner, usually on the bottom or top left-hand side. Triangulation measurements are distances to an item of interest within a room from two fixed points. Angular displacement uses one fixed point and a compass direction and is the most appropriate for large outdoor scenes (Wilkinson 2001).

A grid system is well suited for large-scale scenes such as explosions where shrapnel and other evidence are hurled away from a central location. A grid system can also be used in small scenes, for example, quadrants within a vehicle. Squares within the grid typically use the alphabet along one coordinate and numerals along the other. Using both alpha and numeric coordinate identifiers simplifies and reduces the possibility of inadvertently switching the coordinates. Figure 4.21 is an example of a grid system at a fire scene. A baseline and distance system can also be used for large outdoor scenes, particularly explosions, where there are no "natural" rectilinear baselines.

While most scenes can be layered and examined room by room, some scenes require more carefully controlled examination. This is particularly true in fire death scenes where the location of small items is very important, particularly in the vicinity of the body. Using techniques developed in archeology, the scene is divided into grid squares using the walls as reference base lines. Rope, string, or even chalk can be used to mark out grids, numbered in one direction, lettered in the other. These grids may be 2–3 ft (0.5–0.8 m) in critical areas and as large as 10×10 ft (3×3 m) in surrounding areas with lighter debris concentrations.

Debris is removed from each grid square, layer by layer, for manual/visual search and sieving. Evidence (or unknown materials) recovered from each grid is kept in a bag, can, or envelope designated by corresponding number/letter. This ensures that at a later stage the evidence can be placed back to within 1 ft of its original location. While time-consuming and labor-intensive, this is the best way of finding and documenting evidence that permits its physical reconstruction after the scene has been completely searched.

FIGURE 4.21   ◆   Methodical techniques using gridding assist in forensic evidence collection and documentation. *Courtesy of Lamont "Monty" McGill, McGill Consulting, Gardnerville, NV, by permission.*

Archaeologists use a systematic approach of coordinate systems, which account for not only the location but also the depth of where an object is found. This approach can be helpful when layering debris at fire scenes that has fallen onto an area of fire origin. Such cases include multistory buildings, where collapsed floors bury important evidence. In some cases fallen debris often preserves evidence on lower floors.

"Iso-damage" curves have been used by archaeologists in documenting the impact of fires on historic monuments. One study placed more intense damage within the interior of a section of the Parthenon due to a fire set by the Celts in 267 ACE (Tassios 2002).

As previously noted, it is important to place these layers spatially in the documentation process. If forensic evidence collection and documentation are viewed as a scientific endeavor, the use of archaeological techniques will only enhance this effort.

DOCUMENTATION OF WALLS AND CEILINGS

While most investigators are diligent about identifying floor coverings and assessing their potential contributions to fire spread, the same cannot be said of wall and ceiling materials and coverings. As can be seen from the compartment fire modeling survey form shown in Figures 4.15 and 4.16, the nature and thickness of walls, ceilings, and their coverings are important for accurate reconstruction of the event.

In some cases, noncombustible walls of concrete, masonry, or stucco will not contribute to the fire spread, but their low thermal conductivity and large thermal mass may affect some stages of the fire's development. Plaster (whether with metal or wood lath) will dehydrate and fail, allowing fire to penetrate into ceilings or wall cavities. Modern gypsum board will resist fire spread for some time if properly installed (typically 15–20 min of direct fire exposure) before it collapses. "X" or fire-rated gypsum wallboard is $\frac{5}{8}$ in. or thicker and contains fiberglass fibers to strengthen the gypsum. As a result, it will withstand 30 min or longer of direct fire contact.

High temperatures from fires have an effect on sandstone surfaces and materials. Sandstones with different cement types exposed to fire have been studied and documented for changes in color, spalling, cracking, rounding-off of edges, and disintegration. Monominerallic sandstones with quartz grain and silica cement show thermal expansion cracks at 575°C. Similar behavior is seen in mica and feldspar sandstones. Sandstones containing carbonate and clay materials disintegrate between 450 and 750°C (Hajpál 2002).

Walls and ceilings can also be made of solid wood, plywood paneling, fiberboard (high density like Masonite or low density like Celotex), particle board, OSB (oriented-strand board), or even metal. Each has its own contribution to fire spread. Noncombustible walls can be covered with combustible materials. Thin plywood paneling or low-density cellulose can contribute enormously to fire spread. Karlsson and Quintiere (1999) estimated that lining a room or covering a ceiling with low-density fiberboard would cut the development-to-flashover time in half compared to that for the same room with noncombustible walls or ceilings. These materials will almost guarantee a flashover fire (if adequately ventilated), and can be destroyed so completely as to make their detection difficult.

The investigator must be familiar with these materials and how they are installed. Irregular squiggles or zigzag lines of charred adhesive on cement or plasterboard walls are a sure sign that paneling was glued there at one time. Remnants of the paneling will usually survive at the toe plate or behind baseboards, plumbing, or electrical fixtures. The small nails used to secure paneling or cellulose tiles are very different from those used to secure gypsum board.

Wood or metal furring strips may be used to install tiles on walls or ceilings and their presence should be taken as a cue to search out and identify the sometimes fragmentary remains of combustible walls or ceilings. Wall or ceiling tiles are sometimes installed only with dabs of cement on the back side, so patterns of large dark dots on remaining surfaces should be carefully examined. Samples of unburned wall covering should be measured, identified, and retained for later confirmation.

In one case of the author's experience, postfire scene photos showed only bare studs, with gypsum board on only one (exterior) face of a common wall. Careful searching revealed that the wall facing the room of origin was covered only in thin plywood veneer paneling that had burned nearly completely and contributed to an externally intense and fast-growing fire.

The interior dimensions of all rooms involved in the fire (by flame or smoke penetration) must be recorded (preferably to the nearest ±2 in. [50 mm]). This includes the heights of various rooms (not assuming all rooms in a structure to be the same). Height is often overlooked as a measurement, but it is critical when evaluating the size of fire required for flashover or effects of a given size of initial fire, estimating smoke filling rates, visibility, and tenability, and the like. Unusual features such as skylights should be measured and photographed.

The nature of the ceiling itself will affect the spread of smoke and fire. A flat (level), smooth ceiling will allow the smoke and hot gases from the ceiling jet to spread uniformly in all directions, a pitched (slanted) ceiling, of course, will direct the majority of the spread upwards. Headers above doors, exposed structural beams, even decorative "add-on" beams, and ceiling-mounted ductwork will limit spread dramatically, in some cases allowing the buildup of a sufficient hot gas layer in the side of the room containing the initial fire to the point of transition to flashover. Open ceiling joists (common in unfinished basements) will direct most of the accumulating hot gases along their length while dramatically reducing (if not preventing) transverse spread into adjacent joist spaces. Such features must be measured and documented.

The action of fire and subsequent extinguishment and overhaul activity may have obliterated signs of wall and ceiling covering so efforts should be made to find prefire photos or videos that show the character of the ceiling and walls. Failing that, interviews should be taken of owners, occupants, guests, customers, visitors, maintenance personnel, etc., asking them to describe ceiling and wall finish (as well as type and placement of furniture).

LAYERING

At scenes where there has been significant destruction and collapse of furniture, walls, or ceilings, investigators would be well advised to process at least the most critical portions in layers. Most evidence of the critical early stages of a fire will be beneath the sometimes overwhelming overburden of collapsed ceiling and roof structure.

Undamaged areas of the building may be surveyed to establish what sorts of materials to expect. The roof structure can be photographed and its direction of collapse noted and then removed. Roof or ceiling insulation can then be removed—noting whether it is loose (blown-in) fiberglass, mineral wool or cellulose, or "batts" or roll insulation. Samples should be taken for later identification if ignition or spread through the insulation is considered possible.

The ceiling material can then be identified: gypsum wallboard, lath-and-plaster (wire or wood lath makes a difference in manner and time of collapse), ceiling tile, plywood, or wood plank. An investigator should never assume that all insulation, ceiling, or lining materials are consistent throughout even a small residence since repairs, renovations, or additions will usually be made with materials available at the time.

Although not common in modern structures, the presence of combustible ceilings should not be dismissed. Low-density cellulose ceiling tiles were widely used for many years prior to about 1960. They can be 12 × 12-in. tiles or 2 × 4-ft panels. Decorative-edged tongue-and-groove pine or fir boards (sometimes called beadboard) were widely used in the late 19th to early 20th centuries for walls and ceilings. Such linings will add dramatically to the fuel load and increase the rate of fire development in the room (cutting time to flashover by as much as half [per Karlsson and Quintiere 1999]).

The nature and thickness of ceiling materials must be noted. Samples are strongly recommended for later testing. Suspended ceilings are very common in commercial structures, as they can lower ceilings and offer concealment for electrical, plumbing, and HVAC services. These ceilings, while usually noncombustible, allow fire gases to penetrate the large plenum area they offer and spread throughout the concealed space. They typically fail as the steel wires or lightweight steel grid members reach their annealing temperatures and lose their tensile strength. (This can occur in as little as 10 min once the flames reach the ceiling.)

Under the ceiling material light fixtures, furnishings, and victims will be found. Treatment of fatal fire scenes will be covered in a later chapter. Between the ceiling and the floor covering most of the critical evidence will be found, but even here, the sequence of position may be of importance.

Window glass will tend to break and collapse when the temperatures or heat fluxes become high enough during the fire's development. This usually occurs after the smoke products have condensed on the glass, and the fracture pattern would be expected to be thermal rather than mechanical (unless the glass is toughened safety glass). Glass falling inward may well fall on furnishings or floors that are already fire damaged. If a window is broken before the fire reaches it, the fracture pattern will be "mechanical" in appearance and the glass, bearing little or no soot deposits, may fall and protect unburned materials beneath (keeping in mind that radiant heat in a subsequent postflashover may melt and char protected material beneath the glass).

The location of first collapse of ceiling or wall covering may be indicated by careful examination of the debris at this stage, which should be photographed before further excavation. As the furnishings are documented and removed, the nature of floors and floor coverings can be observed. The distribution of carpet, tile, bare wood, composite floor coverings, and vinyl or asbestos (tile or sheet) should be noted. Comparison samples of carpet, floor covering, and underlayment should be taken in areas of suspected origin.

Testing has shown that carpet fiber content cannot be accurately estimated from observation alone, and even using a match or lighter flame will reveal only whether the face yarn is synthetic or natural fiber. Fire tests have further demonstrated that a carpet may not support flame spread alone, but if provided with a particular type of pad, it may burn readily. It is always best to recover and preserve a small (at least 6 × 6-in.) sample of unburned carpet and pad for later identification. Such samples can be recovered in most scenes from under large furniture or appliances or in protected corners of the room. Such samples can also help forensic testing by their use as a comparison sample to establish what volatiles they contain or may yield upon burning.

SIEVING

In the ashes, all evidence seems to be the same shades of white, gray, or black, and not readily distinguishable from fire debris. Searching by hand and visual examination may result in evidence critical to a complete reconstruction being overlooked. Wet sieving of the debris affords a better chance of detecting small items such as bone fragments, glass, keys, or jewelry that may go unnoticed in dry ash.

Sieving is most effective when done with at least three sieve frames stacked together using 1-, 0.5, and 0.25-in. mesh (some examiners will use a fourth sieve with window screen if searching for tiny tooth or bone fragments). Debris is placed in and a garden hose or hose reel line is used to wash debris through. The debris should never be pushed through by hand since that can shatter bones and teeth. Figure 4.22 is an example of a sieve.

Dry sieving should be avoided since the water washes the gray ash off and makes small objects visible by color or reflectance. If the objects recovered can be visually identified, they can be placed directly into evidence bags. Items that cannot be readily identified should be kept in a separate container, labeled as to grid of recovery, until they can be properly analyzed.

IMPRESSION EVIDENCE

Impression evidence is a general term for the transfer of pattern or contour information from one surface to another. This transfer may take the form of one harder material deforming a softer material on contact. A shoe leaving an impression in soft clay outside a window and the ragged edge of a hatchet blade leaving a striated rough cut

FIGURE 4.22 ◆ Sieving assists the investigator in collecting evidence that is not readily distinguishable from fire debris. Searching by hand and visual examination may result in evidence critical to a complete reconstruction being overlooked. *Courtesy of Lamont "Monty" McGill, McGill Consulting, Gardnerville, NV, by permission.*

on a piece of wood are examples of this. This transfer can also occur when there is a transfer medium that preserves the shape of contoured surfaces that come into contact. For example, a dusty shoe can leave an identifiable print on a clean surface, while a clean shoe can remove dust from a dirty surface and leave the same information. Photography is the primary means of documenting most impressions, but care must be taken that the image is not distorted by the photograph. As shown in Figure 4.23, the "L-shaped" American Board of Forensic Odontology (ABFO) No. 2 scale, measuring 15 × 30 cm, is the appropriate scale for shoeprints since it helps prevent distortion (LeMay 2002).

Fingerprints and shoeprints are commonly found forms of impression evidence. When impression evidence is found, it should be documented in place by photograph and note, and then, whenever possible, the object bearing the impression should be recovered and submitted to the lab. The object is always preferable to a photo or cast, but if not removable, photos or casting or lifting by adhesive or electrostatic lifter will be necessary.

In addition to identification, shoeprints can offer information about where someone entered or left a room or building, where he/she went, and in what sequence. If two or more people have been present, shoeprints have been used to establish which of the perpetrators went where in the building.

Items bearing tool marks should also be individually wrapped in packages to protect their surfaces. When dealing with toolmark impressions, similar collection

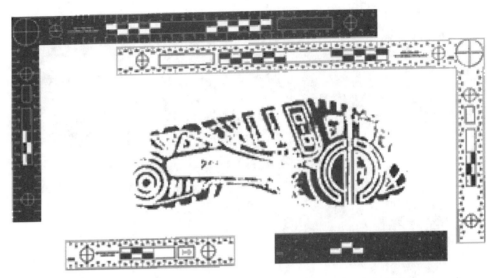

FIGURE 4.23 ◆ An example of a forensic photographic scale alongside impression evidence for a shoe by the transfer of pattern or contour information from one surface to another. *Courtesy of Armor Forensics, by permission.*

guidelines apply to the suspected tools. Wrap each tool separately to prevent shifting during submission to the laboratory. Place each tool in a separate envelope or box with a folded sheet of paper over the end of the tool to minimize damage and loss of trace evidence adhering to the surfaces and to prevent rusting.

Fingerprints.   A finger touching a clean surface like glass or metal can leave a reproduction of its friction ridge contours on the surface by the transfer of the nearly invisible oils, fats, and sweat secretions normally found on the skin. Such patterns are often called "latent" because they are not readily visible to the unaided eye and require some sort of physical, chemical, or optical treatment to make them visible and recordable (and comparable to "inked" record prints).

The skin can also be contaminated with blood, food, grease, or paint that leaves behind a visible or "patent" impression. Even though the skin is pliable and deforms on contact with most other materials, it can deform soft materials such as cheese, chocolate, solvent- or heat-softened plastic or paint, and window putty and leave a three-dimensional or molded "plastic" impression of its ridge detail. The same considerations apply to other deformable materials like rubber shoe soles, gloves, tires, or cloth, where each may deform softer materials or leave a pattern on harder ones through transfer of some intermediate medium or even residues of itself.

Impressions of friction ridge features from fingers, palms, and feet can and do survive fires. The multitude of chemical, physical, and optical techniques available today to enhance fingerprints (both latent and patent) has made it possible to recover prints from difficult, textured or contaminated surfaces. Heat alone can cause the constituents in skin oils to darken or even react with the surface beneath, producing a "patent" impression from a latent one.

Fingerprints are unique to an individual, they are permanent, everyone has them, and chances for transfer or contact are good (with something at the scene, with an incendiary device, or with a container used to transport or distribute ignitable liquids).

In addition, there are vast and long-term repositories of reference prints (which include noncriminals as well as criminals), and the advent of powerful and fast computers has made it possible to scan millions of record prints in a short time to establish a short list of candidate matches. These databases are designed to run single prints and even partial prints and some systems can now process palm prints.

If fire exposure has not been so severe as to melt or severely char the base material, it is worth considering examination for fingerprints. Handling must be very careful and kept to an absolute minimum. Transport to the lab must be done carefully (preferably carry by hand) using containers or devices that minimize contact with potential print-bearing surfaces.

Water contamination from condensation, hose stream, or environmental exposure once meant that prints on paper or cardboard would be impossible to develop (since the amino acids detected by ninhydrin processing are water-soluble). Now the advent of physical developer (which reacts with the fatty constituents that are not water-soluble) makes it possible to process wet paper or cardboard. Small particle reagent (SPR) allows processing on nonporous surfaces like metal, glass, and fiberglass even while still wet. Forensic light sources (multiple wavelength, high intensity) and a variety of chemicals and powders make it possible to recover prints from textured or contaminated surfaces.

Soot can often be washed off smooth metal or glass surfaces with running water, leaving behind prints "developed" by the soot carbon. Fingerprints in blood can be enhanced with amido black or leucocrystal violet (LCV) sprays, which react chemically with blood to form a dark-colored product. The solvent for the LCV has been seen to help rinse overlying soot away from the surface. The print turns a dark blue–purple and is easily distinguished. In one case where two fires had been set in a house, LCV washed soot away, revealing blood spatters on walls where a resident had been bludgeoned to death some 2 years previously (the fires having been subsequently set to simulate drug gang activity).

Charred or burned paper documents are important to a fire scene examination, particularly when business records and important documents are involved. Due to their fragile condition, pack fire-damaged documents on soft cotton sheets and hand-carry them to the laboratory for examination. Do not treat the papers with any lacquer or coating if they are to be processed for identification or comparison.

Shoeprints.   Often compromised by the foot traffic of emergency personnel, tires of vehicles, water, or structural changes, shoeprints can, however, survive fires, if they are left on surfaces that have not been destroyed by fire or heat. Shoeprints left behind on doors that are forcibly "kicked in" may actually be enhanced by the action of the fire. In the example case shown in Figure 4.24, the dusty shoeprint was lightened by heat exposure and the wood around it scorched, increasing its contrast and readability.

Shoeprints are second only to fingerprints in their potential value for linking a person with a scene in that shoes are often (although not always) individualized by the accidental features they acquire through use and damage, they tend not to be changed or discarded for prolonged periods of time, and they are likely to be left behind at many scenes since entry (if not approach) to a building is nearly always accomplished on foot. Paper or cardboard can bear latent shoe impressions (developable by physical developer chemical treatment even after water exposure) or patent impressions in soil, dust, or blood.

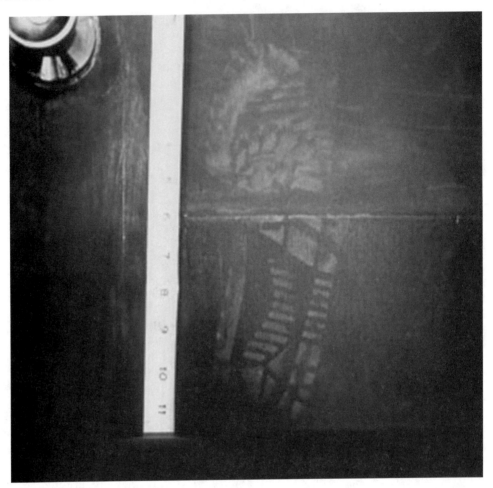

FIGURE 4.24 ◆ Shoeprints left behind on doors that are forcibly "kicked in" may actually be enhanced by the action of the fire. In this case the dusty shoeprint was lightened by heat exposure and the wood around it scorched, increasing its contrast and readability. *Photo by Joe Konefal, by permission.*

TRACE EVIDENCE FOUND ON CLOTHING AND SHOES

Trace evidence such as soil, glass, paint chips, metal fragments, chemicals, and hairs and fibers may cling to the shoes or clothing of a suspect. Liquids may also soak into the clothing or remain on the bottom of the individual's shoes.

Glass, plasterboard, sawdust, and metal shavings have all been found embedded in shoes to offer a link with a scene (as long as comparison samples are recovered). Footwear and clothing can also absorb residues of flammable liquid or chemical accelerants used in arson attacks.

A New Zealand study illustrates the necessity of analyzing clothing and shoes of suspected arsonists for the presence of ignitable liquids. The study measured the approximate amount of petrol transferred to the clothing and shoes of a person during the action of pouring it around the room. It addressed the pouring heights and floor surfaces. Results of the study showed that petrol was always transferred to the shoes and often transferred to both the upper and the lower clothing (Coulson and Morgan-Smith 2000).

In tests using a carbon strip extraction method, it was found that 10 ml of gasoline on clothing worn continuously evaporated much more quickly than on the same clothing left on the lab bench at 20–25°C and was barely detectable after only 4 hr of wear (Morgan-Smith 2000). Clothing must be recovered and packaged soon after exposure.

The investigator must be aware that the solvents used in glues holding footwear together (particularly athletic shoes) can interfere with identification of ignitable liquid residues, as they can produce false positives (some manufacturers have been known to use gasoline as a substitute glue solvent). Preservation of both shoes separately from other evidence and each other in vapor-tight containers is essential (Lentini, Dolan, and Cherry 2000).

Shoes can also pick up material from scenes. One notable example is the case where a wad of charred drapery fabric was found melted into the bottom of an athletic shoe. The wearer of the shoe was dumped in the parking lot of a county hospital suffering from extensive second-degree burns from head to foot with his clothing badly burned and still smoldering. His claim of having injured himself when the carburetor of his car engine malfunctioned was not supported by the distribution of burns to himself or his clothing. The charred drapery remnant was matched to drapery material recovered from a nearby apartment fire that had been ignited with a substantial quantity of gasoline poured through several rooms. Clearly the gasoline had ignited prematurely and the suspect had had to escape through the burning structure. The observation that the synthetic uppers of the shoes were melted and scorched and the drapery fragment was melted into the bottom of the fire-damaged shoe precluded the excuse that he was passing by the fire scene and trod in the debris well after the event.

The general forensic guidelines for handling items of clothing stipulate that they should be marked directly on the waistband, pocket, and collar with the investigator's initials and date found. Pack these items separately into clean paper bags, wrapping each item separately in clean paper.

When dealing with glass, collect as many fragments as possible and place them in paper bags, boxes, or envelopes for later physical reconstruction or reassembly. Package the fragments so that their movement within the containers is kept to a minimum. Keep all evidence separated from any control or comparison samples to avoid cross-contamination. Paint chips, particularly those at least 0.5 in.^{2} should be placed in pillboxes, paper envelopes, cellophane, or plastic bags and carefully sealed. Smaller glass or paint chips should be placed in a folded paper bindle and then sealed in a paper envelope.

Trace evidence sometimes consists of hairs and fibers. Comparison samples from victims or suspects may reveal a direct or indirect relationship. Loose hair combings and clumps cut from various areas should be packaged separately in pillboxes, paper envelopes, cellophane, or plastic bags. The outside of the container should be sealed and labeled.

DEBRIS CONTAINING SUSPECTED VOLATILES

Items containing suspected ignitable liquids should be sealed in clean metal, glass, or special polymer containers. Avoid the use of common polyethylene plastic bags since they are porous and allow evaporation or cross-contamination and may contain volatile chemicals that may contaminate the evidence, resulting in false-positive or -negative results when subjected to laboratory analysis.

Investigators should seek to collect and preserve uncontaminated flammable and combustible liquids from the scene. Seal samples of suspected liquids measuring up to 1 pint in all-glass bottles, jars with Bakelite or metal tops, or metal cans. Such comparison samples should be carefully labeled as to their source if they are not submitted in their original containers.

Samples of fire debris should be sealed in clean empty paint cans or glass jars to prevent evaporation of volatile liquids or in specialized polymeric bags that have been developed for fire debris preservation. The container should be filled no more than three-fourths full using clean tools to prevent contamination. Clean disposable plastic gloves should be worn when collecting the debris and then discarded after each sample is taken.

The collecting officer should label the container with the location at which the sample was collected. Collectors are cautioned not to use rubber stoppers or jars with rubber seals since gasoline, paint thinners, and other volatile liquids contain hydrocarbons, which can compromise the sample by dissolving the seals. Also note prior to sealing the container if an odor was present or a reading was obtained on a hydrocarbon detector. Plastic bottles are inappropriate containers for debris, liquid, or comparison samples. Various advanced techniques are available for collecting liquids from concrete floors using adsorbents such as calcium carbonate. Further details are included in *Kirk's Fire Investigation,* 5th ed. (DeHaan 2002).

◆ 4.8 PHOTOGRAPHY

Fire investigators should not fail to photograph and sketch carefully even the seemingly least significant fire scene detail. Recent benchmark estimates place the average direct costs of performing an investigation in the range of $500 to $1000 (Icove, Wherry, and Schroeder 1998). With the total cost per color slide amounting to approximately $0.25 in quantity, photographs and sketches become the least expensive investigative tools from a cost-effective standpoint.

Fire investigators are not usually required to qualify as expert photographers, yet they must be able to comprehend and utilize their equipment to the maximum benefit (Berrin 1977). With this in mind, the following recommended items should be included as minimum equipment for forensic fire scene photography. This discussion is limited to still photography and does not include video photography.

DOCUMENTATION AND STORAGE

The "Photograph Field Notes" form (906-8; Figure 4.11) is used to record the description, frame, and roll number of each photograph taken at the fire scene. The form is designed to be filled out as the photographs are taken. The frame and roll numbers are later used on the fire scene sketch to indicate the location and direction from which they were taken. The documentation should include a brief caption that permits an independent reviewer correctly to identify the object of interest and the position from which the photos were taken. Documentation of each roll of film should appear on a separate form. The "remarks" field is used to document the disposition of the film.

At the time of the investigation, a descriptive photographic index should be completed using NFPA 906-8 or a similar photo log from the appendix of *Kirk's Fire Investigation,* 5th ed. (DeHaan 2002). After processing, the original negatives and prints should be stored in a specifically marked envelope. The photographic index, complete with narrative, should be included in the investigation report.

FILM CAMERAS

The most versatile cameras available to fire investigators are 35-mm single-lens reflex, with focal plane or between-the-lens shutter systems (Berrin 1977). Single-lens reflex cameras are high cost but allow for close-ups and specialized scene photography and introduce little distortion. Viewfinder cameras are low cost but introduce parallax errors and cannot perform many useful photographic functions.

With the introduction of electronically operated shutter systems, the 35-mm camera has become the recommended standard for fire investigation photography (NFPA 2001; Peige and Williams 1977). Competitive pricing has placed the purchase cost of many acceptable units below $100.

FILM FORMATS

Economic considerations play a key role when deciding upon a film format. Print film requires development and costly printing of photographs. Some investigators may choose to use color slide films having an ASA rating between 100 and 500 to maintain resolution. In some processing laboratories, the customer can choose prints, negatives, and slides at development time. Black-and-white photography of fire scenes has been all but abandoned due to its inability to record important colored fire and burn patterns (Eastman Kodak 1968).

For purposes of economy and portability, some investigators choose slides over negatives. Investigators are free to choose only selected slides for printing and inclusion in the report. Advantages of slide film include its ability to display details during courtroom presentations. With the advent of digital slide scanners, the images of selected photographs can also be imported into word-processed documents and electronic presentation programs.

DIGITAL CAMERAS

Newer and higher-resolution digital cameras are being introduced each year. Their resolution of image quality is often measured in millions of pixels (megapixels). The present minimum acceptable resolution is 2 megapixels, which is still inferior to the resolution available using standard film photography. The higher the number of pixels, the better the quality. Many cameras today will capture 6 megapixels per image on a flash card or similar downloadable medium.

The improper use of digital cameras in investigative photography may undermine the viability of a case, due mainly to evidentiary considerations. Until general acceptance of digital photographs becomes a standard, investigators are advised to use them sparingly and rely upon film cameras as the primary image documentation tool.

As with all photographs, the courtroom tests for their authentication must be met, each image used being a "true and accurate representation" and "relevant to the testimony" to the fire investigation, particularly under Federal Rules of Evidence 403 (Lipson 2000). Systematic handling and processing of digital images is the only method for assuring their long-term acceptability (NFPA 2001, sect. 13.2.2.3).

It is impractical and costly to store high-resolution images permanently on some media or flash cards and they must be copied onto another medium. Two copies of the original electronic/magnetic medium should be made and placed in the case file in the same manner as an audio or video recording. A third copy is used as a "working copy." Images should not be compressed, as this causes loss of detail and clarity.

However, digital cameras and their electronic images do have their place in the production of preliminary photographs of fire scenes, as the images can be easily processed to lighten, darken, or enhance features and easily transferred to printed documents. They are also helpful in producing panoramic views of fire scenes, a technique discussed in a later section. However, file compression and other manipulations may reduce the quality of the images and introduce suspicions of "manipulation" of the photo's content.

DIGITAL IMAGING GUIDELINES

The first draft of "Definitions and Guidelines for the Use of Imaging Technologies in the Criminal Justice System" was published in October 1999 in *Forensic Science Communications* (FBI 1999). The guidelines were prepared by the Scientific Working Group on Imaging Technology (SWGIT) and cover the "documentation of policies and procedures of personnel engaged in the capture, storage, processing, analysis, transmission, or output of imagery in the criminal justice system to ensure that their use of images and imaging technologies are governed by documented policies and procedures."

Agencies or individuals using photography, both simple and complex, should develop a standard operating procedure and departmental policy. The SWGIT procedures recommend preserving original images by storing and maintaining them in an unaltered state and in their native file formats. Duplicates or copies should be used for working images. Original images should be preserved in one of the following durable formats: silver-based (noninstant) film, write-once compact recordable disks (CDR), or digital versatile disk recordable (DVD-R).

If image processing techniques are used on the original image, they should be documented with standard operating procedures. These procedures should be visually verifiable and include cropping, dodging, burning, color balancing, and contrast adjustment. Advanced techniques include those that increase the visibility of the image through multi-image averaging, integration, or Fourier analysis. Other techniques include cropping, overlaying, and creating panoramic views from individual series of photographs. An image processing log is recommended so that the process could be later replicated.

A chain of custody should be maintained for the media on which original images are recorded. This chain of custody should document the identity of the personnel who had custody and control of the digital image file from the point of capture to archiving.

When compressing images, lossy algorithms should be avoided. Image capture devices should render an accurate representation of the images. Different applications will dictate different standards of accuracy.

Training is important when using imaging technologies. Formal uniform training programs should be established, documented, and maintained. Proficiency testing will assure that camera equipment, software, and media keep pace with hardware or software updates.

LIGHTING

No matter what format camera is used, lighting plays an important role in fire scene photography. The intensely burned areas tend to absorb light from natural or artificial sources. Powerful electronic flashes are useful for lighting char details in dark scenes.

There are instances where burn patterns can be best photographed using an oblique flash to light the area of interest. Therefore, a camera should be selected that allows for an externally connected flash unit for oblique or remote illumination of the scene (NFPA 2001, part 13).

ACCESSORIES

Many attachments are available for cameras. However, for the sake of simplicity, nonexpert investigators are often recommended to limit their use. The maximum effectiveness of the single lens provided with the camera (typically 50- to 55-mm focal length) is demonstrated later. Furthermore, the use of filters and interchangeable lens configurations might disqualify photographs or slides used in courtroom presentations when investigators cannot authoritatively qualify their uses, advantages, deficiencies, or effects (Icove and Gohar 1980). A clear filter to protect the lens is the only recommended filter for fire scene photography (NFPA 2001, part 13.2.2.5). Some photographers prefer using polarizing or UV filters in place of the clear filter.

The single most recommended accessory for photographing fire scenes is a sturdy tripod. This simple device enables the photographing of scenes with clear detail when low-lighting conditions command longer shutter exposures. Also, a tripod allows the optimization of depth of field according to the available lighting conditions. For example, an aperture setting of f/16 will usually produce a depth of field ranging from 3 ft (0.91 m) to 10 ft (3.05 m) when the camera is focused on a point 6 ft (1.8 m) from the lens. As demonstrated later, panoramic scenes are most easily recorded when using a tripod.

MEASURING AND IMAGE CALIBRATION DEVICES

Measuring and image calibration devices are used to provide additional information in both processing and interpretation. These devices should be used uniformly from case to case. Even though present fire and explosion investigation guidelines in *NFPA 921* recommend the use of an 18% gray scale calibration card (NFPA 2001, part 13.2.2.6), there are many advantages of using a color calibration card. With the move toward the universal use of color photography for fire scenes, the investigator should consider the use of a combined color and gray scale calibration card.

A color and gray scale calibration card bearing the agency name, case number, date, and time should be photographed on the first frame of each roll of film. Professional film processing laboratories will calibrate their equipment using these cards to ensure that the best color calibrations and gray scales are met. Figure 4.25 is an example of a commercially available card, which bears the essential calibration and documentation information.

Calibration cards also have measuring devices (rulers) on their edges to document distances and sizes of objects photographed. Other measurement devices include longer folding rulers and yellow numbered tent cards to identify the location of individual

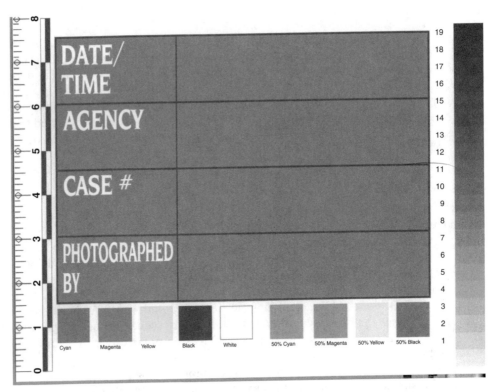

FIGURE 4.25 ◆ Commercially available gray and color calibration chart suitable for use in forensic fire scene documentation. *Courtesy of Armor Forensics, by permission.*

points of interest or forensic evidence, as shown in Figure 4.26. Bright-colored pointers can be used to indicate important features or document direction of fire travel.

AERIAL PHOTOGRAPHY

In rural areas, the U.S. Soil Conservation Service has orthophotoquads of a majority of the farmlands within its area of responsibility. These photos are at various resolutions. Aerial photographs do not always need to be taken from aircraft. Perspective photographs, such as in the previous example, can be taken from higher floors of adjacent buildings. Photos taken from the elevated platform of a fire truck can be useful, providing a closer look at the overall pattern of damage.

PHOTOGRAMMETRY

Photogrammetry, the science of extracting data from photographs, was once out of the range of expertise of the typical fire investigator but has become more accessible in recent years through affordable, user-friendly computer software. When using photogrammetry, two or more photographs are taken and used to extract absolute coordinates and distance measurements from key features in those images (see Figure 4.27). Close-range photogrammetry programs remove the shortening of object lines due to perspective that is present in nearly all photographs, thus allowing users to obtain accurate measurements and create realistic models (Eos Systems Inc. 2003).

For example, the actual walking distance from a deceased victim in a bed to the doorway threshold may be crucial in an investigation. Photogrammetry can capture measurement data in large or complex scenes where tape measure and note pad are

(a)

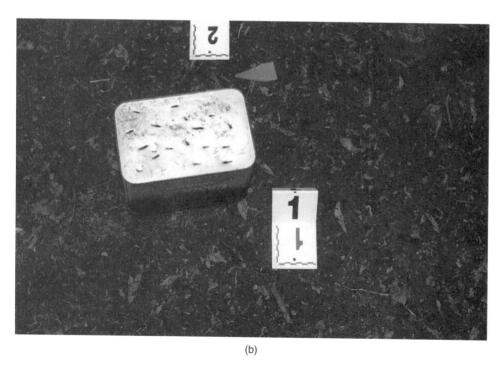

(b)

FIGURE 4.26 ◆ (a) Crime scene measurement devices include yellow numbered tent cards to identify the location of individual points of interest or forensic evidence, such as ignitable liquid containers. (b) Underside of can bears repeated stabbing penetrations from a knife blade, indicating that it was intentionally made to spread flammable liquid contents. Tool marks may link it to the offender.

(c)

FIGURE 4.26 ◆ *continued* (c) Numbered evidence markers illustrate the position of evidence. Plastic pointers or arrows indicate positions of positive canine alerts at the scene. They may also be used to indicate the direction of fire spread in the fire scene. *Courtesy of John DeHaan.*

inadequate or too time-consuming. It also allows investigators to measure evidence after the fact that may have been moved or cleaned up or no longer exists at the scene.

Photogrammetry has been in use in North America, Europe, and Australia for many years for accident scenes and major crimes and has been proposed for advanced fire investigations. In one case in the United States, the technique was used to reconstruct forensically a model of a kitchen area to enhance a "V" burn pattern (King and Ebert 2002).

Some photogrammetry programs allow users to map two-dimensional images over three-dimensional surfaces that have been created, to provide photorealistic models with perspective-corrected phototextures. A wide range of software programs exists that calculate these measurements and construct three-dimensional models from two-dimensional photographs. Figure 4.28 is an example of the use of photogrammetry to construct a three-dimensional view of the exterior of a building using actual two-dimensional photographs.

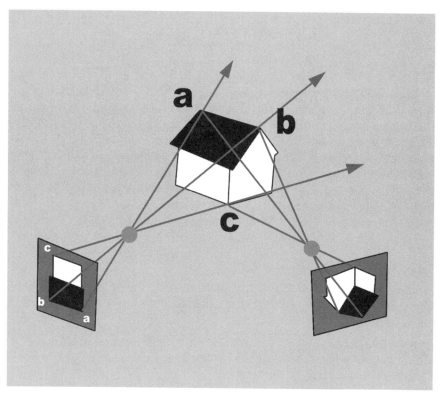

FIGURE 4.27 ◆ In photogrammetry, two or more photographs are used in such a way as to extract absolute coordinates and measurements from key points in those images. *Photomodeler image courtesy of Eos Systems, Inc.*

FIGURE 4.28 ◆ An example of the use of photogrammetry to combine several two-dimensional images to construct a three-dimensional view. *Photomodeler image courtesy of Eos Systems, Inc.*

161

◆ 4.9 SKETCHING

Fire scene sketches, whether or not to scale, are important supplements to photographs. A sketch graphically portrays the fire scene and items of evidence as recorded by the investigator. The "Sketch Field Notes" form (906-9; Figure 4.12) is for use by investigators when drawing simple two-dimensional sketches of fire scenes. These sketches may be rough exterior building outlines or detailed floor plans. Sketches allow the investigator to illustrate relationships between objects that cannot be captured via photography such as those in separate rooms, under or behind large furniture, or visible only from overhead or via cross section. See the Appendix in *Kirk's Fire Investigation,* 5th ed., for additional details (DeHaan 2002).

GENERAL GUIDELINES

Good scene sketches do not require highly artistic work. Forensic fire scene sketches serve many purposes and should depict the following information (DeHaan 2002).

- The outline and dimensions of the building, room, vehicle, or area(s) of interest
- The locations of pertinent evidence or critical features, such as fire patterns and plume damage
- The locations and dimensions of all major fuel packages involved in the fire
- Locations and travel distances of possible points of entry and exit of victims and suspects using Global Positioning System (GPS) data
- Conditions and dimensions of windows, doors, floors, ceilings, and wall surfaces including sill and soffit heights for later use in fire scene reconstruction and analysis

In some instances, multiple sketches using the same overall dimensions are necessary. Several separate types of evidence can be preserved in this manner including isochars, plume damage, and sample (evidence collection) sites. A sketch can often reveal characteristics not readily obvious in a photograph.

The necessary elements of a good forensic sketch are as follows (DeHaan 2002).

- The investigator's full name, rank, and agency, the case number, and the date/time when the sketch was prepared
- The full names of other individuals involved in constructing the sketch
- The location and geographic orientation of the fire scene. Latitude and longitude using GPS data may be desired for rural scenes to determine jurisdiction boundaries (state, county, township, etc.) or relocate the scene accurately
- A legend containing descriptions of all symbols used and their meanings, scale, and other significant information

The area of surface demarcations, thermal effects, penetrations, and loss of material can also be outlined using colored markers on sketches or photographs. One system would be to use yellow lines to outline areas of surface deposits, green lines for thermal effects, blue lines for penetrations, and red lines for areas where the material has been consumed completely.

A sketch is usually initially drawn by hand, sometimes followed up by a computer-assisted drawing. An effective hand sketching tool is a grooved gridpad, where the pencil lead follows a slight indentation on a plastic surface (Accu-Line 2003).

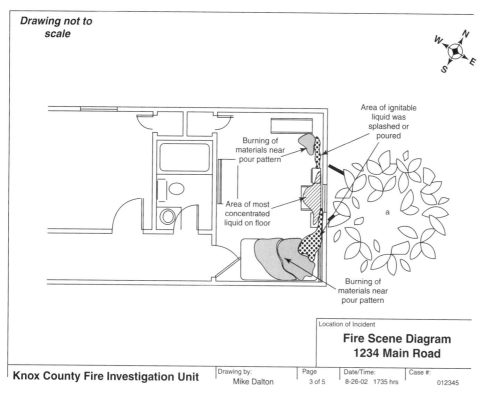

Drawing not to scale

Area of ignitable liquid was splashed or poured

Burning of materials near pour pattern

Area of most concentrated liquid on floor

a

Burning of materials near pour pattern

Location of Incident
**Fire Scene Diagram
1234 Main Road**

| Knox County Fire Investigation Unit | Drawing by: Mike Dalton | Page 3 of 5 | Date/Time: 8-26-02 1735 hrs | Case #: 012345 |

FIGURE 4.29 ◆ A typical fire scene diagram using a sketching program sufficient to produce accurate representations of burn patterns, location of evidence, and other details. *Courtesy of Michael Dalton, Knox County Sheriff's Office, by permission.*

A reliable fire scene sketching program called Visio by Microsoft Corporation is sufficient to produce accurate representations of structures. Microsoft provides an add-on crime scene package at no charge. Experienced investigators have developed templates for use in fire scene analysis (Microsoft 2002). Figure 4.29 shows an example of a typical fire scene diagram produced using the Visio.

A special application of sketching is to copy an overhead or plan view of the scene onto several clear (transparency) sheets. Each sheet can be used to record a different indicator such as fire travel vectors, char depths, calcination, and furniture patterns. The overlays then can be compared to one another or to the original floor plan to illustrate convergence on a possible area of origin. Putting them on separate sheets avoids the problem of too much detail in a single diagram.

TWO- AND THREE-DIMENSIONAL SKETCHES

With the advent of computer-aided drafting (CAD) tools, two- and three-dimensional sketches are now a reality for courtroom exhibits. Figure 4.30 shows an example of this technique. Plastic overlays can also be placed over the large sketches, which can be annotated in court using either prepared overlays or impromptu markings on the overlays. Using a convenient coordinate system will allow the accurate placement of critical evidence such as the orientation of torsos of victims, fire plumes, and evidence locations.

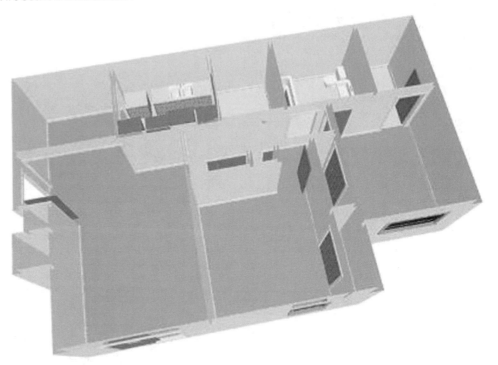

FIGURE 4.30 ◆ Construction of a three-dimensional view using an architectural rendering program. *Courtesy of D. J. Icove.*

TOTAL STATION SURVEY MAPPING

A new generation of computer-driven surveying technologies has come into use for mapping of large interior and exterior scenes. A single reference location is selected and located using a GPS integrated with a Geographic Information System (GIS). The laser sighting unit is then trained on individual features—corners of rooms, curb lines, evidence locations. The computer notes each feature's direction, angle of elevation, and distance. The program then draws a plan of the scene with extremely accurate dimensions (error of ±1 cm at 150 m is typical) even without reflector posts. Systems made by Leica and Topcon that have been in use by traffic accident investigation teams in many police jurisdictions for years are being used by some major arson investigation agencies.

◆ 4.10 ESTABLISHMENT OF TIME

Several natural processes can be used to establish one of the most difficult factors in scene investigation—the passage of time. These processes include evaporation, warming, cooling, drying, and melting.

EVAPORATION

One of the most critical factors in fire reconstruction is the use of accelerants. Estimating the evaporation of accelerants may provide important insight as to the passage of time and the dynamics of the resulting fire. The evaporation rate of a liquid depends on its vapor pressure (which is temperature dependent), the temperature and nature of the surface on which it is deposited, its surface area, and the circulation of air around it.

Vapor pressure is a fundamental physical property of all liquids and solids. It is the pressure that could be generated by the material if placed in a vacuum and allowed to come to equilibrium. It is also a measure of how volatile the material is (i.e., how easily it vaporizes). Being temperature dependent, the higher the temperature, the higher the vapor pressure. Note that it is the temperature of the liquid (or of the surface on which it is spread) that is critical, not necessarily the ambient temperature of the room. A pan of acetone on a hot stove will evaporate very quickly even in a cold room.

It is also highly dependent on molecular weight—the lower the molecular weight of a substance, the higher its vapor pressure and the faster it will evaporate. The physical form of the material is also critical. A very thin film of volatile liquid on a surface will evaporate more quickly than a deep pool. Volatile liquids on porous surfaces like cloth or carpet will evaporate more quickly than they will from a free-standing pool of the same size. The larger the size of the pool and the more air movement around the pool, the faster the evaporation (DeHaan 1999a).

A complex mixture like gasoline contains more than 200 compounds, some of which, like pentane, are very volatile and evaporate very quickly. Others, like trimethylbenzene, evaporate very slowly and will persist for a long time. When such a liquid is poured out, the lightest, most volatile compounds will represent the entire bulk of the vapors generated. So it is the presence of toluene, pentane, and other volatile components that controls the ignitability of the vapors being generated. Components like octane or heavier hydrocarbons hardly evaporate at all at room temperatures. The vapors created by evaporation are easily distinguished from the liquid residues based upon their gas chromographic profiles.

This process means that partially evaporated gasoline will have a very different gas chromographic profile than fresh gasoline. Simple evaporation and burning of gasoline result in the same disproportionate loss of the lighter and more volatile components. The process happens more quickly with the added heat flux from the flames. This property can be used to estimate the relative time of exposure. Gasoline that is present in fire debris that is a contaminant from firefighting gas-powered equipment or intentionally added after the fire will have far more volatiles than would be expected from normal evaporation or combustion.

Changing the physical form of a liquid by placing it on an absorbent wick or aerosolizing it (changing it to a mist by releasing it under pressure) will increase its vapor pressure and make even its heavier components evaporate more quickly or make them more readily ignitable. If evaporation is a concern, from the standpoint of either creating a risk by evaporation of a liquid fuel or reconstructing time factors (time since release), several factors must be documented. These factors include accurate identification of the material itself; the temperature of ambient air, surfaces, or the liquid itself at the time of release; and the nature of surfaces on which it was spread (liquids evaporate much more quickly from thin materials like clothing with free air movement than from solid porous materials like shoes).

The predominant air conditions are also critical: High temperatures, direct sun, wind, mechanical movement by fans, HVAC systems, and movement of equipment, vehicles, or people all increase the evaporation rate. Evaporation of flammable liquids from evidence that could be significant in establishing the cause of the fire or linking a person with the scene needs to be arrested by sealing the evidence in an appropriate container and keeping it as cool as possible to reduce further losses. The time at which it was sealed should be noted on the container.

DRYING

Drying is usually closely related to evaporation (as in the drying of water from cloth-ing) but when blood or other complex liquids are involved, there may be other processes that occur. Blood, for instance, changes chemistry, color, and viscosity as it dries, becoming darker and stickier (if it is drying as a pool on a nonporous surface). Documentation of the drying state of blood involves more than simple photography, as the texture needs to be assessed, usually using a sterile cotton swab (that would be used in its eventual collection in any event). The time at which this observation is made must also be recorded. The same considerations apply to paint, adhesives, plastic resins, or other materials that change mechanically as they dry. Also, the ambient tem-perature must be noted if drying is a critical concern.

COOLING

Cooling and melting are time indicators that play roles in crime scene investigation (such as ice cream on the counter still partly frozen) but occasionally are useful in fire reconstruction. The cooling (or lack thereof) of materials directly exposed to fire can be assessed by direct contact (ouch!), thermal imaging, or temperature probe (inex-pensive digital thermometers that are good from −45 to 200°C (−50 to 400°F) are available from kitchen and housewares stores. Once again, ambient conditions of wind, rain, sun, standing or running water, etc., need to be recorded at the same time.

Cooling of a body is a routine estimate in death investigations but is often ignored in fires. That is unfortunate because the death and the fire need not have occurred at the same time. An adult human body, once cooled to room temperature, must be ex-posed to heat (even from a fire) for a long time before its internal core temperature rises (ask anyone who's ever tried to cook a 25-lb roast). The thermal inertia of human tissue is similar to that of pine wood or polyethylene plastic. The internal temperature (rectal or liver) of a body at a fire scene should be recorded, not for estimation of the time of death interval as much as for exclusion of situations where the victim has been dead for many hours prior to the fire.

◆ 4.11 SPOLIATION

Of major concern to fire investigators is the preservation of evidence to prevent its de-struction or alteration before it is submitted for competent examination and analysis. Destruction or alteration of evidence, particularly when it will be the subject of pend-ing or future litigation, is often referred to as *spoliation*.

The impact of failing to prevent spoliation can result in testimony being disal-lowed, sanctions, and potentially civil or criminal remedies (Burnette 2000). Standards presently establish practices for examining and testing items of evidence that may or may not be involved in product liability litigation (ASTM E860). In laboratory exam-inations where the evidence will be altered or destroyed, all persons involved in the present or potential cases should be given the opportunity to make their opinions known and be present at the testing.

A list of common testing standards is shown in Table 1.3. Legal experts provide the following practical advice and pointers to avoid spoliation issues in potential or pending cases (Hewitt 1997).

TABLE 4.3 ◆ Summary of Case Citations and Opinions on Spoliation of Evidence

Principle	Case Citation	Discussion
Definition	*County of Solano v Delancy,* 264 Cal. Rptr. 721, 724 (Cal. Ct. App. 1989) *Miller v Montgomery County,* 494 A.2d 761 (Md. Ct. Spec. App. 1985)	• Failure to preserve property for another's use in a pending or future litigation • Destruction, mutilation, or alteration of evidence by a party to an action
Dismissal	*Allstate Insurance Co. v Sunbeam Corp.,* 865 F. Supp 1267 (N.D. Ill. 1994) *Transamerica Insurance Group v Maytag, Inc.,* 650 N.E. 3d 169 (Ohio App. 1994)	• Dismissal of action as result of spoliation by deliberate or malicious conduct • Dismissal of action from failure to preserve all evidence prior to suit
Duty to preserve evidence	*California v Trombetta,* 479, 488–89 (1984)	• Duty to preserve exculpatory evidence that will play a role in suspect's defense
Exclusion of expert testimony	*Bright v Ford Motor Co.,* 578 N.E. 2nd 547 (Ohio App. 1990) *Cincinnati Insurance Co. v General Motors Corp.,* 1994 Ohio App. LEXIS 4960 (Ottawa County Oct. 28, 1994) *Travelers Insurance Co. v Dayton Power and Light Co.,* 663 N.E.2nd 1383 (Ohio Misc, 1996) *Travelers Insurance Co. v Knight Electric Co.,* 1992 Ohio App. LEXIS 6664 (Stark County Dec. 21, 1992)	• Exclusion of expert testimony for failure to protect evidence prejudicial to innocent party • In products liability actions, the court may preclude expert testimony as a sanction for spoliation of evidence • Negligent or inadvertent destruction of evidence sufficient for sanctions excluding deposition testimony of expert witness • Court struck opinion evidence of plaintiff's expert based upon physical evidence no longer available
Evidentiary inferences	*State of Ohio v Strub,* 355 N.E.2d 819 (Ohio App. 1975) *U.S. v Mendez-Ortiz,* 810 F.2d 76 (6th Cir. 1986)	• Attempts to suppress evidence indicating consciousness of guilt • Intentional spoliation or destruction of evidence infers that it was unfavorable to spoliator's cause
Independent torts	*Continental Insurance Co. v Herman,* 576 S.2d 313 (Fla. 3rd DCA 1991) *Smith v Howard Johnson Co., Inc.,* 615 N.E.2d 1037 (Ohio 1993)	• Independent actions for intentional or negligent spoliation of evidence • Cause of action exists in tort for interference with or destruction of evidence
Criminal statutes	Ohio Statute, Section 2921.32	• "Obstructing justice" by destroying or concealing physical evidence of the crime or act

Source: Derived from information given by Burnette (2000).

- ◆ Recognize the duty to other interested parties.
- ◆ Keep current on your field of expertise and track the law on spoliation.
- ◆ Retain only properly qualified experts and seek their guidance.
- ◆ Have regard for industry standards and recommendations and develop standard procedures for the storage of evidence.
- ◆ Put those in control of evidence on notice of your rights.
- ◆ Raise the opposing party's standard of care if necessary and build a record with the party in control of evidence.
- ◆ Notify other interested parties and provide them with an opportunity to examine evidence.
- ◆ Seek either a voluntary undertaking or a formal agreement to preserve evidence.
- ◆ Consider also obtaining a preservation or protection order.
- ◆ If compelled to destroy or damage evidence, first document it thoroughly.
- ◆ Consult a lawyer.

Summarized in Table 4.3 are relevant fundamental case citations and opinions. Fire investigators should keep up to date on legal issues surrounding the problem of spoliation and how to guard against violating its principles and practices. When questions arise, always contact your legal advisor or prosecutor for clarification or advice. Thorough documentation by photo and note is the best protection against problems arising from accidental damage during storage and transport.

◆ 4.12 SUMMARY

Fire scenes often contain complex information that must be thoroughly documented, since a single photograph and scene diagram are often not sufficient to capture vital information on fire dynamics, building construction, evidence collection, and avenues of escape for the building's occupants.

Through a comprehensive effort of forensic photography, sketches, drawings, and analysis, fire scene documentation is accomplished. Various computer-assisted photographic and sketching technologies now can help to ensure accurate and representational diagramming.

The next chapter relies upon accurate fire scene documentation to assist in the analysis of intentionally set fires. From visual observations of the manner of ignition of the fire, the presence or absence of accelerants, and the area of origin, much can be learned about the arsonist. Later chapters rely upon accurate documentation to construct fire models, design tests, and compare case studies.

Problems

4.1. Find an example of a public fire report on the U.S. Fire Administration's Web site. What is your assessment of the completeness of this report? What information would you add to the report?

4.2. Visit the scene of a fire and take exterior survey photographs of the building without crossing onto the property. What information can you glean from the exterior visual information?

4.3. Review the list of case citations on spoliation of evidence. What guidance would you provide to ensure that your local community investigator meets or exceeds this professional conduct?

■ ■

Suggested Reading

Cooke, R. A., and R. H. Ide. 1985. *Principles of fire investigation,* chaps. 8 and 9. Leicester, U.K.: Institution of Fire Engineers.

DeHaan, J. D. 2002. *Kirk's fire investigation,* 5th ed., chap. 7. Upper Saddle River, N.J.: Prentice Hall.

National Institute of Justice. (June 2000). *Fire and arson scene evidence: A guide for public safety personnel.* Washington, D.C.: NIJ.

CHAPTER 5 ◆ Arson Crime Scene Analysis

You know my methods in such cases, Watson: I put myself in the man's place, and having first gauged his intelligence, I try to imagine how I should myself have proceeded under the same circumstances.

—Sir Arthur Conan Doyle,
"The Adventure of the Musgrave Ritual"

Based upon published statistics from the National Fire Protection Association (NFPA), in 2001 United States fire departments responded to 1,734,500 fires, or one every 18 s. Of these incidents, 30 percent occurred in structures, 20.3 percent in vehicles, and 49.7 percent in outside properties. Loss of life is always a grave concern in fires. The NFPA estimates that 3745 civilians died in 2001, excluding those in the 9/11 disaster. These statistics come from fire departments that responded to the NFPA's national fire experience survey (Karter 2002).

A leading cause of fire in the United States, arson continues to be both an urgent national problem and truly a contemporary crime. Often characterized as a clandestine tool for criminals, the true arson picture is not clearly known. The NFPA statistics for 2001 estimate that arson caused 45,500 structure fires and 39,500 vehicle fires.

The NFPA estimates that in 2001, arson fires caused over $1 billion in damage, excluding the events of 9/11. Economically, arson impacts insurance premium rates, removes taxable property assets, and degrades our communities. Historically, the inner areas of our large cities often are most hard-hit, and the result is that much of the cost of this destructive crime is placed on those who can least afford it. The correlation between a healthy economy and a decline in business failure equates to fewer arson-for-profit cases. However, historical trends show repeatedly that a decline in the economy results in an increase in arsons (Decker and Ottley 1999).

If an investigator understands the crime of arson based upon its motivations, it may enhance investigative efforts and provide a focus for intervention efforts. Examination of the fire scene and reporting of the results may facilitate dialogue between the various disciplines and investigative units involved in arson study and investigation.

It is intended that the information supplied in this chapter will assist arson investigators in developing skills for reading and interpreting the characteristics of crime scene evidence and applying that evidence to behavior and patterns of thinking on the part of the arsonist.

The definitions for arson range widely, due primarily to differing statutory terminology. The most accepted definition is that ***arson*** is the willful and malicious burning of property (Icove et al. 1992).

The criminal act of arson is usually divided into three elements (DeHaan 2002).

- **There has been a burning of property.** This must be shown to the court to be actual destruction, not just scorching or sooting (although some states include any physical or visible impairment of any surface).
- **The burning is incendiary in origin.** Proof of the existence of an effective incendiary device, no matter how simple it may be, is adequate. Proof must be accomplished by showing specifically how all reasonable natural or accidental cases have been considered and ruled out.
- **The burning is shown to be started with malice.** This act of burning requires a specific intent of destroying property.

An arsonist is usually a person apprehended, charged, and convicted of one or more arsons. Arsonists commonly use accelerants, which are any type of material or substance added to the targeted materials to enhance the combustion of those materials and to accelerate the burning (Icove et al. 1992).

DEVELOPING THE WORKING HYPOTHESES

Chapter 1 explored in depth the development of a working hypothesis using the scientific method. Several factors that may contribute to a working hypothesis for an incendiary fire are discussed in *NFPA 921* (NFPA 2001, chap. 19).

Displayed in Figure 5.1 is a mosaic of several factors that may contribute to a working hypothesis in an incendiary fire or explosion, derived from *NFPA 921*. Note that these factors should not be considered all-inclusive; other indicators may exist. However, one should be cautioned that one or more of these factors is not necessarily sufficient to constitute a finding of an incendiary fire.

MULTIPLE FIRES

When an offender is involved with three or more fires, particular terminology is used to describe that crime (Icove et al. 1992).

- ***Mass arson*** involves an offender who sets three or more fires at the same site or location during a limited period of time.
- ***Spree arson*** involves an arsonist who sets three or more fires at separate locations with no emotional cooling-off period between the fires.
- ***Serial arson*** involves an offender who sets three or more fires with a cooling-off period between the fires.

This chapter includes a detailed discussion of how arson crime scenes are analyzed. The analysis often leads to the identification of offenders, their method of operation, and their areas of frequent travel.

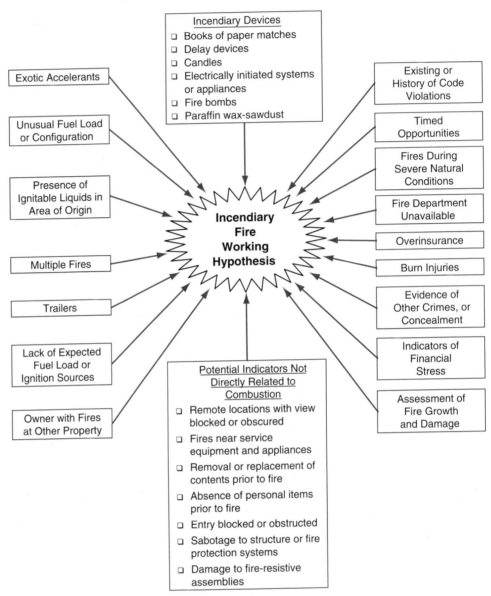

FIGURE 5.1 ◆ Several factors that may contribute to a working hypothesis for an incendiary fire or explosion. *Derived from* NFPA 921 *[NFPA 2001, chap. 19].*

◆ 5.2 CLASSIFICATION OF MOTIVE

A jury often wants to have a demonstrable ***motive*** considered with the evidence so that they can justify the verdict to themselves, even though it is not a legal requirements. Although motive is not essential to establish the crime of arson, and need not be demonstrated in court, the development of a motive frequently leads to the identity of the offender.

Establishing motive also provides the prosecution with a vital argument when presented to the judge and jury during trial. It is thought that the motive in an arson case often becomes the mortar that holds together the elements of the crime.

It is in the area of motives that most of the literature on fire setting and arson is concentrated, and this research offers a number of classification schemes and typologies, most often based on motives. Several of the earlier typologies contributed significantly to the current understanding of the motives and psychological profiles of arsonists (Hurley and Monahan 1969; Icove and Estepp 1987; Inciardi 1970; Levin 1976; Lewis and Yarnell 1951; Robbins and Robbins 1967; Steinmetz 1966; Vandersall and Wiener 1970; Wolford 1972). Geller (1992) offers an exhaustive review of that literature and identifies 20 or more attempts to classify arsonists into typologies.

For the purposes of classification, FBI behavioral science research defines *motive* as an inner drive or impulse that is the cause, reason, or incentive that induces or prompts a specific behavior (Rider 1980). A motive-based method of analysis can be used to identify personal traits and characteristics exhibited by an unknown offender (Douglas et al. 1992; Icove et al. 1992). For legal purposes, the motive is often helpful in explaining why an offender committed his or her crime. However, motive is not normally a statutory element of a criminal offense.

The motivations discussed in this chapter are also outlined and described in the *NFPA 921—Guide for Fire and Explosion Investigations* (NFPA 2001) and the FBI's *Crime Classification Manual* (Douglas et al. 1992; Icove et al. 1992).

OFFENDER-BASED MOTIVES

Law enforcement-oriented studies on arson motives are ***offender-based***; that is, they look at the relationship between the behavioral and the crime scene characteristics of the offender as it relates to motive. One of the largest present-day offender-based studies consists of 1016 interviews of both juveniles and adults arrested for arson and fire-related crimes, during the years 1980 through 1984, by the Prince George's County Fire Department (PGFD), Fire Investigations Division (Icove and Estepp 1987). These offenses include 504 arrests for arson, 303 for malicious false alarms, 159 for violations of bombing/explosives/fireworks laws, and 50 for miscellaneous fire-related offenses.

The overall purpose of the PGFD study was to create and promote the use of motive-based offender profiles of individuals who commit incendiary and fire-related crimes. Prior studies failed to address completely the issues confronting modern law enforcement. Of primary concern were the efforts to provide logical, motive-based investigative leads for incendiary crimes.

The study was conducted primarily because fire and law enforcement professionals were entitled to take upon themselves the task of conducting their own independent research into violent incendiary crimes. The PGFD study determined that the following motives are most often given by arrested and incarcerated arsonists (Icove and Estepp 1987):

- vandalism
- excitement
- revenge
- crime concealment
- profit, and
- extremist beliefs.

Scientific literature and research on arsonists have historically been conducted largely from the forensic psychiatric viewpoint (Vreeland and Waller 1978). Many forensic researchers do not necessarily assess the crime from the law enforcement perspective. They may have limited access to complete adult and juvenile criminal records and investigative case files and must often rely upon the self-reported interviews of the offenders as being totally truthful. The researchers do this without the capabilities and time to validate the information through follow-up investigations. Other researchers have cited that methodological difficulties including small sample sizes of interviews and skewed databases may also bias the previous studies (Harmon, Sosner, and Wiederight 1985).

VANDALISM-MOTIVATED ARSON

Vandalism-motivated arson is defined as malicious or mischievous fire setting that results in damage to property (see Table 5.1 and Figure 5.2). Some of the most common targets of usually juvenile arsonists are schools, school property, and educational facilities. Vandals also frequently target abandoned structures and combustible vegetation. The typical vandalism-motivated arsonist will use available materials to set fires with book matches and cigarette lighters as the ignition devices for the fires.

Usually vandalism-motivated arsonists will leave the scene and not return to it. Their interest is in setting the fire, not watching it or the fire-fighting activities it generates. On average, vandalism arsonists will be questioned twice before being arrested and charged. They will offer no resistance when arrested but will qualify and minimize their responsibility. After an initial not guilty plea, vandalism-motivated arsonists will typically change the plea to guilty before trial.

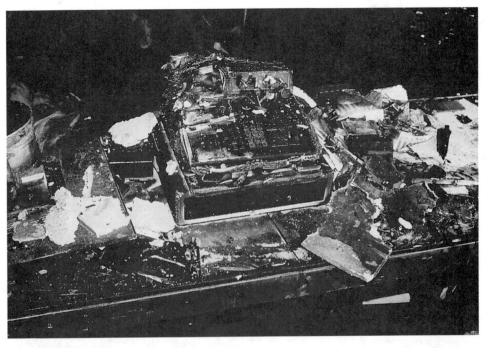

FIGURE 5.2 ◆ Vandalism-motivated arson showing a superficially damaged cash register. *Courtesy of D. J. Icove.*

TABLE 5.1 ◆ Vandalism-Motivated Arson

Characteristics
  Victimology:
    targeted property
	Educational facilities common target
	Residential areas
	Vegetation (grass, brush, woodland, and timber)

Crime scene indicators frequently noted	Multiple offenders acting spontaneously and impulsively Crime scenes reflect spontaneous nature of the offense (disorganized) Offenders use available materials at the scene and leave physical evidence behind (footprints, fingerprints, etc.) Flammable liquids occasionally used Entrance may be gained through windows of secured structures Matchbooks, cigarettes, and spray-paint cans (graffiti) often present Materials missing from scene and general destruction of property
Common forensic findings	Flammable liquid analysis Presence of fireworks Glass particles on suspect's clothing if entered by breaking a window
Investigative considerations	Typical offender is a juvenile male with 7–9 years of formal education Records of poor school performance Not employed Single and lives with one or both parents Alcohol and drug use usually not associated Offender may already be known by police and may have arrest record Majority of offenders live more than 1 mi from the crime scene Most offenders flee from scene immediately and do not return If the offenders return, they view the fire from a safe vantage point Investigators should solicit assistance from school, fire, and police
Search warrant suggestions	Spray-paint cans Items from the scene Explosive devices Flammable liquids Clothing: evidence of flammable liquid, glass particles Shoes: footprints, flammable liquid traces

Source: Updated from Icove et al. (1992).

EXAMPLE 5.1   Vandalism—Case Study of a Vandalism-Motivated Serial Arsonist

A 19-year-old high-school dropout was responsible for a series of arsons in a northeastern city that were set using available materials and lit by a cigarette lighter. He admitted to setting 31 fires in vacant buildings and garages as well as dumpsters and old vehicles. He was questioned twice before being arrested and formally charged with nine of the house fires. When interviewed he stated, "I just burned them for the hell of it. You know, just to have something to do. Those old houses and that other stuff was not worth anything anyway."

His mother and father had divorced when he was 2 years old and he had lived alternately with his mother and grandmother. He had no contact with his father and said that his mother had remarried several times. After dropping out of school in the 10th grade, he had worked sporadically at unskilled jobs and continued to live with his grandmother.

Several times, he reported just striking a match, tossing it into dry leaves or grass, and walking away without even seeing if it ignited. "I did not care nothing about watching no fire. It was just something to do. There was not much going on most of the time, you know. Just hanging out. It was just kid stuff, just for the hell of it. Half of the guys I knew that hung out would set a fire for the hell of it" (Sapp et al. 1995).

EXCITEMENT-MOTIVATED ARSON

Offenders *motivated by excitement* include seekers of thrills, attention, recognition, and, rarely but importantly, sexual gratification (see Table 5.2 and Figure 5.3). The arsonist who sets fires for sexual gratification is quite rare.

FIGURE 5.3 ◆ Excitement-motivated arsonists sometimes target garages, which contain and have easily accessible all the materials and fuels needed for setting the fire. *Courtesy of D. J. Icove.*

TABLE 5.2 ◆ Excitement-Motivated Arson

Characteristics
  Victimology:
    targeted property
	Dumpsters
	Vegetation (grass, brush, woodland, timber)
	Lumber stacks
	Construction sites
	Residential property
	Unoccupied structures
	Locations that offer a vantage point to observe fire
	  suppression and investigation safely

Crime scene indicators
  frequently noted
	Often adjacent to outdoor "hangouts"
	Often uses available materials on hand to start
	  small fires
	If incendiary devices are used, they usually have time-
	  delayed triggering mechanisms
	Offenders in the 18–30 age group more prone to use
	  accelerants
	Match/cigarette delay device frequently used to ignite
	  vegetation fires
	Small group of offenders is motivated by sexual
	  perversions, leaving ejaculate, fecal deposits,
	  pornographic material

Common forensic findings	Fingerprints, vehicle and bicycle tracks Remnants of incendiary devices Ejaculate or fecal materials

Investigative considerations	Typical offender is a juvenile or young adult male with
	  10+ years of formal education
	Offender unemployed, single, and living with one or
	  both parents from middle- to lower-class bracket
	Offender generally is socially inadequate, particularly
	  in heterosexual relationships
	Use of drugs or alcohol limited usually to older offenders
	History of nuisance offenses
	Distance offender lives to the crime scene determined
	  through a cluster analysis
	Some offenders do not leave, mingling in crowds to
	  watch the fire
	Offenders that leave usually return later to assess the
	  damage and their handiwork

Search warrant suggestions	Vehicle: Material similar to incendiary devices, floor
	  mats, trunk padding, carpeting, cans, matchbooks,
	  cigarettes
	House: Material similar to incendiary devices, clothing,
	  shoes, cans, matchbooks, cigarettes, lighter, diaries,
	  journals, notes, logs, records and maps documenting
	  fires, newspaper articles, souvenirs from the
	  crime scene

Source: Updated from Icove et al. (1992).

Potential targets of the excitement-motivated arsonist run the full spectrum from so-called nuisance fires to fires in occupied apartment houses at nighttime. A limited number of firefighters have been known to set fires so they can engage in the suppression effort (Huff 1994). Security guards have set fires to relieve boredom and gain recognition. Research into this motive category further categorizes excitement-motivated arsonists into several subclassifications, including thrill-, sex-, and recognition/attention-motivated arsonists (Icove et al. 1992).

EXAMPLE 5.2 Excitement—Case History of an Excitement/Recognition-Motivated Serial Arsonist

A 23-year-old volunteer fireman was charged with setting a series of fires shortly after joining the fire department. The fires initially consisted of trash cans and dumpsters and then progressed to unoccupied and vacant structures.

He became a suspect in the fires when he consistently arrived first on the scene and frequently reported the fire. He stated that he set the fires so that he could get practice and others would view him as a good fireman. He commented that his father was "real proud of his firefighter son" (Sapp et al. 1995).

REVENGE-MOTIVATED ARSON

Revenge-motivated fires are set in retaliation for some injustice, real or imagined, perceived by the offender (see Table 5.3 and Figure 5.4). Often revenge is also an element of other motives. This concept of mixed motives is discussed later in this chapter.

FIGURE 5.4 ◆ A revenge-motivated fire set on a bed, which is characteristic of a focused target having personal significance. *Courtesy of D. J. Icove.*

TABLE 5.3 ◆ Revenge-Motivated Arson

Characteristics Victimology: targeted property	Victim of revenge fire generally has a history of interpersonal or professional conflict with offender (lover's triangle, landlord/tenant, employer/employee) Tends to be an intraracial offense Female offenders usually target something of significance to victim (vehicle, personal effects) Ex-lover offender frequently burns clothing, bedding, and/or personal effects Societal revenge targets displace aggression to institutions, government facilities, universities, corporations
Crime scene indicators frequently noted	Female offenders usually burn an area of personal significance, using victim's clothing or other personal effects Male offenders begin with an area of personal significance, tend to overkill by using more accelerants or incendiary devices than are necessary
Common forensic findings	Laboratory tests for accelerants, pieces of incendiary bomb, cloth, fingerprints
Investigative considerations	Offender is predominantly an adult male with 10+ years of formal education If employed, the offender is usually a blue-collar worker in the low socioeconomic status Resides in rental property, loner, unstable relationships Event happens months or years after precipitating incident Most often have some periodic law enforcement contact for burglary, theft, and/or vandalism Use of alcohol more prevalent than drugs, with possible increase after the fire Usually alone at scene and seldom returns after fire started to establish an alibi Lives in affected community, with mobility an important factor Revenge-focused analysis assists in determining true victim Investigative investment significant
Search warrant suggestions	If accelerants suspected: shoes, socks, clothing, bottles, flammable liquids, matchbooks

Source: Updated from Icove et al. (1992).

What may be of concern to investigators is that the event or circumstance that is perceived as unjust may have occurred months or years before the fire-setting activity (Icove and Horbert 1990). For threat assessment purposes, this time delay may not be readily recognized, and investigators are urged to pursue the historical incidents in which a person or property has been targeted.

Revenge and spite-motivated fires account for more of the serious arson cases due to the "overkill" reflected in the actions of the offender. In these cases, when one canister of a flammable liquid would suffice in setting a building on fire or incinerating a body, the offender may use much larger quantities, reflecting the "rage" or "vengeance" of the act.

The broad classification of revenge-motivated arsonists is further divided into subgroups based on the target of the retaliation. Studies show that serial revenge-motivated arsonists are more likely to direct their retaliation at institutions and society than at individuals or groups (Icove et al. 1992).

EXAMPLE 5.3   Case History of an Institutional Retaliation "Revenge" Serial Arsonist

John (not his real name) is 31 years old and claimed to have set more than 60 fires at various local government facilities in the city where he lives. His fire-setting activities have taken place since he was 19 years old.

John started setting fires after he was sentenced to 180 days in the local jail facility for a petty theft. He claimed to have set 5 fires while in the jail and then later 20–25 trash can fires in the local city hall. His method of operation consisted of simply walking through and dropping lighted matches into the trash containers.

His stated motive for setting the fires was "to cost the city some trouble and money. They did not treat me fair and I will get even." When asked when his revenge against the city would be satisfied, his response was "when the whole damn city hall burns down and the jail too" (Sapp et al. 1995).

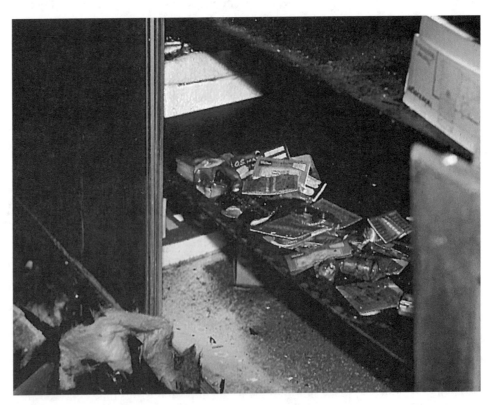

FIGURE 5.5 ◆ A crime concealment-motivated arson where more valuable inventory was removed prior to the fire. *Courtesy of D. J. Icove.*

TABLE 5.4 ◆ Crime Concealment-Motivated Arson

Characteristics Victimology: targeted property	Dependent on the nature of the concealment
Crime scene indicators frequently noted	Murder: attempts to obliterate forensic evidence of potential lead value or conceal victim's identity, commonly using an accelerant in a disorganized manner Burglary: uses available materials to start the fire, characteristic of involving multiple offenders Auto theft: strips and burns vehicle to eliminate fingerprints Destruction of records: sets fire in area where records usually kept
Common forensic findings	Determine if victim is alive at time of fire and why he/she did not escape Document injuries, particularly if concentrated around genitals
Investigative considerations	Common to find alcohol and recreational drug use Offender expected to have history of contacts of arrests by police and fire departments Offender is likely a young adult who lives in the surrounding community and is highly mobile Crime concealment suggests accompanied to scene by coconspirators Murder concealment is usually a one-time event
Search warrant suggestions	Refer to other category of primary motive Gasoline containers Clothing, shoes, glass fragments, burned paper documents

Source: Updated from Icove et al. (1992).

CRIME CONCEALMENT-MOTIVATED ARSON

Arson is the secondary criminal activity in this motivational category of ***crime concealment-motivated arson*** (see Table 5.4 and Figure 5.5). Examples of crime concealment-motivated arsons are fires set for the purpose of covering up a murder or burglary or to eliminate evidence left at a crime scene.

Other examples include fires set to destroy business records to conceal cases of embezzlement and fires set to destroy evidence of an auto theft. In these cases, the arsonist may set fire to the structure to destroy the evidence of the initial crime, to obliterate latent fingerprints and shoeprints, and sometimes to attempt to render useless DNA or serological evidence linking them to a victim left behind to die in the fire.

EXAMPLE 5.4 Case History of a Crime-Concealment Serial Arsonist

A 31-year-old unemployed laborer, 3 weeks after being released from prison, admitted responsibility for burglarizing and setting fires to 12 houses in a residential area over a period of 7 months. All but one of the fires occurred while the owners were out of town.

The offender stated, "I just drove through and looked for newspapers and other things where the people were gone for a while. I just took money and jewelry and stuff that would get lost or burn up in the fire. I would pour gasoline on everything and use a candle to light it off."

The fires were all identified as arson fires because of their rapid development and the detected presence of flammable liquids in the debris. The extensive damage made it difficult for the owners to determine if valuables in the homes had been removed. The offender stated that he was identified when, "One of the ladies spotted a ring in a pawn shop in another town. I would always take stuff somewhere else to sell it. What happened was she recognized the ring and called the cops."

His criminal history record consisted of a burglary conviction when stolen property was recovered and traced back to him. The offender's rationalization was, "I figured it out that if the fire burned everything nobody would know anything was gone. Last time, I got caught on a serial number, so this time I decided I would not leave anyway for them to know what was gone" (Sapp et al. 1995).

PROFIT-MOTIVATED ARSON

Arsonists in this category of ***profit-motivated arson*** expect to profit from their fire setting, either directly for monetary gain or more indirectly to profit from a goal other than money (see Table 5.5 and Figure 5.6). Examples of direct monetary gain include

FIGURE 5.6 ◆ Profit-motivated arsons to leverage overinflated insurance sometimes involve well-planned fires set in vacant dwellings fueled with excessive amounts of accelerants. *Courtesy of D. J. Icove.*

TABLE 5.5 ◆ Profit-Motivated Arson	
Characteristics Victimology: targeted property	Property targeted includes residential, business, and transportation (vehicles, boats, etc.)
Crime scene indicators frequently noted	Usually well-planned and methodical approach, with crime scene more organized by containing less physical evidence With large businesses, multiple offenders may be involved Intent is the complete destruction with excessive use of accelerants, incendiary devices, multiple points of fire origin, trailers Lack of forced entry Removal or substitution of items of value prior to fire
Common forensic findings	Use of sophisticated accelerants (water-soluble) or mixtures (gasoline and diesel fuel) Components of incendiary devices
Investigative considerations	Primary offender is an adult male with 10+ years of formal education Secondary offender is sometimes the "torch," who is usually a male, 25–40 years of age, and unemployed Offender generally lives more than 1 mi from the crime scene, may be accompanied to the scene, leaves, and usually does not return Indicators of financial difficulty Decreasing revenue with increasing and unprofitable production costs Technology outdates processes Costly lease or rental arrangements Personal expenses paid with corporate funds Hypothetical assets, overstated inventory levels Pending litigation, bankruptcy Prior fire losses and claims Frequent changes in property ownership, back taxes, multiple liens
Search warrant suggestions	Check financial records If evidence of fuel/air explosion at scene, check local emergency rooms for patients with burn injuries Determine condition of utilities as soon as possible

Source: Updated from Icove et al. (1992).

insurance fraud, liquidation of property, dissolution of businesses, destruction of inventory, parcel clearance, and gain employment. The latter is exemplified by the case of a construction worker wanting to rebuild an apartment complex he destroyed or of an unemployed laborer seeking employment as a forest firefighter or as a logger to salvage burned timber.

Arson-for-profit may have interesting twists where the offender benefits directly or indirectly. Arsonists have set fire to western forests in order to have their equipment rented out to support part of the suppression effort. What may be the most disturbing of all are cases of parents murdering their own children for profit, with fire used to cover the intentional death of the child. While this motive is uncommon, it is by no means rare (Huff 1994). Cases are documented where an insured child is murdered, but more commonly the parents wish to profit from getting rid of a perceived nuisance or hindrance—their own child. This is particularly true in a single-parent or divorce situation where the child is viewed as an impediment to freedom or marriage. Filicide by fire is discussed in a later section of this chapter.

Other nonmonetary reasons from which arsonists may profit range from setting brush fires to enhance the availability of animals for hunting to burning adjacent properties to improve the view. Also, fires have been set to escape an undesirable environment, such as in the case of seamen who do not wish to set sail (Sapp et al. 1993, 1994).

EXAMPLE 5.5   Case History of a Profit-Motivated Serial Arsonist

Arnold (not his real name) was a professional arsonist (a "torch") and was serving 2 years in prison for one of his arsons. He admitted having burned down 35 to 40 vacant houses over his career. He became involved in setting fires for profit when talking with a real estate agent who was trying to find him an apartment. "He asked me if I would find somebody to burn a place. Did I know someone who was involved in demolition? I told him yeah and it started from there."

Arnold then partnered with the real estate agent and the demolition specialist. The real estate agent identified the targeted properties and made sure they were vacant. The demolition specialist taught Arnold how to set the fires, even accompanying him on his first job.

Arnold claimed that the fire-setting technique he used never resulted in a confirmed arson. "They could not tell. They would say it is under investigation. When they say it is under investigation, they are pretty sure it is arson but they cannot prove it." His method of operation was to pour 5 to 10 gal of odorless white gas in the attic and leave a time-delayed chemical timer. The ignited gasoline resulted in a fast-developing fire that collapsed the roof and overwhelmed the suspicions of the fire investigators, who found it hard to determine that an accelerant was used.

Arnold was concerned that no one would be injured in the fires he set. "I never burned any place where anybody was inside. I made sure I was not hurting anybody. That was important, it really was."

After being implicated with the real estate agent for an attempted arson, Arnold served 2 years in prison. "It was wrong, but at the time I was making good money and I was not hurting anybody. It was easy, but it is a dangerous job. I was afraid every minute while I was doing it."

As an afterthought, Arnold claimed that he was also a victim of his arsons. "I am the one that got hurt. I was underpaid. I was making $700 or $800 a fire, the real estate man was making about $10,000 on his share of the insurance policy and I am the one that got sent away and everything" (Sapp et al. 1995).

EXTREMIST-MOTIVATED ARSON

Arsonists involved in *extremist-motivated arson* may set fires to further social, political, or religious causes (see Table 5.6). Examples of extremist-motivated targets include abortion clinics, slaughterhouses, animal laboratories, fur farms, furrier outlets, and even, now, sport-utility vehicle (SUV) dealers. The targets of political terrorists reflect the focus of the terrorists' wrath. Random target selection also simply generates fear and confusion. Self-immolation has also been an extremist act.

TABLE 5.6 ◆ Extremist-Motivated Arson

Characteristics	
Victimology: targeted property	Analysis of targeted property essential in determining specific motive and represents the antithesis of the offender's belief
	Targets include research laboratories, abortion clinics, businesses, religious institutions
Crime scene indicators frequently noted	Crime scene reflects organized and focused attack by the offender(s)
	Frequently employ incendiary devices, leaving a nonverbal warning or message
	Overkill when setting fire
Common forensic findings	Extremist arsonists are more sophisticated offenders and often use incendiary devices with remote or time-delay ignition
Investigative considerations	Offender is frequently identified with cause or group in question
	May have previous police contact or an arrest record with crimes such as trespassing, criminal mischief, or civil rights violations
	Postoffense claims should undergo threat assessment examination
Search warrant suggestions	Literature: writings, paraphernalia pertaining to a group or cause, manuals, diagrams
	Incendiary device components, travel records, sales receipts, credit card statements, bank records indicating purchases
	Flammable materials: materials, liquids

Source: Updated from Icove et al. (1992).

EXAMPLE 5.6 Extremist-Motivated Arson

This is a fictional account of a serial extremist-motivated arson attack based upon news accounts (Associated Press 2001).

Federal and local authorities stepped up efforts to investigate a series of arson attacks across several states. The fires are thought to be extremist-motivated, an attempt to lash out against companies, universities, and institutions that threaten the environment.

The group claiming responsibility for the attacks actually posted a manual on its Web site that explains to novice arsonists how to build and use simple incendiary devices. The instruction manual gives guidance on how to maximize damage by arson fires. Targets of the group have included luxury homes, logging trucks, and an agricultural research center. The arsonists do little to cover up their crimes, and in one case several ignitable liquid-filled containers were found beneath a targeted property.

Little is known about the group, only that authorities are concerned that as the arson activity escalates, public safety officers are placed at greater risk of getting injured or killed fighting or responding to these fires.

◆ 5.3 OTHER MOTIVE-RELATED CONSIDERATIONS

Other factors involving motives appear in the literature and are important to address. This information is provided in order to clarify inaccuracies and misconceptions.

PYROMANIA

Perhaps most conspicuous by its absence is any mention of *pyromania* in this discussion of motivations. For an authoritative definition of the term, refer to the American Psychological Association's (1994) *Diagnostic and Statistical Manual of Mental Disorders,* 4th ed. (DSM-IV), the standard for psychological and psychiatric diagnoses for over 40 years.

A review reveals that each edition of the DSM has treated this topic differently. The current edition, DSM-IV, does not list pyromania as a diagnosed personality disorder. The definition of this disorder cycles through the years due to varying opinions and the lack of a solid definition.

DSM-IV's diagnostic criteria for pyromania include one or more of the following conditions.

A. Deliberate and purposeful fire setting on more than one occasion.
B. Tension or affective arousal before the act.
C. Fascination with, interest in, curiosity about, or attraction to fire and its situational contexts (e.g., paraphernalia, uses, consequences).
D. Pleasure, gratification, or relief when setting fires, or when witnessing or participating in their aftermath.
E. The fire setting is not done for monetary gain, as an expression of sociopolitical ideology, to conceal criminal activity, to express anger or vengeance, to improve one's living circumstances, in response to a delusion or a hallucination, or as a result of impaired judgment (e.g., in Dementia, Mental Retardation, Substance Intoxication).
F. The fire setting is not better accounted for by Conduct Disorder, a Manic Episode, or Antisocial Personality Disorder.

Researchers ask if the fire-setting impulses, characteristic of the various definitions of pyromania, could be a manifestation of some other disorder. Some report that the "irresistible impulse" to set fires may actually just be an impulse not resisted (Geller, Erlen, and Pinkas 1986).

Sexual fantasies or desires have often been linked to pyromania in the popular literature, but this is not borne out by the interview research. Sex as a motive is highly overrated.

The so-called motiveless arsonist in many cases knows his or her motive for setting fires but it does not necessarily make sense to normal outsiders. The arsonist may also lack the capacity to express such concerns.

Fire investigators are cautioned not to label a subject a pyromaniac since this is a diagnosis to be made by a mental health professional. They are reminded that each field brings a bias or different slant on how pyromania is viewed, with investigators, criminologists, psychologists, and psychiatrists each having a slightly different definition.

MIXED MOTIVES

Interviews conducted with incarcerated arsonists underscore the complexity of human behavior, particularly when ***mixed motives*** arise. When questioned on the motives for their arsons, they gave responses indicating that, while there was always a primary motive (vandalism, excitement, revenge, crime concealment, profit, or extremism), there were also often secondary and supplementary motives.

It should also be noted that arson can be used directly as a weapon, with the intent only of killing the "target." Such homicides can be linked to a wide variety of motives, including self-defense. A thorough reconstruction of the entire incident, rather than just the fire-setting event, may reveal the offender's actual intent for the homicide.

Researchers, most notably Lewis and Yarnell (1951), assert that revenge is present as a motive in all arsons to a greater or lesser degree. Motives for arson, like other aspects of human behavior, often defy a structured, unbending definition. Another rationale is that arsonists may lack the social and communicative skills necessary to articulate their motives clearly. Embarrassment about the true motive may also lead to providing an alternative or false motive to investigators or mental health professionals.

Add the element of power and revenge, and one can see the problem of strict, unyielding classification. Fire investigators also should be aware that since arson is a criminal tool, motivations may change based upon the target, situation, or motivation.

The best examples of many of the factors that arise in a serial arson case are given in an actual account of an arson from an offender's point of view: the well-articulated account by a female serial arsonist who used the pen name Sarah Wheaton. Included in her published article are edited excerpts from her discharge summary, which included a diagnosis of borderline personality disorder. Her treatment therapy included biofeedback, social skills training, and clomipramine, a psychotropic drug. At the time of writing the article, Ms. Wheaton (2001) stated that she was fire-thinking-free for 8 months.

EXAMPLE 5.7   Memoirs of a Serial Fire Setter

Sarah Wheaton (not her real name), a former self-admitted compulsive serial fire setter, is now working on a master's degree in psychology. At the end of her college freshman year, during the summer of 1993, she was involuntarily admitted to a psychiatric hospital for 2 weeks for treatment for fire setting.

The following material in italics is a direct quote from Ms. Wheaton's (2001) description of her treatment and perception of what she experienced as a serial arsonist. Her viewpoint was published as "Personal Accounts: Memoirs of a Compulsive Firesetter" in *Psychiatric Services*, Volume 52, pages 1035–1036, 2001. Copyright 2001, the American Psychiatric Association; http://PS.psychiatryonline.org. Reprinted by permission.

Reason for admission: *This 19-year-old single female was living in a dorm at the University of California at the time of admission. The patient had been involved in bizarre activity, including lighting five fires on campus that did not remain lit.*
History of present illness: *This young lady is a highly intelligent, highly active young woman who had been the president of her class in her high school for all four years. She had been going full-time to the university and working full-time at a pizzeria. She had called the police threatening suicide. When she was brought in, she was in absolute and complete denial. Patient claimed that everything was wonderful and she did not need to be here. She was admitted for 72-hour treatment and evaluation.*
Hospital course: *The hospital course initially was quite tempestuous. It was difficult to determine what the diagnosis was, as the patient was on the one hand very bright and*

very endearing to the staff and on the other hand was very unpredictable. She jumped over the wall on the patio (AWOL). The police were called and ultimately brought her back. She was subsequently certified to remain in custody for up to 14 days for intensive treatment because she attempted to cut herself with plastic and/or glass. She became more open after this, more tearful and at times more vulnerable. She tended to run from issues, to try to help everyone else, and not look at herself. Her father was seen in family therapy with her by a social worker.

Mother is reported by father to be both an alcoholic and have a history of bipolar illness. The patient herself reported sexual abuse by an older stepbrother when she was about age nine to age 11.

Initially I intended to use antimanic medication with her, either carbamazepine or lithium, but opted not to as she was adamant against medication. The patient did seem to stabilize without medication. She showed dramatic improvement, although the staff and I still have concern. The stable environment she has been able to pull together may not exist after discharge. She is scheduled to go to Washington, D.C., on July 1 to work as an intern in the office of one of the congressional representatives. She had done this two years ago working as a page.

Aftercare instructions: *No follow-up appointment is scheduled because she is discharged today and will be leaving for Washington on the first.*

Prognosis given by psychiatrist: *Prognosis is very guarded given the severity of her condition.*

Mental status: *She firmly denies…destructive ideation, including firesetting at this time.*

Discharge diagnosis *[33 hospitalizations after the initial hospitalization]: Axis I. Major depressive disorder, recurrent, with psychosis. Axis II. Obsessive-compulsive personality disorder; history of pyromania; Borderline personality disorder. Axis III. Asthma. Axis V. GAF 45.*

As a student of psychology Ms. Wheaton's article accurately traces many of the traits and characteristics of serial fire setters, particularly those reported in the literature and supported by interviews of law enforcement agents specializing in arson. Quoted below are her observations on how fire dominated her life from preschool to college.

Fire became a part of my vocabulary in my preschool days. During the summers our home would be evacuated because the local mountains were ablaze. I would watch in awe.

Below I have listed some of my thoughts and behaviors eight years after the onset of deviant behavior involving fire. I have also included suggestions for helping a firesetter.

Firesetting behaviors on a continuum. *Each summer I look forward to the beginning of fire season as well as the fall—the dry and windy season. I set my fires alone. I am also very impulsive, which makes my behavior unpredictable. I exhibit paranoid characteristics when I am alone, always looking around me to see if someone is following me. I picture everything burnable around me on fire.*

I watch the local news broadcasts for fires that have been set each day and read the local newspapers in search of articles dealing with suspicious fires. I read literature about fires, firesetters, pyromania, pyromaniacs, arson, and arsonists. I contact government agencies about fire information and keep up-to-date on the arson detection methods investigators use. I watch movies and listen to music about fires. My dreams are about fires that I have set, want to set, or wish I had set.

I like to investigate fires that are not my own, and I may call to confess to fires that I did not set. I love to drive back and forth in front of fire stations, and I have the desire to pull every fire alarm I see. I am self-critical and defensive, I fear failure, and I sometimes behave suicidally.

Before a fire is set. *I may feel abandoned, lonely, or bored, which triggers feelings of anxiety or emotional arousal before the fire. I sometimes experience severe headaches, a rapid heartbeat, uncontrollable motor movements in my hands, and tingling pain in my* *right arm. I never plan my fire, but typically drive back and forth or around the block or park and walk by the scene I am about to light on fire. I may do this to become familiar with the area and plan escape routes or to wait for the perfect moment to light the fire. This behavior may last anywhere from a few minutes to several hours.*

At the time of lighting the fire.   *I never light a fire in the exact place other fires have occurred. I set fires at random, using material I have just bought or asked for at a gas station—matches, cigarettes, or small amounts of gasoline. I do not leave signatures to claim my fires. I set fires only in places that are secluded, such as roadsides, back canyons, cul-de-sacs, and parking lots. I usually set fires after nightfall because my chances of being caught are much lower then. I may set several small fires or one big fire, depending on my desires and needs at the time. It is at the time of lighting the fire that I experience an intense emotional response like tension release, excitement, or even panic.*

Leaving the fire scene.   *I am well aware of the risks of being at the fire scene. When I leave a fire scene, I drive normally so that I do not look suspicious if another car or other people are nearby. Often I pass in the opposite direction of the fire truck called to the fire.*

During the fire.   *Watching the fire from a perfect vantage point is important to me. I want to see the chaos as well as the destruction that I or others have caused. Talking to authorities on the phone or in person while the action is going on can be part of the thrill. I enjoy hearing about the fire on the radio or watching it on television, learning about all the possible motives and theories that officials have about why and how the fire started.*

After the fire is out.   *At this time I feel sadness and anguish and a desire to set another fire. Overall it seems that the fire has created a temporary solution to a permanent problem.*

Within 24 hours after the fire.   *I revisit the scene of the fire. I may also experience feelings of remorse as well as anger and rage at myself. Fortunately, no one has ever been physically harmed by the fires I have set.*

Several days after the fire.   *I revel in the notoriety of the unknown firesetter, even if I did not set the fire. I also return again to see the damage and note areas of destruction on an area map.*

Fire anniversaries.   *I always revisit the scene on anniversary days of fires that I or others set in the area.*

Fires not my own.   *A fire not my own offers excitement and some tension relief. However, any fire set by someone else is one I wish I had set. The knowledge that there is another firesetter in the area may spark feelings of competition or envy in me and increase my desire to set bigger and better fires. I am just as interested in knowing the other firesetters' interests or motives for lighting their fires.*

Suggestions for helping a firesetter.   *The likelihood of recidivism is high for a firesetter. The firesetter should be able to count on someone always being there to talk to about wanting to set fires. Firesetting may be such a big part of the person's life that he or she cannot imagine giving it up. This habit in all aspects fosters many emotions that become normal for the firesetter, including love, happiness, excitement, fear, rage, boredom, sadness, and pain.*

A firesetter should be taught appropriate problem-solving skills and breathing and relaxation techniques. Exposure to burn units and disastrous fire scenes may be therapeutic and may enable the firesetter to talk openly about physical and emotional reactions. Doing so will not only help the firesetter but also give mental health professionals a deeper understanding of the firesetter's obsession.

The self-reported insight of this individual clearly confirms the findings of scientific, psychiatric, and law enforcement research. Of particular note are the pre- and postoffense behaviors at the fire scene, the significance of anniversaries, and the highly suggestive impact of the publicity of fires set by other arsonists. Astute fire investigators can use this information to assist them in identifying and solving cases set by compulsive serial arsonists.

FAKED DEATHS BY FIRE

Insurance fraud by burning a substituted body in an attempt to obliterate the true identity is a crime unique to profit-motivated arson. This form of arson often involves advanced planning, particularly in acquiring the body and evading suspicion by the investigating authorities.

Case studies have shown that the warning signs for faked deaths include, but are not limited to, the following (Reardon 2002).

- The death occurs shortly after filing of the insurance application and/or during the contestability period.
- The death occurs abroad.
- The insured has financial problems.
- Large policies or multiple small policies that do not require medical examinations exist. The policies have misrepresentations or omissions.
- The insured uses aliases.
- Previous policies have been canceled.
- The levels of insurance are inappropriate for the actual earned income.
- There is no body, or the body is in an unidentifiable condition.

Reardon emphasizes that the fundamental component in the payment of any life insurance claim is proving that the insured is actually dead. He also recommends thoroughly investigating these types of claims and litigating where appropriate to reduce the likelihood of paying the claimant.

EXAMPLE 5.8 Out-of-Country Faked Death by Fire Scheme

In 1998, Donny Jones (not his real name) completed an application for a term life policy bearing a face value of $4 million. Jones already had a $3 million policy with another insurer. Six months after the $4 million policy was written, the insurance company received the news that Jones had died in a car accident outside the United States.

The insured prepared his plan carefully. While out of the country with a friend, Jones rented a large SUV, placed his bicycle in the car and, drove away at 10 PM, supposedly to take a 3-hr drive through the desert to an adjoining town. Early the next morning, the SUV was found burning at the side of a desert highway. Local authorities found no traces of the mountain bike in the SUV, no signs of collision, and, initially, no body. Later, a body was found in the vehicle at an impound yard, but the remains were quickly claimed by a traveling companion of Mr. Jones and cremated.

Due to the unique circumstances, the insurer initiated an immediate investigation into the claim by arranging for a forensic anthropologist to examine the skeletal bones and for a fire protection engineer to document and impartially determine the cause of the vehicle fire. The forensic anthropologist determined that the remains were those of an elderly man of Indian heritage, not of the insured, who was a 33-year-old Caucasian. The engineer also interpreted for the insurance company technical information on the incendiary nature of the fire, the staged accident, and forensic evidence collected during the investigation by the foreign authorities.

The insurance company denied the claim and filed suit in the U.S. District Court for declaratory judgment, seeking rescission of the policy. By initiating the litigation, this permitted the insurance company to begin to gather and preserve evidence from the authorities, police, and public and private sectors. The investigation outside the United States required the involvement of the U.S. consulate followed by litigation, where the Hague Convention generally requires approval from a U.S. court and the appropriate foreign authority.

Eventually, the insured was discovered working in the United States through a routine background investigation by a firm for which the insured had been working, under an alias. The insurance policy was rescinded, and the insured later pled guilty in Federal court to criminal charges of wire fraud and was ordered to pay full restitution (Reardon 2002).

FILICIDE

Filicide by fire is a crime where the murderer is the parent of the victim and has used fire to mask the death, often making it appear to be an accident. Even though cases of a parent killing a child appear to be rare, these events are becoming more frequently studied in academics (Stanton and Simpson 2002).

Studies by the FBI, although based upon limited case samples, still provide an insight into this crime. They speculate that due to the thoroughness needed to conduct such investigations, this phenomenon may be widespread and underreported (Huff 1997).

Multiple motives are found for filicide, and as for any crime, there are various motives for these murders, including the following examples.

- ◆ **Unwanted child.** Falsely believing that she and her spouse or lover can then live unburdened, a mother commits child murder to remove the perceived obstacle or nuisance child(ren).
- ◆ **Acute psychosis.** The parent is psychotic, such as the severely depressed single mother who stabbed her three small children to death and then set their apartment on fire.
- ◆ **Spousal revenge.** An estranged husband cruelly decides to deprive his wife of her most cherished treasure, her child.
- ◆ **Murder-for-profit.** Parents take out large life insurance policies on their children shortly before killing them by fire.

In any of these cases, the authorities may not be suspicious initially, for various reasons, including the lack of a thorough investigation of the fire; the presumption that the fatal fire was accidental, caused by the child playing with matches; masking of overwhelming compassion for the parents; and sadness for the children. This reaction may override "red flags" of suspicion (Huff 1997).

While there is no single indicator, several common factors appear in cases of filicide, including the following.

Victimology

The children are often young, preschool age; have little training of how to behave in a fire; and are thus less likely to escape. The perception is that younger children are more disposable and more easily "replaced" later.

Preoffense behavior

Look for unusual behavior just before the fire, such as preparation of the children's favorite meal or a visit to a special place.

Temporal

The fires take place at night or early in the morning when the children are most likely to be in bed asleep. This gives the parents time to plan the event and, in some cases, lock the children in their rooms during the fire setting.

Crime scene characteristics

Children are asked to change their sleeping arrangements the night of the fire and sleep in a room not normally used. Staging of scenes takes place in some cases where the children have already been shot, stabbed, or strangled to death and a fire set to cover the crime, frequently with a flammable liquid such as gasoline. In some cases escape routes are blocked and doors locked. Drugs may be administered to make the child sleepy or unconscious.

Offender characteristics

Half-hearted rescue attempts by adults are reported, with no appreciable element of danger encountered. Therefore, they do not display smoke-filled clothing, burns, or watery eyes as would be expected. The parents most commonly live in manufactured housing and are no older than their mid-30s.

Postoffense behavior

After the incident, the adults exhibit inappropriate behavior, such as little or no grief, seldom speak about the victims, favoring the discussion of material losses (including insurance coverage), and are too quick to return to life as usual.

In totality, there will be a combination of one or more of the above indicators. However, investigators are cautioned not to allow one factor alone necessarily to raise suspicion. They should move forward in a cautious, yet professional manner.

EXAMPLE 5.9 Mother Involved in Filicide

Firefighters responding to a late-summer fire found a 22-month-old female hidden under a pile of clothes in a locked bedroom closet, dead from smoke inhalation. The child's mother told firefighters that her young daughter had been known to play with the cigarette lighter, which was found in the kitchen.

Firefighters doubted the story, but it became believable when the deceased child's brother, who was about the same age, demonstrated that he was able to operate the lighter. Investigators then ruled the fire accidental, initially concluding that the deceased girl had lit some paper napkins in the kitchen, taken them into the bedroom, and gone into the closet, with the door locking behind her.

All the facts surrounding this fire were consistent with a child playing with a cigarette lighter and causing her own death. There was no reason to suspect foul play until 5 years later, when the son, now 9 years old, suffered third-degree burns in another fire. When firefighters arrived at the home, they found the child unconscious behind a locked bedroom door. The mother fled immediately after the fire.

One week later, fire investigators located and interviewed the mother, who allegedly confessed to setting both fires, telling them that she was angry with her husband. In each of these fires, there were individual factors indicating filicide by arson (Huff 1997).

◆ 5.4 GEOGRAPHY OF SERIAL ARSON

Offender-based classification of arson extends beyond merely identifying the motive. Investigators should also closely examine the actual geographical locations selected by the arsonist. An analysis of these sites may reveal much about the intended target, add insight into the arsonist's motive, and provide a potential surveillance schedule for attempting to identify and apprehend the arsonist.

The results of a joint research effort on the geography of serial violent crime illustrate the patterns found in serial arson cases. Results indicate that criminal offenders who repeatedly set fires exhibit temporal, target-specific, and spatial patterns that are often associated with their modus operandi (method of operation; m.o.) (Icove, Escowitz, and Huff 1993).

Many geographers, criminologists, and law enforcement professionals are concerned with the rising tide of serial violent crimes, including arson, that plagues the United States and other Free World countries. Many of these violent offenders purposely use jurisdictional boundaries to evade detection from law enforcement officials.

Serial arsonists often create a climate of fear in entire communities. Community leaders compound this problem by pressuring law enforcement agencies to identify and apprehend the fire setter quickly. Often the arsonist evades apprehension for months, frustrating even the most experienced investigators. Unpredictable gaps often occur between incidents, leaving law enforcement authorities questioning whether the arsonist has stopped his or her fire setting, left the area, or been apprehended for another offense.

PREVIOUS RESEARCH FINDINGS

Few studies have been conducted on the geographic distribution of serial violent crimes, including arson. The studies that do exist focus on specific characteristics found in particular crimes (Icove 1979; Rossimo 1999).

A keynote analysis of violent crime measured it in terms of a spatial and ecological perspective (Georges 1978). The social-ecological view is that the environment has a direct relationship with crime. In this view, an absence of community controls, such as found in inner cities, correlates with high crime rates. Several historical approaches trace methods to interpret the geography of crime, particularly as it impacts arson and fire-related crimes.

Transition Zones.   The Chicago School of Sociology approach first proposed that crime was related to the environment (Park and Burgess 1921). This ecological approach assumes that high crime rates flourish in urban areas called *transition zones*. These zones have mixed land uses, high population fluxes, and a lack of opportunities for middle and upper economic classes to succeed.

In many cities, encircling concentric rings form zones of transition, representing diffusion between the central business district and residential housing. These zones often contain rooming houses, ghettos, red-light districts, and diverse ethnic groups.

An example of the transition zone theory is given in a study of 15 fire bombings in Knoxville, Tennessee, during 1981 (Icove, Keith, and Shipley 1981). An analysis of the fire bombings investigated by the Knoxville Police Department's Arson Task Force showed that 66 percent of the incidents occurred within specific census tracts having significant population decline or growth.

The Knoxville study also examined a census tract in a low-income area where three fire bombings occurred during the reporting period. U.S. Census data indicated that this tract had the highest minority population, the highest percentage of poverty, and the second-lowest mean income of all the tracts where bombings occurred.

Centrography.   The technique of *centrography* uses descriptive statistics to measure the central tendency of crime on a two-dimensional spatial plane. In certain crimes, the offender's residence may be close to the site of the crime commission. Another hypothesis is that the older the offender, the greater the mean distance from the crime scenes to his or her residence. This mobility is due, in part, to the offender's access to bicycles and vehicles, as well as being over the curfew age.

A study for the U.S. Department of Justice explored the concept of centrography and described geographic **anchor points** for criminals (Rengert and Wasilchick 1990). Examining the geographic locations of buildings targeted by burglars, the study concluded that criminals select targets close to an anchor point that minimizes the travel and time required to commit their crimes. Often, the dominant anchor point is close to the offender's residence. Other anchor points away from the criminal's residence include bars and video arcades.

Other geographic analysis approaches take place on both the micro and the macro levels. The micro-level approach examines the exact physical location of the crime, such as the type of building, vehicle, or street. The macro-level approach tends to aggregate the data into zones. These zones can be census tracts, police beats, or other geographic areas. This approach tends to increase the scale of the analysis.

Centrography has been used to examine spatial–temporal relationships of other violent crimes (LeBeau 1987). Observations of the spatial distribution of crime use the x and y axes of a Cartesian coordinate system overlain on a street-block map. The locations of numerous incidents are analyzed using the **mean center** approach to measure the central tendency by tracking the movement of these mean centers and explore the standard deviational ellipse to describe the distribution of incidents.

The spatial and temporal trends of serial arsonists and bombers have been explored on a local level (Icove 1979; Rossimo 1999). Patterns not previously documented among serial incendiary crimes were found using cluster analysis techniques—individual and group serial arsonists and bombers tend to be geographically bounded by natural and artificial boundaries. When the offender moves his or her residence, the cluster of activity normally follows.

Temporal trends have also been examined by Icove (1979) using two- and three-dimensional cluster analysis techniques. This form of analysis detects subtle time-of-day and day-of-week trends correlated with geographic changes in the cluster centers. Other temporal trends such as lunar activity are also documented.

EXAMPLE 5.10 Incendiary Crime Detection

This is an example of the application of geographic analysis to detect local clusters of activity of serial arsonists. After plotting the locations of arsons in an Eastern city from February through September 1974, a major cluster center of activity was detected. Three comparative techniques are shown in Figures 5.7 and 5.8, illustrating the use of two-dimensional gridded incident, shaded, and three-dimensional contour maps by Icove (1979).

A further analysis showed that a major cluster center was formed by two grids totaling 12 arson fires. Investigation revealed that during that time period, a gang of juvenile boys engaged in starting incendiary fires within a section of the large city. Properties targeted by these youths included garages and wooded areas.

When plotted on a map along with other arsons in the area, the incidents formed a major cluster. Data from these fires were collected and analyzed to provide a surveillance schedule for local law enforcement. A temporal analysis of the time of day and day of the week for the incidents contained in this major cluster was performed and revealed that the juveniles set these incendiary fires during breaks in the school day. In the summer months, fires occurred in the evening hours, but the trend reverted to the original pattern at the start of school in the fall. With few exceptions, the fires were on weekdays. During this step, lunar patterns are often detected. In this example, 75 percent of the fires occurred within 4 days of either the new or the full moon phase.

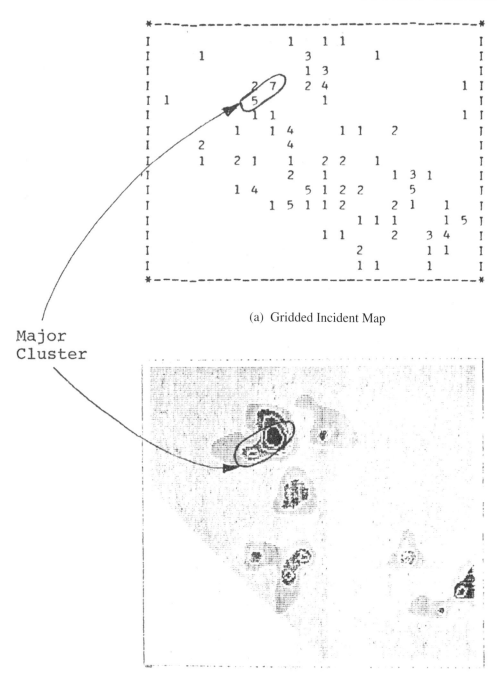

(a) Gridded Incident Map

Major
Cluster

(b) Contour Map of Incidents

FIGURE 5.7 ◆ Gridded incident and contour maps with major cluster centers of arson fires.
Source: Icove (1979). *Courtesy of D. J. Icove.*

Major
Cluster

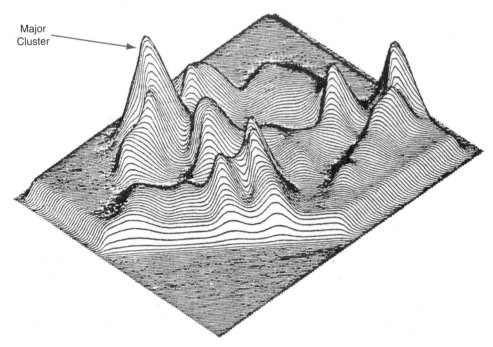

FIGURE 5.8 ◆ The detection of geographic clusters of arsons in three dimensions with a 35° counterclockwise rotation. Source: Icove (1979). *Courtesy of D. J. Icove.*

Urban Morphology. Georges (1978) examined the concept of ***urban morphology***, where crime tends to concentrate within certain zones or within areas with close access to transportation routes. For example, in his earlier study on the geographic distribution of arsons during the 1967 Newark riots, fires were found to be set along main commercial routes (Georges 1967). The study further documented earlier theories that as cities grow, crime tends to spread along main transportation routes (Hurd 1903).

CRIME PATTERN ANALYSIS

Crime pattern analysis is a methodology for detecting recurring patterns, trends, and periodic events in the times, dates, and locations of incidents. Once patterns are detected in a serial arson case, much can be learned to help predict the next event and classify the behavioral patterns exhibited by the offender.

A criminal investigative analysis (also referred to as a "profile") is usually prepared by an expert, based upon exposure to numerous case histories, personal experiences, educational background, and research performed. Suspects developed by crime analysis do not always fit each and every prediction since the predictions are made based upon similar historical cases, and the offender may often modify his or her m.o. in response to an active investigation.

Several key points are developed during a criminal investigative analysis involving the geography of serial incidents: temporal analysis, target selection, spatial analysis of the cluster centers, and standard distance.

Temporal Analysis. The first type of pattern analysis in serial arson cases is temporal analysis, which determines if any time-of-day or day-of-week trends exist. These trends may vary due to the offender modifying his or her m.o. in response to an active investigation. This modification may be a conscious act by the suspect.

Target Selection.   Another form of pattern analysis deals with the type and location of the target selected by the offender. A commonly observed characteristic in cases involving serial arsonists is that the offender often escalates the targets selected. This escalation is reflected in more life-threatening situations as time progresses. A typical scenario is that the arsonist starts with small grass and brush fires, then moves to outbuildings and vacant structures and, finally, to occupied structures.

Spatial Analysis.   Using a technique called spatial analysis, an investigator can determine a series of arson fires set repeatedly within the same geographic area to form a cluster of activity. Finding this center of activity often reveals information about future target selections or where the offender may live or work. As indicated previously, serial arsonists and bombers often tend to be geographically bounded by natural and artificial boundaries. Therefore, the offenders either consciously or unknowingly maintain their activities within a bounded area, not often crossing major highways, rivers, or railroad tracks.

Long-term analysis of these cluster centers over time is useful to determine if the offender has changed his or her residence or place of work. In cases spanning 3 years or more, a cluster center should be calculated for each year and plotted on a map (Icove 1979). If the cluster centers do not move significantly, or if they hover in a specific area, the offender has not likely changed residence or job. When a marked change is detected, this information is conveyed to the law enforcement agency, which compares this knowledge with the potential movement of the suspects.

Cluster Centers.   The mean center is the primary measurement of a spatial distribution of a cluster of data points. Given that the columns on a map equate to x values on a Cartesian coordinate system, and the rows to y values, the equations for calculating the mean center (Ebdon 1983) are

$$\bar{x} = \frac{1}{n} \sum_{i=1}^{n} x_i,$$

$$\bar{y} = \frac{1}{n} \sum_{i=1}^{n} y_i.$$

For the general case of cluster center z, the formula is

$$z_i = \frac{1}{n} \sum_{i=1}^{n} x.$$

A more suitable center of activity for crime pattern analysis is known as the center of minimum travel or what is often referred to as the ***centroid***. This center is the point or coordinate at which the total sum of the squared Euclidean distances to all other points on the map is the lowest.

The centroid is more meaningful than a mean center when examining the geographic locations of crimes committed by a single offender. The reasoning behind this assumption is that an offender who walks to his or her crime scenes will choose the minimum travel distance. The mean center does not always agree with the centroid.

The calculation of the centroid is not a simple formula but requires an iterative process of minimizing a performance index. The best clustering procedure using this technique is referred to as the *K*-means algorithm (MacQueen 1967). The application of this algorithm (Tou and Gonzalez 1974) is best broken down into logical steps.

Standard Distance. Once a cluster center is calculated for the geographic locations of a series of crimes, law enforcement officials can use this knowledge to concentrate their investigative and surveillance efforts within that neighborhood. For the purposes of geographic crime analysis, the ***standard distance*** refers to the area of study bounded by a radius of one standard deviation from the cluster center.

This search radius from the cluster center is useful in focusing a criminal investigation within a reasonable distance from the cluster center of activity. For example, in a detailed crime analysis of a serial arson case profilers often recommend a surveillance schedule and area of concentration within the area bounded within one standard deviation from the cluster center (Icove 1979).

Several case examples are presented to illustrate the phenomena associated with the geography of crime. The following cases were submitted for crime pattern analysis after an intensive investigation by a local law enforcement agency had exhausted all logical leads (Icove, Escowitz, and Huff 1993).

EXAMPLE 5.11   Shifting Cluster Centers

During the fall and winter months in a Southwestern state, an unknown arsonist was suspected of setting 24 fires involving fields, vehicles, mobile homes, residences, and other structures. Some of the fires were set after the offender broke into the structures.

TABLE 5.7 ◆ A Month-to-Month Temporal Analysis (Time of Day, Day of Week, and Selected Target) Should Be Performed When Examining the Patterns of Geographically Shifting Cluster Centers

Variable	Aug.	Sept.	Oct.	Nov.	Dec.	Total
8 AM–4 PM					1	1
4 PM–12 AM	1		4	3		8
12 AM–8 AM		1	4	6	3	14
Unknown			1			1
Monday					1	1
Tuesday	1					1
Wednesday					2	2
Thursday			1	2	1	4
Friday				2		2
Saturday		1	4	6		11
Sunday				2	1	3
Field				1	1	2
Vehicle			2			2
Mobile home				2	1	3
House, vacant	1	1	1			5
House, occupied					1	1
Structure		1	3	5	2	11
Total	1	1	9	9	4	24

Source: Icove, Escowitz, and Huff (1993).

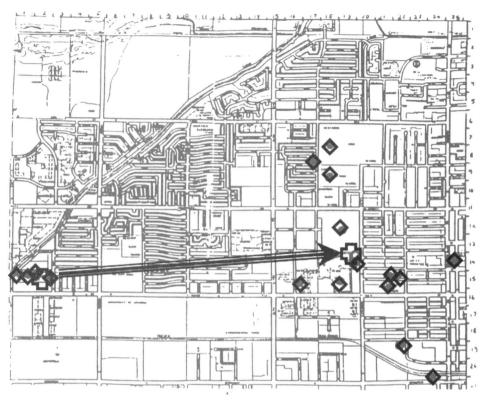

FIGURE 5.9 ◆ Geographic display of 24 fires showing the shift from west to east of two clusters' centers of activity. Source: Icove, Escowitz, and Huff (1993). *Courtesy of D. J. Icove.*

A temporal analysis of the 24 incidents revealed that the arsonist favored the late evening and early morning hours. The majority of the fires occurred on the weekends, as shown in Table 5.7. An analysis of the targets selected by the arsonist showed that the structures were either unoccupied, vacant, or closed for business. The arsonist escalated the severity of his fire setting, first using available materials and later turning to flammable liquids to accelerate the fires.

A geographic cluster analysis showed that the first three fires clustered tightly on the west side of the town. The remaining fires concentrated in a larger cluster of activity on the east side (Figure 5.9).

A crime analysis of this case was conducted and the results were forwarded to the agency that requested the assistance. The investigators were told to concentrate their efforts on any suspects who first lived near the west cluster and then changed their residence to close to the center of the east-side cluster.

Armed with this information, the law enforcement agency later charged and convicted a 19-year-old, single white male who matched the characteristics set forth by the crime analysis. A key element in the case was that the offender lived near both the cluster centers of the arsons. The geographic centers of the activity shifted when the offender moved from the west to the east side of town (Icove, Escowitz, and Huff 1993).

EXAMPLE 5.12   Temporal Clustering

A Southeastern city was plagued with 52 arson fires in vacant buildings, vehicles, commercial businesses, residences, and garages over a 2-year period. The law enforcement agency investigating the case was stumped and turned to crime analysis after exhausting all traditional leads.

A crime analysis revealed temporal and geographic patterns in the time and locations of the arsons. The majority of the fires were set during the afternoon and evening weekday hours. There were two 6-month gaps in activity. An overwhelming majority of the arsons occurred within a 1-mi radius of the downtown area of the city, with two additional clusters adjacent to two lakes. The crime analysis returned to the agency stated that the offender lived close to the cluster center of activity of his fires. When later apprehended, the arsonist admitted in most cases walking to the scenes of his fires and using available materials and matches or a lighter carried to the scenes. Even though it is a high-risk scenario for the offender to set the fire in the afternoon and evening hours, his intimate knowledge of the geographic area diminished his possibility of being detected and followed.

Four months after the department received the crime analysis, a 29-year-old white male was arrested after fleeing the scene of a fire he had just set. The suspect later told police that he first started setting fires when his relationship with a girlfriend failed. Further investigation revealed that the offender set all the fires within walking distance of his residence. The different clusters of fires were the result of the subject changing his residence. Periods of inactivity in his fire setting were correlated directly with times when the offender had ongoing social relationships. Only when these relationships failed did the offender return to setting fires.

When arrested, the suspect had in his possession a map that marked the locations of the fires he had set. The offender's prior arrest history included charges for disorderly conduct, criminal mischief, harassment, and filing a false report to law enforcement (Icove, Escowitz, and Huff 1993).

◆ 5.5 SUMMARY

Arson is defined as the willful and deliberate destruction of property by fire. Although establishing a motive for fire setting is not a legal requirement of the elements of the criminal offense, it can help focus investigative efforts and aid in the prosecution of the arsonist. Motives for fire setting usually include one or more of the following general categories: vandalism, excitement, revenge, crime concealment, profit, and political terrorism. Motiveless arson or pyromania is not considered an identifiable classification.

Arson crime scene analysis is still in its infancy. Problems associated with its use and application include the broad degree of knowledge required to assess and interpret the actual scenes, particularly using a motive-based approach.

A combined background in the principles of fire protection engineering as well as in behavioral and forensic sciences can certainly bring more to the analysis. Clearly, the concept that the "ashes can speak" is paramount in the application of this technique.

Future work in this area should combine the knowledge of fire pattern analysis with advanced photographic techniques to assist in the visualization of fire scenes long after they occur and deteriorate.

Problems

5.1. Research a local serial arson case that appears in the media. Plot the locations of these fires and conduct a temporal analysis. What can you learn from this analysis?

5.2. For the case from problem 5.1, determine what motive was developed. Is the assessment by the media correct? What was the motive stated by the prosecution? By the defense?

Suggested Reading

Danforth, J. C., Interim report to the Deputy Attorney General concerning the 1993 Confrontation at the Mt. Carmel Complex, Waco, Texas. U.S. Government Publication. July 21, 2000.

Huff, T. G. 1993. Filicide by Fire, *Fire and arson investigator.* June pp. 24–26.

Icove, D. J., V. B. Wherry, and J. D. Schroeder. 1998. *Combating arson-for-profit: Advanced techniques for investigators.* Columbus, Ohio: Battelle Press.

Rossimo, D. K. 2000. *Geographic profiling.* Boca Raton, Fla: CRC Press.

Sapp, A. D., T. G. Huff, G. P. Gary, and D. J. Icove. 1995. *A motive-based offender analysis of serial arsonists* (monograph). Quantico, Va: National Center for the Analysis of Violent Crime, FBI Academy.

Stauss, E. 1993. Paul Kenneth Keller. *National Fire and Arson Report* II: 4–6.

Wood, B. 1997. Arson profiling. *Fire Engineering Journal* 60:208.

CHAPTER 6 ◆ **Fire Modeling**

Data! Data! Data! I can make no bricks without clay.

Sir Arthur Conan Doyle,
"The Adventure of the Copper Beeches"

The concept of fire modeling as it applies to forensic fire investigation is unique to the last decade, although models have existed since the 1960s. Previously, fire modeling was centered on explaining the physical phenomena of fires, particularly when applied to verifying existing experimental data.

It was the effort of a few fire scientists and engineers working at and associated with the National Institute of Standards and Technology (NIST) and the Fire Research Station (BRE, FRS, UK) that pushed the acceptability and application of fire modeling out of laboratory conditions and into the world of forensic fire scene reconstruction. These early successes of fire modeling applied to the field of fire litigation and reconstruction further underscore its usefulness (Babrauskas 1996). Several of the keynote studies that contributed significantly to this effort are described in this textbook.

The purpose of this chapter is not to make the reader an expert in fire modeling, but to allow him/her to gain a better appreciation for its added value to an investigation. Ample references are provided should more information be needed.

This chapter also answers the following questions that often arise when fire investigators are confronted with a fire of exceptional scale or impact: (1) What exactly is a fire model? (2) In which aspect of my investigation can fire modeling help? (3) What are the realistic and reliable results of a fire model? (4) Should more than one type of model be run to increase confidence in the results? and (5) What is the future of fire modeling?

◆ 6.1 HISTORY OF FIRE MODELING

Research by Mitler (1991) at NIST best establishes the historical framework for the application of fire scene modeling. Work at NBS(NIST) NIST in 1927 was the first attempt to understand and explain in scientific terms the issues surrounding post-flashover compartment fires by relating the room gas temperature to the available mass of fuel being consumed (Inberg 1927).

The first fire model was developed through the work in Japan in 1958 to relate the ventilation factor to steady-state fire development (Kawagoe 1958). The second model was constructed in Sweden (Magnusson and Thelandersson 1970), followed by NIST (Babrauskas 1975).

Historical reasons for developing and applying mathematical fire modeling were also accurately articulated and predicted by NIST (Mitler 1991). Suggested are six major areas where a good mathematical fire model of a structure can render assistance:

- avoid repetitious full-scale testing,
- help designers and architects,
- establish flammability of materials,
- increase the flexibility and reliability of fire codes,
- identify needed fire research, and
- help in fire investigations and litigation.

These NIST areas of concentration are as on point today as they were over a decade ago. With the impact of environmental limitations on testing, fire modeling can certainly assist in avoiding full-scale testing. The impact of the placement and combination of certain fuel items, changes in venting, and thickness of materials can now be evaluated without the need for repeated full-scale testing.

Designers and architects have already found modeling to assess the flammability of materials as they specify new construction. New flexible fire performance codes are becoming more generally accepted as modeling introduces alternatives or new designs (such as atrium construction) that traditionally have not been addressed in historical prescriptive code. Fire modeling can now also address the optimum designs for evacuation of people from buildings during emergencies.

As the science of fire dynamics advances, computational modeling identifies new areas of fire research, into the phases of growth, flame spread, and flashover. Modeling can aid in gap assessments of certain areas that previously were not fully understood by designers, engineers, researchers, and even fire investigators.

Finally, forensic fire investigations and litigation have become one of the greatest areas of advancement that fire modeling has impacted. More detailed information on these impacts is given in this chapter.

◆ 6.2 FIRE MODELS

Computer fire models have a broad range of applications in the area of fire science and engineering. The fire model works to supplement the information found in conjunction with forensic evidence, witness interviews, media film footage, and preliminary fire scene examinations. Historically, these fire models, as shown in Table 6.1, have included eight major overlapping categories (Hunt 2000).

Fire models usually emulate the impact of fires, not the physical fire itself. There are two recognized approaches to modeling fires: probabilistic and deterministic. ***Probabilistic*** models usually center on the application of stochastic mathematics to estimate or predict an outcome (within certain likelihoods), such as human behavior (SFPE 2002a, chap. 3-12). In ***deterministic*** models, the investigator relies upon mathematical relationships that form the underpinnings of the physics and chemistry of fire science. Deterministic models can range from a simple straight-line approximation to correlating test data with complex fire models solving hundreds of simultaneous equations.

TABLE 6.1 ◆ Classes of Computer Fire Models

Class of Model	Description	Example(s)
Spreadsheet	Calculates mathematical solutions produced through interpretation of actual case data	FiREDSHEETS, NRC Spreadsheets
Zone	Calculates fire environment through two homogeneous zones	CFAST, ASET-B, BRANZFIRE, FireMD
Field	Calculates fire environment through solution of conservation equations, usually with finite-element mathematics	FDS, JASMINE, FLOW3D, PHOENICS, SOFIE
Postflashover	Calculates time–temperature history for energy, mass, and species and is useful in evaluating structural integrity in fire exposure	COMPF
Fire protection performance	Calculates sprinkler and detector response times for specific fire exposures based on the response time index (RTI)	DETACT-QS, DETECT-T2, LAVENT
Thermal and structural response	Finite-element calculations as to the structural fire endurance of a building	FIRES-T3, HEATING7, TASEF
Smoke movement	Calculates the dispersion of smoke and gaseous species	CONTAM96, Airnet, MFIRE
Egress	Calculates the evacuation times using stochastic modeling using smoke conditions, occupants, and egress variables	EXITT, EXIT89, EVACNET

Source: Updated from Hunt (2000).

Mathematical models are usually in the form of a mathematical relationship that is produced from the interpretation of actual test data. These types of models are usually done through "hand calculations" using the formula and a scientific calculator, a spreadsheet, or a simple computer program. Smoke filling rates, flame heights, virtual origin, and other approximations can usually be calculated by hand in several minutes. When based upon sound mathematical representations and properly applied, fire models can often assure that the scientific method has been satisfied.

Since mathematical models essentially are discussed throughout this textbook in the form of quick problem solutions, the zone and field models are emphasized here. The discussion is limited to a survey of the concepts and is by no means intended to serve as an instruction manual for their use. In fact, most guidance issued with fire models cautions that they are for use only by those having significant competency in the fire engineering field and that the results should supplement the user's common judgment.

The use of spreadsheets to solve simple fire protection engineering relationships arose from a May 1992 study, "Methods of Quantitative Fire Hazard Analysis," prepared for the Electric Power Research Institute (EPRI) by the University of Maryland, Department of Fire Protection Engineering. The study was distributed through the Society of Fire Protection Engineers (Mowrer 1992).

The study describes fire hazard analysis models used in the Fire-Induced Vulnerability Evaluation (FIVE) methodology, which is used by the Nuclear Regulatory Commission (NRC). The FIVE analysis uses realistic industrial loss histories to evaluate hazards quantitatively, including those produced in fires resulting from ignitable liquid spills, cable trays, and electrical cabinets. Many of these examples predict the impact of the fire plumes and ceiling jet temperatures, hot gas layers, thermal radiation to targets, and critical heat fluxes. Many of these concepts are applicable to problems previously posed in fire scene reconstruction and analysis (Mowrer 1992).

From this study came ***FiREDSHEETS*** from the University of Maryland, Department of Fire Protection Engineering (Mowrer and Milke 2001). These spreadsheets concentrate on the specifics of compartment fire analysis and have extensive property tables for typical materials first ignited, as illustrated in Figure 6.1.

A later and more industry-oriented set of spreadsheets known as FDS2 was developed for the NRC fire protection inspection program (Iqbal and Salley 2002). The format of the spreadsheet output is more intuitive in its approach, as with, for example, financial spreadsheets.

Materials: Thermal properties of boundary materials

ID	MATERIAL	k [kW/m.K]	ρ [kg/m3]	c_p [kJ/kg.K]	α [m2/s]	kρc [kW/m^2.K]^2.s
1	Aluminum (pure)	2.06E−01	2710	0.895	8.49E−05	5.00E+02
2	Concrete	1.60E−03	2400	0.75	8.89E−07	2.88E+00
3	Aerated concrete	2.60E−04	500	0.96	5.42E−07	1.25E−01
4	Brick	8.00E−04	2600	0.8	3.85E−07	1.66E+00
5	Concrete block	7.30E−04	1900	0.84	4.57E−07	1.17E+00
6	Cement-asbestos board	1.40E−04	658	1.06	2.01E−07	9.76E−02
7	Calcium silicate board	1.25E−04	700	1.12	1.59E−07	9.80E−02
8	Alumina silicate block	1.40E−04	260	1	5.38E−07	3.64E−02
9	Gypsum board	1.70E−04	960	1.1	1.61E−07	1.80E−01
10	Plaster board	1.60E−04	950	0.84	2.01E−07	1.28E−01
11	Plywood	1.20E−04	540	2.5	8.89E−08	1.62E−01
12	Chipboard	1.50E−04	800	1.25	1.50E−07	1.50E−01
13	Fiber insulation board	5.30E−05	240	1.25	1.77E−07	1.59E−02
14	Glass fiber insulation	3.70E−05	60	0.8	7.71E−07	1.78E−03
15	Expanded polystyrene	3.40E−05	20	1.5	1.13E−06	1.02E−03

FIGURE 6.1 ◆ Representative boundary thermal properties table from the FiRED-SHEETS spreadsheet program from the University of Maryland. *From Mowrer and Milke [2001]. Applications of Fire Behavior and Compartment Fire Models Seminar, Tennessee Valley Chapter of the Society of Fire Protection Engineers, Oak Ridge, TN, September 2001.*

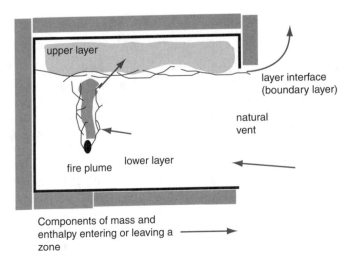

FIGURE 6.2 ◆ Representation of the two-zone fire model showing the location of the upper and lower layers, layer interface, fire plume, and compartment vent. *Courtesy of NIST, from Forney and Moss [1992].*

◆ 6.4 ZONE MODELS

Two-zone models usually emphasize the concept that a fire in a room or enclosure supports two unique zones and predict the conditions within each zone. In these models, the room is divided into an upper (warmer) and a lower (cooler) zone, each having a unique measurable temperature, concentration of by-products of combustion, and intersection height. Complex two-zone models of up to 10 rooms typically take a few minutes to run using Pentium-class computer workstations.

TABLE 6.2 ◆ **Numerical Limits in the Software Implementation of the CFAST Model**

Feature	Maximum
Simulation time (s)	86,400
Compartments	30
Object fires	30
Material thermal property definitions	56
Slabs in a single surface material	3
Fans in mechanical ventilation systems	5
Ducts in mechanical ventilation systems	62
Connections in compartments and mechanical ventilation systems	62
Independent mechanical ventilation systems	10
Targets (includes 1 per compartment and 1 per object)	30
Data points in a history or spreadsheet file	900

Source: Peacock et al. (2000).

Figure 6.2 shows a schematic representation of the concepts behind a two-zone model. A survey published by the Society of Fire Protection Engineers has shown ASET-B and DETACT-QS to be the most widely used zone models (Friedman 1992). Another widely distributed zone model is CFAST, the Consolidated Model of Fire Growth and Smoke Transport, which combines the engineering calculations from the FIREFORM model and the user interface from FASTlite to construct a unique comprehensive zone modeling and fire analysis tool. (Peacock et al. 2000). Table 6.2 summarizes the present limitations of the CFAST model, version 3.1.7.

OVERVIEW

The CFAST two-zone model is a finite-element model based upon not only the physics and chemistry, but also the results of experimental and visual observations in actual fires. CFAST computes

- production of enthalpy and mass by burning objects;
- buoyancy and forced transport of enthalpy and mass through horizontal and vertical vents; and
- calculations of temperatures, smoke optical densities, and species concentrations.

Two zones actually refer to two individual control volumes known as the upper and lower zone. The upper zone of heated gases and by-products of combustion form an upper layer that is penetrated only by the fire plume. These hot gases and smoke fill the ceiling layer, which then slowly descends. A boundary layer marks the intersection of the upper and the lower areas. The only exchange between the upper and the lower layers comes with the action of the fire plume.

Zone models use a set of differential equations to solve within each zone for pressure, temperature, carbon monoxide, oxygen, and soot production. The accuracy of these calculations varies with not only the model but also the reliability of the input data describing the circumstances and room dimensions and, very importantly, the assumptions as to the heat release rates of the materials burned in the fire.

CFAST 3.1.7 is the last version to have a graphical user interface (GUI) that allows the user to enter and modify characteristics about the physical form of the rooms, the fire signature, and graphics output. Enhancements in CFAST 5.0.1 allow an interface with NIST's Fire Dynamics Simulator program.

Since CFAST does not have a pyrolysis model to predict fire growth, a fuel source is entered or described by its fire signature (heat release rate). The program converts this fuel information into two characteristics: enthalpy (heat) and mass. In an unconstrained fire, the burning of this fuel takes place within the fire plume. In constrained fires, the unburned fuel seeks out portions of the fire plume where there is sufficient oxygen. These portions may be the upper or lower layer of the room of fire origin, the plume in the doorway leading to an adjoining room, or even the layers or plumes in adjacent rooms.

FIRE PLUMES AND LAYERS

As in previous mathematical models where calculations of heat flux used the fire plume's virtual origin, the CFAST model instructs the fire plume to act as a "pump" to move enthalpy and mass into the upper layer from the lower layer. Mixed with this enthalpy and mass are in-flows and out-flows from horizontal or vertical vents (doors, windows, etc.), which are modeled as plumes.

Some assumptions of the model do not always hold in real fires. Some mixing of the upper and lower layers does take place at their interfaces. At cool wall surfaces, gases will flow downward as they lose heat and buoyancy. Also, heating and air-conditioning systems will cause mixing between the layers.

Interroom horizontal flow of the fire from one room to the next occurs when the upper layer descends below the opening of an open vent. As the upper layer descends, pressure differentials between the lower layer may cause air to flow in the opposite direction, resulting in the two flow conditions. Upward vertical flow may occur when the roof or ceiling of a room is opened.

HEAT TRANSFER

In the CFAST model, unique material properties of up to three layers can be defined for surfaces of the room (ceiling, walls, flooring). This design consideration is useful, since in CFAST the heat transfer is to the surface in the form of convective heat transfer and through the surfaces by conductive heat transfer.

Radiative heat transfer takes place among three areas: fire plumes, gas layers, and surfaces. In radiative heat transfer, emissivity is dominated primarily by species contributions (smoke, carbon dioxide, and water) within the gas layers. CFAST applies a combustion chemistry scheme that balances carbon, hydrogen, and oxygen in the room of fire origin among the lower-layer portion of the burning fire plume, the upper layer, and lower layer-entrained air being absorbed into the upper layer of the next connecting room.

LIMITATIONS

Zone models have their limitations. Six specific aspects of physics and chemistry either are not included or have very limited implementations in zone models: flame spread, heat release rate, fire chemistry, smoke chemistry, realistic layer mixing, and suppression (Babrauskas 1996). Some errors in species concentrations can result in errors in the distribution of enthalpy among the layers, a phenomenon that impacts the accuracy of temperature and flow calculations. CFAST has also been known to over-predict the upper-layer temperatures.

The absence of a fire growth model in CFAST can be compensated for by the careful selection of information gleaned from historical calorimeter testing, where rates of energy and mass release by burning objects are collected.

Even with these known limitations, zone models have been extremely successful in forensic fire reconstruction and litigation. CFAST zone models have been successfully introduced in federal court litigation (see chapter 1, example 1.2). These successes are a result of their correct application by knowledgeable scientists and engineers, continued revalidation in actual fire testing, and applied research efforts.

◆ 6.5 FIELD MODELS

Field models, the newest and most sophisticated of all deterministic models, rely upon computational fluid dynamics (CFD) technology. These models are attractive, especially as a tool in litigation support and fire scene reconstruction, due to their ability to display the impact of information visually in three dimensions.

However, field models have their downside since typically the creation of input data is very time-consuming and often requires powerful computer workstations to compute and display the results. The computation time to run field models may be days or weeks depending upon the complexity of the model.

COMPUTATIONAL FLUID DYNAMICS

CFD models estimate the fire environment by dividing the compartment into smaller cells instead of two zones. The program then numerically solves simultaneously the conservation equations for combustion, radiation, and mass transport to and from each cell surface. The present downfalls are the lack of an intuitive graphical interface for encoding information and the extreme amount of time sometimes needed to input data and to run the model.

There are many advantages of using CFD models over zone models such as CFAST. With higher geometric resolution of CFD models, the solutions are more refined. Higher-speed workstations allow CFD models to be run, where in the past larger mainframe and minicomputers were required.

Since CFD models are used in companion areas such as fluid flow, combustion, and heat transfer, their underlying technology is generally more accepted. Once these models become more broadly validated, they will gain further acceptance in situations where scientifically based testimony is sought, such as in forensic fire scene reconstructions.

CFD has a long history of use in forensic fire scene reconstructions, and it is gaining in popularity due to its intuitive graphical interface. CFD programs now outpace the capabilities of the zone models, which, for the most part, operate under maintenance modes only. Moreover, CFD programs such as FDS make the technology economically feasible due to its wide use, testing, and growing acceptance.

NIST CFD TECHNOLOGY-BASED MODELS

The recommended field model using CFD technology, due to its availability at no cost, focus of research, and acceptance in the community is the Fire Dynamics Simulator (FDS; version 3). This model is available from NIST's Building and Fire Research Laboratory (McGrattan et al. 2002). The first version of FDS was publicly released in February 2000, and it has become a mainstay of the fire research and forensic communities.

FDS is a fire-driven fluid flow model and numerically solves a form of the Navier–Stokes equations for thermally driven smoke and heat transport. The companion program to FDS is Smokeview, which allows the visualization of data produced by FDS. User guides for both FDS and Smokeview are available at NIST's Web site (fire.nist.gov).

Traditional usage of FDS to date has been divided among the evaluation of smoke handling systems, sprinkler and detector activation, and fire reconstructions. It is also being used to study fundamental fire dynamics problems encountered in both academic and industrial settings. FDS uses a combination of models to handle the computational problems: a hydrodynamic model to solve the Navier–Stokes equations, a Smagorinsky form of large eddy simulation (LES) for smoke and heat transport, a numerical grid simulation to handle turbulence, a mixture fraction combustion model, and a finite volume method approach for radiation heat transfer.

Unfortunately, FDS uses a rectilinear grid describe the computational cells. This makes it slightly difficult to model sloping roofs, rounded tunnels, and curved walls without some form of approximation. The boundary conditions for material surfaces consist of assigned thermal constants that include information on its burning behavior.

The latest release of FDS at the time of this publication is version 3, which is a significant enhancement over previous versions. These enhancements include the ability to evaluate heat transfer through walls, impact of water suppression and initial conditions and to add image texturing to solid objects and surfaces. The latest companion release of Smokeview is version 3.0. Its enhanced features include controls over viewing the rectilinear mesh used in constructing the FDS model. Scene clipping allows the visualization of obstructed boundary surfaces in complex models having numerous walls. A more robust feature allows more control over the movement and orientation of the scene as rendered by Smokeview.

Other features include several visualization modes such as tracer particle flow, animated contour slices of computed variables such as vector heat flux plots, animated surface data, and multiple isocontours. The use of new visualization tools can allow for the comparison of fire burn patterns with computed uniform surface contour planes. The option for animated flow vectors and particle animations allows for the comparison of observed heat flux vectors with computed vectors.

A horizontal time bar displays the duration of the run while a vertical color-coded bar displays other variables, such as temperature and velocity. Graphic options include the ability to capture the output of the Smokeview program on a screen-by-screen basis. An enhanced dialogue box gives greater control and navigation to ensure the most accurate representation for comparison with actual fire case histories.

VALIDATION

All fire models must be validated by comparing their predictions against the results of real fire tests if their results are to be relied upon. This testing will sometimes reveal flaws or "blind spots" where some types of enclosures or conditions produce predicted values that differ from observed real fire behavior. If a model has not been shown to give valid results for a particular type of fire, its predictions in an unknown scenario should not be taken as gospel. For instance, FDS has been shown to be very accurate in predicting growth of fires in large compartments but data on how FDS predictions will relate to real fires in very small compartments are just now being gathered and published.

◆ 6.6 CASE HISTORIES

Over the years, the traditional usage of fire models has clustered into two separate yet important areas of concentration:

- building fire safety and codes analysis and
- fire reconstruction and analysis.

Specific fire modeling evaluations have looked at the impact or performance of smoke handling systems, sprinkler and detector activation, and fire/burn patterns in reconstructions. Fire modeling found its way into forensic reconstruction through a long history of applications to real-world events.

Science and litigation are the two driving factors behind the appearance of these applications. Fire protection engineering principles developed from correlating tests to actual case histories stretch the knowledge of fire science as we know it. Attorneys seeking answers to liability for damages due to fires also have a reason to seek the answers to similar questions. Therefore, science and law meet in the courtroom.

Moreover, the public pursues answers to why and how incidents occur and whether fire codes perform as anticipated during actual fires. This is due not only to their curiosity, but to a legitimate concern for answers to questions such as, "Is it safe for my child to live in a nonsprinklered dormitory?"

Several keynote case histories are worth mentioning to show the development of fire modeling in forensic evaluations of fire scenes. These cases are listed below along with the important concepts gleaned from the referenced and published studies.

U.K. FIRE RESEARCH STATION

The Fire Research Station (FRS) located in Garston, U.K., has a 50-year history of assisting with fire investigations for the U.K. government. FRS investigations and reconstructions have included fire incidents at the Stardust Disco in Dublin, at Windsor Castle, in the Channel Tunnel, and on the Paddington–Ladbroke Grove Train. The FRS (2002) uses a combination of a statistical database of loss histories, scene investigations, a fire testing laboratory, and fire models to test hypotheses or investigate unusual phenomenon (FRS 2002).

Early work at the FRS developed a compartment fire model to address smoke and heat venting when its researchers examined the similarities of the Livonia, Michigan, and Jaguar auto plant fire losses (Nelson 2002). These fires at two separate parts manufacturing centers had devastating impacts on the production of automobiles. What began as a search for means to vent hot by-products of combustion through the ceiling formed the basis for the zone fire model and was based upon actual fire testing.

DUPONT PLAZA HOTEL AND CASINO FIRE

The December 31, 1986, fire at the *Dupont* Plaza Hotel and Casino, San Juan, Puerto Rico, killed 98 persons. Due to the extreme and needless loss of life, this case has become a textbook example of the application of fire reconstruction of the initial development of what became a large complex fire. NIST fire researcher Harold "Bud" Nelson worked in cooperation with ATF investigators to collect and analyze data during the scene investigation. This led to a successful prosecution of the individuals who started the fire simply to harass hotel management. The analysis combined technical data and fire growth models to explain the dynamics exhibited by the fire. The analysis looked at mass burning rate, heat release rate, smoke temperatures, smoke layer, oxygen concentration, visibility, flame extension and spread, sprinkler response, smoke detector response, and fire duration. Fire models used in this analysis included FIRST, ROOMFIR, ASETB, and HOTVENT.

The analysis confirmed the area of origin to be the south ballroom. The fire spread next to a foyer, then to the lobby and casino area. Figure 6.3 documents the conditions at 60 s after the fire started.

KINGS CROSS UNDERGROUND STATION FIRE

One of the first important cases using CFD models for fire investigation was the analysis of the Kings Cross Underground Station fire, which occurred on November 18, 1987. The CFD model FLOW3D tracked the fire upward along a 30° incline of the wooden sides and treads of the escalators to the ticket concourse level. The theory of the "trench effect" was verified by the CFD model that described this unique

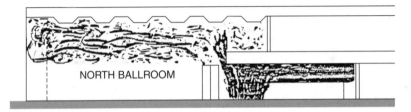

NORTH BALLROOM

FIGURE 6.3 ◆ The analysis combined technical data and NIST fire growth models to explain the dynamics exhibited by the fire at the *Dupont* Plaza. The analysis looked at mass burning rate, heat release rate, smoke temperatures, smoke layer, oxygen concentration, visibility, flame extension and spread, sprinkler response, smoke detector response, and fire duration. *Courtesy of NIST, from Nelson [1987].*

phenomenon. Thirty-one persons died in the fire, including a senior station officer (Moodie and Jagger 1991).

In determining the area of origin for this fire, investigators homed in on the area where the fire damage first started, some 21 m (68 ft) below the top of escalator No. 4. This area was at the lowest point of fire pattern damage to wooden handrails, treads, and risers. An analysis of the paint blistering and ceiling damage was also used to locate the fire origin. The physical indicators were corroborated by witnesses who saw the fire partway up the escalator from below.

The investigation concluded that the fire started beneath the wood steps on the moving staircase. The fire consumed much of the available combustible material in the escalator shaft. Fire damage also extended to the other escalators and the general ticket hall.

In addition to a CFD model, government health and safety personnel obtained ignition characteristics of the samples taken from nearby the area of fire origin. Investigators were surprised to determine that escalator lubricating grease was able to be ignited only if combined with fibrous debris (paper, hair, lint) that acted as a wick.

To assess and validate the results of the CFD model, three types of tests were run. Full-scale fire growth tests determined that the initial fire was about 1 MW in size. Small-scale models implied that above a certain fire size, the flame in the channel of the escalator did not rise vertically but was pushed down into the trench by entrained air, spreading rapidly. Scaled open-channel tests examined the 30° inclined plane and used plywood with various surface coverings to evaluate flame spread.

Lessons learned from this fire include the result of a fire under the influence of the trench effect and the speed at which such a fire can develop and spread. Witnesses underestimated the speed of the smoke spread, the growing intensity of the fire, and the need for rapid evacuation. Smoking on underground rail services has since been banned, and combustible escalators, enclosures, and signage eliminated.

FIRST INTERSTATE BANK BUILDING FIRE

Another NIST engineering analysis was conducted on the May 4, 1988, fire at the 62-story, steel-framed high-rise known as the First Interstate Bank Building, Los Angeles, California (Nelson 1989). The fire involved an afterhours accidental ignition in an office cubicle on the 12th floor that spread across the floor of origin. The fire propagated to the 13th, 14th, and 15th and a section of the 16th floors as exterior windows failed, allowing large flame plumes to entrain against the sheer sides of the building, causing windows above to fail. The fire burned for over 2 hr before being suppressed by firefighters' hose streams inside the building. One of the maintenance personnel was killed after taking an elevator to the fire floor.

The NIST analysis of the fire used the Available Safe Egress Time (ASET) program to predict the smoke layer's temperatures, level, and oxygen content after the calculated time for flashover (smoke temperature of 600°C [1150°F]). The DETACT-QS model was used to estimate the response of smoke detectors and sprinklers to the fire growth. Since there were no sprinklers on the floor of origin, these studies revealed how much damage could have been avoided.

This analysis was one of the first to evaluate burn patterns and flame spread on combustible furniture, glass breakage, flame extension from broken windows communicating the fire to upper floors, and compartment burning rates in reconstructing the fire. The study also advanced the theory of a transfire triangle, which demonstrates the important interdependent relationships among the mass burning (pyrolysis) rate, the heat release rate, and the available air (oxygen) (Nelson 1989).

HILLHAVEN NURSING HOME FIRE

A fire at 10 P.M. on October 5, 1989, at the Hillhaven Rehabilitation and Convalescent Center, Norfork, Virginia, claimed the lives of 13 persons. None of these individuals were burned but all breathed sufficient carbon monoxide (CO) to produce high carboxyhemoglobin (COHb) levels.

NIST fire researcher Harold E. "Bud" Nelson recognized the importance of evaluating the fire dynamics when performing a reconstruction of this incident. His analysis of the room of fire origin in the nursing home showed that the temperature rose to at least 1000°C (1800°F), and the CO concentration of 40,000 ppm.

Using the fire model FIRE SIMULATOR contained in the engineering package, FPEtool predicted the smoke temperatures, smoke layer, toxic gas concentrations, and velocity of the smoke front. An important aspect learned in the NIST analysis was the impact of the "what if" questions that are posed in fire reconstructions (Figure 6.4). These questions centered on the hypothetical impact of different fire activation devices on fire suppression as well as the predicted time of flashover.

PULASKI BUILDING FIRE

A fire in Suite 5127 of the Pulaski Building, 20 Massachusetts Avenue, NW, Washington, D.C. on March 23, 1990, damaged the offices of the U.S. Army Battlefield Monuments Commission (ABMC). A NIST engineering review of the fire was able to apply sound fire protection engineering principles to estimate the rate of fire growth (Nelson 1989).

At 11:24 A.M., an alert ABMC staff member noticed smoke coming from the open door of the unoccupied audio/visual conference room. The room contained an estimated 4530 kg (10,000 lb) of combustible materials consisting of publications and photos stored in corrugated board boxes and cardboard mailing tubes stacked about 1.5 m (5 ft) high. The fire was thought to have started when cartons of mailing tubes were exposed to a frayed energized electrical lamp cord that passed through a small hole from a podium into the projection room.

FPETOOL was used to analyze the fire and arrive at a scenario based upon actual witness testimony and observations. FIRE SIMULATOR estimated the environmental conditions, temperature, depth of smoke, thickness of smoke, and energy vented from the room. Also included was the impact of the failure of the suspended ceiling.

The analysis showed that the fire reached flashover approximately 268 s into the fire. A unique aspect of this fire engineering analysis was the introduction of a time

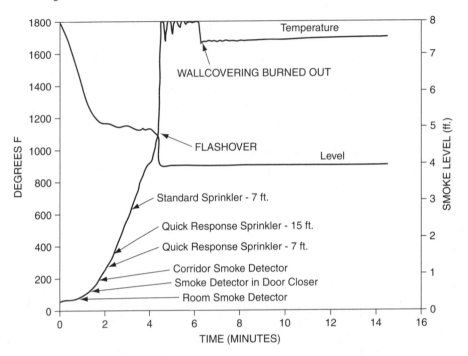

FIGURE 6.4 ◆ The impact at the Hillhaven Nursing Home of the "what if" questions that are posed in fire reconstructions, centering on the hypothetical impact of different fire activation devices on fire suppression as well as the predicted time of flashover. *Courtesy of NIST, from Nelson [1991].*

line of events (Figure 6.5) that compares the approximate time with the observed fire conditions, behavior of the occupants, impact of existing fire protection systems, and "what if" scenarios of the impact of potential fire protection systems.

HAPPYLAND SOCIAL CLUB FIRE

On the morning of March 25, 1990, an arsonist set a fire in the entryway of a neighborhood club using 2.8 liters (0.75 gal.) of gasoline. The fire killed 87 occupants of the two-story club. A NIST fire engineer used the HAZARD I fire model to examine the fire development as well as mitigation strategies that could have influenced the outcome of the fire (Bukowski 1991).

These mitigation strategies included the potential use of automatic sprinkler protection, solid wooden doors at the base of the stairway, a fire escape and enclosed exit stairwell, and a noncombustible interior finish. The analysis also weighed the economic cost estimates to implement these strategies. The study concluded that, due to the combustible wall coverings and the opening of an entry door, allowing the entry of gasoline-fed flames, the fire grew and combustion gases spread so rapidly that everyone was affected within 1–2 min, precluding their escape.

The NIST analysis also took into account occupant tenability at various locations, based upon the output of the HAZARD I fire model (which contains CFAST version 2.0). The tenability study looked at heat flux that could cause second-degree burns, temperatures, and fractional exposure doses based upon both the NIST and the purser toxicity models (Bukowski 1991).

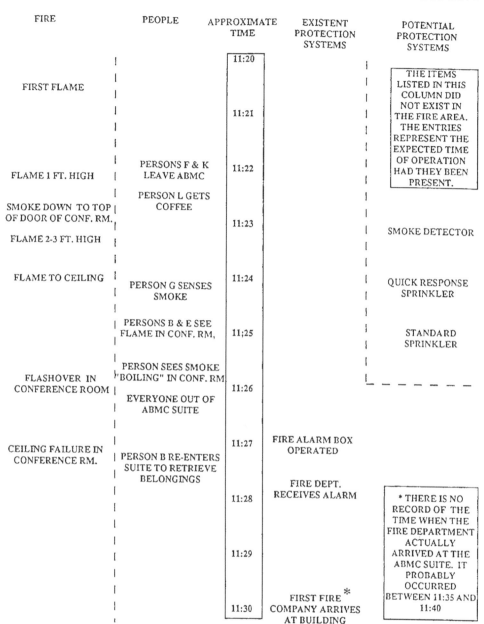

FIGURE 6.5 ◆ The Pulaski time line of events, which compares the approximate time with the observed fire conditions, behavior of the occupants, impact of existing fire protection systems, and "what if" scenarios of the impact of potential fire protection systems. *Courtesy of NIST, from Nelson [1989].*

62 WATTS STREET FIRE

A NIST fire engineer modeled what is now known as the 62 Watts Street Fire, which was reported at 7:36 P.M., March 28, 1994. The fire involved a three-story apartment building in Manhattan, New York. The responding firefighters formed 2 three-person hose teams. One team planned to enter the first-floor apartments, while the second team proceeded up a stairwell to search for fire extension.

Investigation later revealed that the occupant of the first-floor apartment left at 6:25 P.M. A plastic trash bag inadvertently left on top of the kitchen gas range was ignited by the pilot flame. Several bottles of high-alcohol content liquid were left close by, on the adjoining countertop, contributing to the initial fire. The fire burned for approximately an hour, becoming oxygen-deficient and producing large quantities of carbon monoxide, smoke, and other unburned fuels trapped in the hot smoke layer in the small, closed apartment.

As the firefighters in the stairwell forced open the door to the first-floor apartment, a backdraft occurred, where warm fire gases flowing out of the burning apartment were replaced by cool outside air. With the ensuing exchange, the combustible gas mixture ignited, creating a large flame in the stairwell for a period of approximately 6 min, killing the firefighters on the stairs.

Bukowski at NIST modeled this event using CFAST, and the analysis has become a classic study into backdrafts and firefighter safety. Recently, McGill (2003), at Seneca College, Toronto, Canada, modeled this case using the NIST Fire Dynamics Simulator. A view of McGill's Smokeview visualization is shown in Figure 6.6. It took approximately 20 hr to define and construct the model.

CHERRY ROAD FIRE

NIST engineers were called upon to demonstrate the results of calculations using the NIST FDS to evaluate the thermal and tenability conditions of a fire that may have led to the deaths of two firefighters and burned four others (Madrzykowski and Vettori 2000).

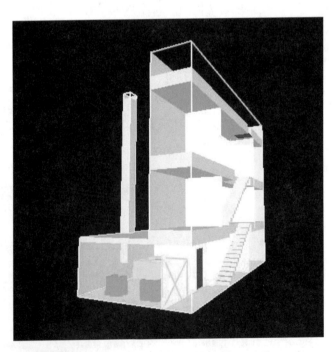

FIGURE 6.6 ◆ Model of the 62 Watts Street fire using the NIST Fire Dynamics Simulator. Room of origin at lower left. The fire plume extending from the door (center) trapped and killed three firefighters on the stairs. *Courtesy of D. McGill [2003], by permission.*

The fire started in a ceiling electrical fixture on the lower level of a townhouse. Responding firefighters entered the first and basement floors of the structure. Both on-scene observations and subsequent modeling showed that the opening of the lower-level sliding glass doors increased the heat release rate of the fire, so that 820°C (1500°F) gases moved up the stairwell. The fire may have smoldered for several hours before ignition, producing fuel-rich gases that ignited while firefighters were inside the building searching for the fire and any victims in the heavy smoke.

The Cherry Road fire FDS model may become one of the best examples of application of the NIST FDS. As the technology expands, so does the ability to address harder questions, concerning transient heat and ventilation, rapid fire growth, and stratified or localized conditions. Furthermore, the fire pattern damage observed correlated with the model's predictions.

◆ 6.7 CASE STUDY 1—WESTCHASE HILTON HOTEL FIRE

A March 6, 1982, fire at the Westchase Hilton Hotel, Houston, Texas, resulted in the deaths of 12 people and serious injuries to 3 others. Hot gases and smoke progressed from a fire in a guest room on the fourth floor down the corridor and, to some degree, throughout the building as illustrated in Figure 6.7. Records indicate that 137 guests were registered at the hotel the evening of the fire.

The fire started in a fourth-floor hotel room (Figure 6.8), with the first odor of smoke being reported at 2:00 A.M. on the tenth floor, followed by light smoke reported at 2:10 A.M. on the eighth floor; the fire was actually discovered at 2:20 A.M. by an occupant returning to his room. The guest attempted to put the fire out himself. Both occupants of his suite survived.

At 2:31 A.M., the district fire chief arrived and saw a fire plume projecting out the exterior window of the room of fire origin (Figures 6.9 and 6.10). Extinguishment

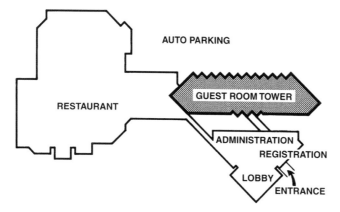

FIGURE 6.7 ◆ The March 6, 1982, fire at the Westchase Hilton Hotel resulted in the deaths of 12 people and serious injuries to 3 others. *Reprinted with permission from NFPA Westchase Hilton Fire Investigation. Copyright © 1982 and 1983, National Fire Protection Association, Quincy, MA 02269.*

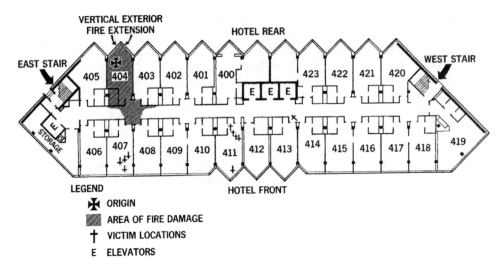

FIGURE 6.8 ◆ The Westchase Hilton Hotel fire started on the fourth floor, Room 404. *Reprinted with permission from NFPA Westchase Hilton Fire Investigation. Copyright © 1982 and 1983, National Fire Protection Association, Quincy, MA 02269.*

of the property started at 2:38 A.M. and ended at 2:41 A.M. Fire investigating officials concluded from these times that flashover in the room occurred between 2:20 and 2:31 A.M.

An investigation revealed that a cigarette smoldered in an upholstered chair in the room of origin. The U.S. Consumer Products Safety Commission (CPSC) later had fire tests performed on the furniture. Heat release rate data from these tests along with assumptions about how wide the door was ajar formed the basis for the input to the model (Table 6.3). Data from previous NIST tests of smoldering ignition of furniture

TABLE 6.3 ◆ Numerical Constants in the Fire Investigation and Reconstruction Model

Feature	
Room floor area	24.51 m² (263.8 ft²)
Room ceiling height	2.44 m (8.00 ft)
Door (vent) soffit	1.99 m (6.56 ft)
Vent width (varied)	0.152–0.457 m (6–18 in.)
Fire height (estimated)	0.914 m (3.0 ft)
Heat loss fraction	0.66
Flashover 0.152-m (0.50-ft) opening[a] 0.076-m (0.25-ft) opening[a]	 199 s 299 s
Untenability, 1.2–1.5 m (4–5 ft)	183°C (361°F)

Source: Derived from Janssens (2000).
[a]Predicted value.

FIGURE 6.9 ◆ At 2:31 A.M., the district fire chief arrived and saw the south view of the Westchase Hilton high-rise tower, with a fire plume projecting out the exterior window of the room of fire origin. Note the areas of external, horizontal, and vertical fire extension. *Reprinted with permission from NFPA Westchase Hilton Fire Investigation. Copyright © 1982 and 1983, National Fire Protection Association, Quincy, MA 02269.*

were also used in evaluating the time factors between detection of smoke odors and observation of a flaming fire.

A zone model called FIRM (Fire Investigation and Reconstruction Model) was later used to evaluate the single compartment of fire origin using data from the CPSC tests (Janssens 2000). The input screen using these data is shown in Figure 6.11. A sensitivity analysis using FIRM was conducted to examine the impact of the position of a door ajar on the room of fire origin.

The model also was used to predict the fire development within the room of origin, tenability conditions, and an evaluation of the time to flashover. Table 6.3 lists the

FIGURE 6.10 ◆ Exterior view, from the hallway, of Room 404, Westchase Hilton Hotel, whose combustible contents were almost totally consumed. *Reprinted with permission from NFPA Westchase Hilton Fire Investigation. Copyright © 1982 and 1983, National Fire Protection Association, Quincy, MA 02269.*

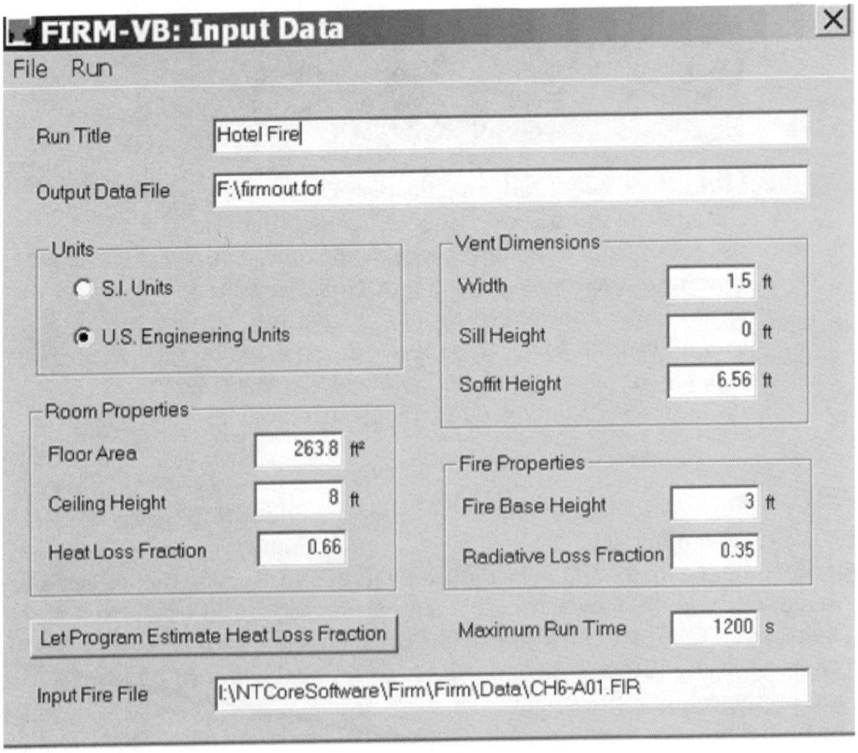

FIGURE 6.11 ◆ Input screen for the Fire Investigation and Reconstruction Model (FIRM).

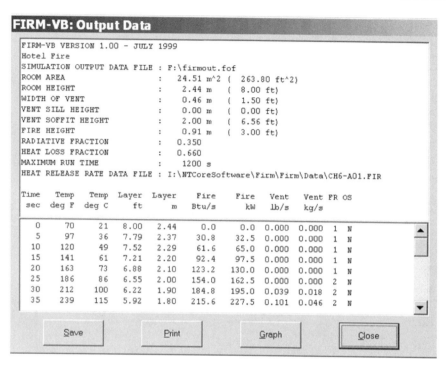

FIGURE 6.12 ◆ Output screen for the Fire Investigation and Reconstruction Model (FIRM).

observed and predicted data for the room of fire origin, and the output file from the FIRM program is shown in Figure 6.12.

Later studies on the behavior of persons in this fire used a data collection interview form reproduced in Figure 6.13 (Bryan 1983). Approximately a month after the fire this form was mailed to 130 guests registered at the time of the fire by the National Fire Protection Association (NFPA) in coordination with the Houston Fire Department. A total of 55 persons responded, which was 27 percent of the registered guests, representing 42 males and 13 females.

The study showed that some of the guests were aware of the fire for as long as almost 2 hr before the fire department was notified. The majority of the notification was by way of people yelling. The most frequent first actions after notification of the fire were dressing, calling the front desk, and attempting to exit. The study failed to reveal any nonproductive behavior by those persons escaping the fire, and some spent time knocking on doors to alert other occupants. Persons escaping the fire did not reenter the building.

Movement through the smoke is always a question in research studies. Over 50 percent of the guests passed through the smoke for distances of from 15 to 600 ft (4.57 to 182 m).

Lessons learned from the analysis of this fire by several sets of experts include the need for better enforcement of the life safety code, education of the public, training of hotel staff, and emergency preparedness. Other lessons showed the utility of scientific review of fire scenes, development of a list of behaviorally related questions and answers, and application of a fire model.

FIRE EXPERIENCE SURVEY
Westchase Hilton Hotel, Houston, Texas
March 6, 1982

We are grateful for your willingness to share your fire experience with us. Your information will help us find ways to avoid recurrences of such tragedies.

Occupation _____ Sex ____ Age ____
Room No. ____

1) How did you first become aware that there was something unusual occurring in the Hotel? ____

2) What time was it? _____ How did you determine the time?

3) When did you realize that what was occurring was a fire? What time was it? _____ How did you determine the time? _____

4) How serious did you believe the fire to be at first?
   ( ) Not at all serious, ( ) Only slightly serious, ( ) Moderately serious, ( ) Extremely serious

5) Were you alone when you became aware of the fire?
   ( ) No ( ) Yes

6) How many persons were with you? ____ They were
   ( ) relatives ( ) others.

7) Were you injured? ( ) No ( ) Yes:
   If Yes, What was the nature and cause of the injury?
   _____

8) What did you do when you realized there was a fire? (state exact sequence of actions)
   First _____
   Second _____
   Third _____
   Fourth _____
   Fifth _____

9) Once aware of the fire, did you:
   (a) Call or attempt to call the hotel operator (switchboard) ( ) No ( ) Yes; If Yes, At what time? _____
   (b) Call or attempt to call the Fire Department directly ( ) No ( ) Yes;
       If Yes, at what time? _____
   (c) Operate a manual fire alarm pull station?
       ( ) No
       ( ) Yes; If Yes, At what time? _____

10) Did you voluntarily leave (i.e., without being requested by hotel or Fire Department personnel)?
    ( ) No ( ) Yes
    If No, Why not? _____
    If Yes, At what time? _____
    How? ( ) Stairway to ground ( ) Stairway to roof
         ( ) Elevator ( ) Window ( ) Exterior Door
         ( ) Other (Specify) _____

11) I left the building: ( ) Unassisted; Assisted by:
    ( ) Hotel staff ( ) Guest ( ) Fire Department
    ( ) Other (Specify) _____

12) After you realized there was a fire, how long did you wait before leaving the building? ____ minutes.
    What were you doing while waiting? _____

13) In your leaving, was there any visible smoke?
    ( ) No ( ) Yes
    Any smoke odor? ( ) No ( ) Yes
    Any flames? ( ) No ( ) Yes
    Did you try to move through smoke? ( ) No
    ( ) Yes: If Yes, How far did you move? ____ feet

(approx.). How far could you see at the time? ____ feet.
Did you turn back? ( ) No ( ) Yes. If Yes, Why did you turn back? _____

14) Did you notice any lighted exit signs? ( ) No
    ( ) Yes

15) During your escape, was there any difficulty in following the direction marked by exit signs?
    ( ) No ( ) Yes:
    If Yes, What was difficulty? _____

16) Any obstructions to escape? _____

17) What aids helped you to escape? _____

18) Did smoke enter your room? ( ) No ( ) Yes:
    If Yes, How? _____
    ( ) Heating/cooling unit ( ) Bathroom vent
    ( ) Around door ( ) Window
    ( ) Don't know ( ) Other (Specify) _____

19) Beginning with your first action, number the sequence of events you took while in your room.
    ( ) Put materials over the heating/cooling unit vent
    ( ) Around door ( ) Around window
    ( ) Turned on TV ( ) Turned off radio
    ( ) Took no action ( ) Other _____

20) Did the smoke detector in your room sound an alarm? ( ) No ( ) Yes

21) Did you hear the building fire alarm?
    ( ) No ( ) Yes: If Yes, At what time? ____
    How long did it operate? _____

22) Did you receive any instructions from hotel staff during the fire emergency? ( ) No ( ) Yes: If Yes, What were they? _____
    Did you receive fire safety instructions from hotel employees prior to the fire? ( ) No ( ) Yes:
    If Yes, What were they? _____
    Did you observe fire safety information within your room? ( ) No ( ) Yes. If Yes, What was the information? _____

23) Did you have previous training on actions to take in a fire?
    ( ) No ( ) Yes: If Yes, Number of times ____
    Type _____
    Given by _____ Last time: Date ____
    Did the training help in this fire?
    ( ) No ( ) Yes

24) Did you receive previous fire safety information from: ( ) Radio ( ) TV ( ) Publication.
    What was the message? _____
    Did it help in this fire? ( ) No ( ) Yes

25) Were you ever involved in a fire before?
    ( ) No ( ) Yes Last time: Date ____

26) Please report any additional comments that you think might help others in a similar fire situation.
    _____

27) Please mark your escape route on the diagram provided. Please identify room number and floor. Thank you.

FIGURE 6.13 ◆ Later studies on the behavior of persons in this fire used this data collection interview form, mailed to 130 guests. A total of 55 persons responded, which was 27 percent of the registered guests, 42 males and 13 females. *Reprinted with permission from Westchase Hilton Fire, National Fire Protection Association, Quincy, MA 02269. This reprinted material is not the complete and official position of the National Fire Protection Association on the referenced subject, which is represented only by the standard in its entirety.*

This chapter has introduced the concept and application of fire modeling to forensic fire investigation. Although models have existed since the 1960s, fire modeling has centered on explaining the physical phenomena of fires, particularly when applied to verifying existing experimental data.

It was the effort of a few fire scientists and engineers working at NIST that pushed the acceptability and application of fire modeling out of the laboratory and into the world of forensic fire scene reconstruction. As indicated, the purpose of this chapter is not to make the reader an expert in fire modeling but to allow him or her to gain a better appreciation of its value in an investigation.

ASTM Subcommittee E05.33 presently oversees four guidelines that standardize the evaluation and use of computer fire models. This committee continues to work on improving the standard. The following is a synopsis of the current standard guides (Janssens 2002).

- ASTM E1355 evaluates the predictive capacity of fire models by defining scenarios, validating assumptions, verifying the mathematical underpinnings of the model, and evaluating its accuracy.
- ASTM E1472 oversees how a fire model is to be documented, including a user's manual, a programmer's guide, mathematical routines, and installation and operation of the software.
- ASTM E1591 covers and documents open literature data that are beneficial to modelers.
- ASTM E1895 examines the use and limitations of fire models and addresses how to choose the most appropriate model for the occasion.

A task group from the Society of Fire Protection Engineers (SFPE) has used ASTM E1355 since 1995 to evaluate several models. The work of this group is ongoing and continues to develop accepted engineering practice guides for each of the models.

The next chapter explores the issues of tenability as it applies to the modeling of toxicity of fires.

■ ■

Problems

6.1. Select one of the cases mentioned in this chapter and model it using the selected program. Obtain a similar fire modeling program and compare and contrast the results derived using it for this analysis.

6.2. Many computer fire models contain a database of materials and their burning properties. Examine the database of two fire models. Compare and contrast the results obtained using each model.

■ ■

Suggested Reading

Hume, B. T. 1993. *Fire models: A guide for fire prevention officers.* Report No. 6/93. London: Home Office Fire Research and Development Group.

Janssens, M. 2000. *Introduction to mathematical fire modeling*, 2nd ed. Lancaster, PA: Technomic.

Madrzykowski, D. 2002. Fire research: Providing new tools for fire investigation. *Fire and Arson Investigator* 52(4):43–46.

Urbas, J. 1997. Use of modern test methods in fire engineering. *Fire and Arson Investigator* 47(December):12–15.

Fire Death and Injuries

7 **CHAPTER**

There's a scarlet thread of murder running through the colorless skein of life, and our duty is to unravel it, and isolate it, and expose every inch of it.

Sir Arthur Conan Doyle,
"A Study in Scarlet"

In every country, particularly in highly industrialized ones, fire kills a significant number of people. In the United States, it is one of the five leading causes of accidental death. The involvement of the investigator or forensic specialist in fatal fires can come in any form, from any sector, and challenge one's talents and knowledge to come to just and accurate conclusions. These cases require the highest degree of cooperation among the investigators, who all have contributions to make toward a successful investigation. When deaths occur in a fire, the event becomes the focus of the press and the public as well as police, fire, insurance, and forensic professionals. When problems occur, they can have far-reaching consequences.

◆ 7.1 PROBLEMS AND PITFALLS

There are several problem areas that can complicate fatal fire investigations and compromise the accuracy and reliability of the conclusions reached.

- **Linkage between the fire and death investigations.** Prejudging the fire and its attendant death as an accident and automatically treating the scene investigation accordingly is a major problem. Fires can be intentional, natural, or accidental in their cause, and deaths can be accidental, homicidal, suicidal, or natural. The linkage between the two events can be direct, indirect, or simple coincidence. The responsibility of the investigation team in fatal fire cases is to establish the cause of both the fire and the death and to determine the connection (if any) between the two.
- **Time interval to death.** Sudden violent deaths are assumed to be instantaneous exposure to insult followed by immediate collapse and death (a shot is fired and the victim collapses to die shortly afterward), and many forensic investigations are considered (and successfully concluded) in this light. Fires, however, occur over a period of time,

creating dangerous environments that vary greatly with time and can kill by a variety of mechanisms. A person may be killed nearly instantaneously by exposure to a flash fire or only after hours of exposure to toxic gases. Investigators must have an appreciation for the nature of fire and its lethal products and not treat the event as a single exposure to a single set of conditions at a precise moment in time that results in instant collapse.

◆ **Understanding heat intensity and duration.** There is little accurate information available to detectives and pathologists about the temperatures and intensities of heat exposure that occur in a fire as it develops. Misunderstandings and misapprehensions can lead investigators seriously astray when they try to assess injuries or postmortem damage.

◆ **Fire-related human behavior.** In most violent deaths, the victim offers a "fight or flight" response to the threat, suffers an injury, collapses, and dies. In fires, the potential responses include going to investigate, simply observing, failing to notice or appreciate the danger, failing to respond due to infirmity or incapacitation from drugs or alcohol, returning to the fire or delaying escape to rescue pets, family, or purses, and fighting or attempting to fight the fire. This variability of responses can vastly complicate the process of solving the critical problem of why the victim failed to escape the fire (and perhaps other people escaped).

◆ **Time interval between fire and death.** Fires can kill in seconds or the death can occur minutes, hours, days, or even months after the victim is removed from the scene. The longer the time interval between the fire and the death, the harder it is to keep track of the actual cause (the fire) and the result (the death). Evidence is lost when a living victim is removed from a scene, and when the victim dies later, away from the scene, it may be too late to recover or document that evidence.

◆ **Conflicts among investigating agencies.** There can be conflicts regarding the perceived or mandated responsibilities of police, fire, medicolegal, and forensic personnel that are often involved in fire death scenes.

◆ **Understanding postmortem effects.** After death, there can be severe postmortem effects on the body that can vastly complicate the investigation by obliteration of evidence. The body can bear fire patterns of heat effects and smoke deposits that can be masked by exposure to fire after death. The body can be incinerated by exposure to flames such that evidence of prefire wounds or even clinical evidence such as blood samples is destroyed. There can be structural collapse and effects of firefighting hose streams and overhaul that induce additional damage.

◆ **Premature removal of the body.** A major problem is the premature removal of a deceased victim from the fire scene. The compulsion to rescue and remove every fire victim is a very strong one, particularly among dedicated firefighters. However, once the fire is under control and unable to inflict further damage to the body of a confirmed deceased, there is nothing to be gained and much to be lost in the way of burn pattern analysis, body fragments (especially dental evidence), projectiles, clothing and associated artifacts (keys, flashlight, dog leash, etc.), and even trace evidence, by the undocumented and hurried removal of the remains.

◆ 7.2 TENABILITY—WHAT KILLS PEOPLE IN FIRES?

Fire investigators should evaluate the common problems and pitfalls when conducting forensic reconstructions, particularly when death occurs. The purpose of this chapter is to review these human tenability factors, present analytical approaches, and provide illustrative case histories.

Structural fires can achieve their deadly result in a number of ways—heat, smoke, flames, soot, and others—but fire conditions change continually as a fire grows and evolves and the conditions of exposure of a would-be victim can vary from "no threat or injury" to almost instantly lethal.

Although they can and usually do act in combination, the major lethal agents of fires are as follows: heat, smoke, inhalation of smoke or toxic gases, anoxia, flames, and blunt trauma. These are discussed in detail later in this chapter.

The ability of humans to escape a fire is measured by the time frame for which their environment remains survivable (tenability). Fires can produce incapacitating effects on humans when they are exposed to heat and smoke. These physiological effects are generally categorized into the following areas (Purser, 2002).

- **Visibility.** Optical opacity of the smoke and irritants produces impaired vision as the distribution of thick smoke descends toward the floor through rooms, stairwells, and hallways.
- **Toxic gases.** Toxic gas inhalation causes confusion, respiratory tract injuries, loss of consciousness, or asphyxiation.
- **Heat transfer.** Heat irritates exposed skin and respiratory tracts, causing pain and varying degrees of burn injuries or hyperthermia.

Of primary concern is the point at which exposure to one or more of the above variables would cause injury or block the individual from successfully escaping the fire, resulting in death. The psychological behavior of people in fires when exposed to these variables affects their decisions and the time required to travel via safe escape routes.

Critical limits to human tenability include a limit of visibility to 5 m, an accumulated dose of carbon monoxide of 30,000 parts per million minute (ppm-min), and a critical temperature of 150°C. The synergistic effects of one or more of these factors may override these individual limits (Jensen 1998).

The major goal of the investigator when conducting a ***tenability analysis*** is to determine how the individual who is escaping a burning structure becomes impaired and how the fire changes his/her environment and perceptions. These techniques are grounded upon both experimental and forensic data, giving a balanced and practical approach.

The physiological and toxicological correlations of how heat transfer and toxic smoke impact animals and humans are based upon experiments. For example, studies correlating ***carbon monoxide*** exposure to ***carboxyhemoglobin*** levels used subjects ranging from laboratory rats to volunteer medical students (Nelson 1998). Some of these data are extrapolated to model the results at higher levels of exposure. These variables must take into account variations in age, health, and stature of the individual. Note that an individual's height affects his/her exposure to the stratified upper smoke layer.

Forensic evaluations of incapacitation also come from forensic data derived from actual case histories and investigations. The major work on the behavior of people in fires came from subject interviews of persons surviving large fires and explosions (Bryan and Icove 1977).

◆ **7.3 VISIBILITY**

The optical opacity of dense smoke and its irritants impairs vision and respiration. This applies to normally sighted people, who find their ability to travel impaired by the smoke. This optically dense smoke affects

+ exit choice, escape decisions, and wayfinding ability;
+ speed of movement; and
+ the eyes and respiratory system.

During structure fires, the occupants often depend on their ability to seek out exit signs, doors, and windows (Jin 1975). *Visibility* of an object depends on several factors such as the smoke's ability to scatter or absorb the ambient light, the wavelength of the light, whether any signs are light emitting or light reflecting, and the individual's visual acuity (Mulholland, in SFPE 2002a, 2-265).

OPTICAL DENSITY

Calculating an estimate for visibility is based upon the term D, the *optical density* per meter (OD/m). Optical density uses a collateral term known as the extinction coefficient, K, which is the product of an *extinction coefficient* per unit mass, K_m, and the mass concentration of the smoke aerosol, m, where

$$K = K_m m, \tag{7.1}$$

$$D = \frac{K}{2.3}, \tag{7.2}$$

with

K = extinction coefficient (m^{-1}),
K_m = specific extinction coefficient (m^2/g),
m = mass concentration of smoke (g/m^3), and
D = optical density per meter (m^{-1}).

The values for K_m are typically 7.6 m^2/g for smoke produced during flaming combustion of wood or plastics and 4.4 m^2/g for smoke produced during pyrolysis (SFPE 2002a).

In terms of the extinction coefficient, K, one problem at a fire is to determine the visibility, S, of light-emitting and light-reflecting exit signs to occupants. S is a measure of how well an individual can see through the smoke. Light-emitting signs are two to four times better than light-reflecting signs (Mulholland, in SFPE 2002a, 2-265), as reflected in equations (7.3) and (7.4),

$$KS = 8 \quad \text{for light-emitting signs,} \tag{7.3}$$

$$KS = 3 \quad \text{for light-reflecting signs,} \tag{7.4}$$

where

K = extinction coefficient (m^{-1}) and
S = visibility (m).

Methods for estimating visibility based upon the mass optical density are considered realistic. In studies, the optical density at which people turned back from a smoke-filled area was a visibility distance of 3 m (Bryan 1977). The study also showed a tendency for women to be more likely to turn back than men. Other factors include the ability of the persons to see exit signs directing them to safe egress from the building. Height of exit signs may be critical to their visibility.

Estimates of visibility are based upon mass optical densities, D_m, derived from test data (Babrauskas 1981). Typical values of mass optical density produced by mattresses in flaming combustion are listed in Table 7.1.

TABLE 7.1 ◆ Mass Optical Density (D_m) for Flaming Mattresses

Type of Material	Mass Optical Density (m^2/g)
Polyurethane	0.22
Cotton	0.12
Latex	0.44
Neoprene	0.20

Source: Derived from Babrauskas (1981).

When estimating the density of the visible smoke, D, use the following equation:

$$D = \frac{D_m \Delta M}{V_c}, \tag{7.5}$$

where

D = optical density per meter (m^{-1}),

D_m = mass optical density (m^2/g),

ΔM = total mass loss of sample (g), and

V_c = total volume of compartment or chamber (m^3).

EXAMPLE 7.1   Visibility

A small, 300-g (0.66-lb), polyurethane mattress cushion on a waiting bench is set afire by a juvenile arsonist and is undergoing flaming combustion. The bench is located in a 6-m-square waiting room with a ceiling height of 2.5 m. Determine the closest visibility of both light-emitting and light-reflecting signs leading to the exit door. Assume that the smoke filling the waiting room is homogeneous.

Total mass loss of mattress	$\Delta M = 300\,g$
Mass optical density	$D_m = 0.22\ m^2/g$ (from Table 7.1)
Volume of compartment	$V_c = (6\,m)(6\,m)(2.5\,m) = 90.0\,m^3$
Optical density	$D = (0.22\ m^2/g)(300\,g)/(90.0\,m^3) = 0.733\ m^{-1}$
Extinction coefficient	$K = 2.3D = (2.3)(0.733\ m^{-1}) = 1.687\ m^{-1}$
Visibility (light emitting)	$S = 8/K = (8)/(1.687\ m^{-1}) = 4.74\,m$
Visibility (light reflecting)	$S = 3/K = (3)/(1.687\ m^{-1}) = 1.77\,m$

The calculations indicated that a light-emitting sign can be seen in this fire at a distance of up to 4.74 m (15.5 ft), compared to 1.7 m (5.58 ft) for an unlit sign.

WALKING SPEED

As previously summarized, optical density impacts a person's decision to choose the closest exit, ability to make proper escape decisions, wayfinding ability, and speed of movement. Experiments with human subjects navigating through nonirritant smoke-filled corridors indicated that speed of movement decreases with increased smoke density (Jin 1975).

Based upon Jin's research, equation (7.6) provides an expression of this relationship.

$$FWS = (-1.738)(D) + 1.236 \qquad (7.6)$$

$$\text{for the range} \qquad 0.13\,\text{m}^{-1} \le D \le 0.30\,\text{m}^{-1} \qquad (7.7)$$

where

FWS = fractional walking speed (m/s) and

D = optical density per meter (m^{-1}).

For this equation, the smoke optical density ranges between 0.13/m (below normal walking speed) and 0.56/m (above walking speed in darkness at 0.3 m/s). The limits on the equation do not allow for delays such as erratic walking and sensory irritation. Jin used wood smoke in his experiments, therefore, his equation should be considered valid only for nonirritant smoke.

WAYFINDING

Jin's expression does not correlate midcourse corrections in **wayfinding**, reduced visibility, and irritability of the smoke. Studies have shown that the average density at which persons turned back was a visibility distance of 3 m ($D = 0.33\,\text{m}^{-1}$ and $K = 0.76$). Poor visibility and irritation of the eyes are the leading factors in reduced wayfinding, followed by irritation of the respiratory system (Jensen 1998).

The term **fractional effective concentration** (FEC) was developed to assess the impact of smoke on visual obscuration on a subject (Purser 2001). The FEC is expressed in general terms as

$$FEC = \frac{\text{dose received at time } t(C_t)}{\text{effective } C_t \text{ dose to cause incapacitation or death}} \qquad (7.8)$$

The FEC, in special situations, is also referred to as the **fractional incapacitation dose** (FID) or the **fractional lethal dose** (FLD).

The following formulas relate to enclosed spaces. As the value of FEC_{smoke} becomes closer to 1, the level of visual obscuration increases and the chance of escape decreases significantly.

$$FEC_{smoke} = \frac{D}{0.2} \qquad \text{for small enclosures.} \qquad (7.9)$$

$$FEC_{smoke} = \frac{D}{0.1} \qquad \text{for large enclosures.} \qquad (7.10)$$

SMOKE

Smoke contains water vapor, CO, CO_2, inorganic ash, toxic gases, and chemicals in aerosol form as well as soot. Soot is agglomerations of carbon from incomplete combustion large enough to produce visible particles. These particles may be very hot and are not cooled readily as they are inhaled so they may induce edema and burns where they lodge in the mucosal tissue of the respiratory system. Soot particles are active adsorbents so they may carry toxic chemicals and permit their ingestion or inhalation (with direct absorption by the mucosal tissues). Soot can be inhaled in quantities sufficient to block airways physically and cause mechanical asphyxiation. Soot in smoke can also obscure the vision of victims and prevent their escape.

Toxic products of combustion can include a wide variety of chemicals depending on what is burning and how efficiently it is burning (temperature, mixing, and oxygen concentration are all important variables in determining what species are created). Material can be classified generally into three basic categories.

- **Toxic:** CO, HCN (hydrogen cyanide), H_2S (hydrogen sulfide), and phosgene.
- **Acidic:** HCl (hydrogen chloride)—produced during the combustion of polyvinyl chloride plastics (PVC); sulfur oxides (SO_x), which form H_2SO_3 (sulfurous acid) and H_2SO_4 (sulfuric acid)—produced by oxidation of sulfur-containing fuels; and nitrogen oxides (NO_x), which form HNO_2 (nitrous acid) and HNO_3 (nitric acid)—from nitrogen-containing fuels.
- **Organic Irritants:** Formaldehyde and acrolein (2-propenal, C_3H_4O)—produced by the combustion of cellulosic fuels; and isocyanates—produced by combustion of polyurethanes.

Acidic gases dissolve in the water of the mucous membranes and generate the corrosive acids described above. These acids cause the epithelial cell membranes to dissolve and release the fluids as edema. Hydrogen sulfide combines with water to form sodium sulfide, which destroys epithelial cell membranes, but also inhibits cytochrome a_3 (Cya_3) oxidase, which is necessary for cellular function (Feld 2002).

For exposure to hydrogen chloride (HCl), walking movement is usually affected starting at concentrations of 50 ppm, with total cessation of movement when approaching a concentration of 300 ppm. The effects of concentrations of HCl above the 1000 ppm level are most likely severe enough to prevent escape (Purser 2001).

Table 7.2 shows the effects of various fire gases on both the escape and the incapacitation of humans.

HCl is a major combustion/decomposition product of vinyl plastics, in both the flaming and the smoldering modes. Hydrogen bromide (HBr) or hydrogen fluoride (HF) occurs when some synthetic rubbers are burned. Acrolein is created when wood or cardboard is burned.

TABLE 7.2 ◆ Irritant Concentrations of Fire Gases Predicted to Cause 50% Impaired Escape or Incapacitation in the Human Population

Common Fire Gases	*Impaired Escape (ppm)*	*Incapacitation (ppm)*
Hydrogen chloride (HCl)	200	900
Hydrogen bromine (HBr)	200	900
Hydrogen fluoride (HF)	200	900
Sulfur dioxide (SO_2)	24	120
Nitrogen dioxide (NO_2)	70	350
Formaldehyde (CH_2O)	6	30
Acrolein (C_3H_4O)	4	20

Source: Derived from Purser (2001).

The previous section concerned a majority of visibility and irritants effects on humans as they try to navigate through fires. ***Toxic gases*** contained within smoke can also have a narcotic effect that asphyxiates victims. The dominant narcotic gases in smoke that impact the nervous and cardiovascular systems are carbon monoxide (CO) and hydrogen cyanide (HCN). Carbon dioxide and reduced oxygen levels, while not toxic themselves, may have severe effects on tenability.

Increased exposure to toxic gases may cause confusion, loss of consciousness, and eventually asphyxial death. The prediction of asphyxiation to incapacitation and death can be modeled (Purser, in SFPE 2002a).

CARBON MONOXIDE

Carbon monoxide is produced in fires by the incomplete combustion of any carbon-containing fuel. It is not produced at the same rate in all fires. In free-burning (well-ventilated) fires, it can be as little as 0.02 percent (200 ppm) of the total gaseous product. The CO concentrations in smoldering, postflashover, or underventilated fires range from 1 to 10 percent in the smoke stream.

When inhaled and absorbed into the bloodstream, it forms a complex with the heme portion of the hemoglobin molecule, carboxyhemoglobin (COHb). CO has an affinity for hemoglobin that is 200–300 times stronger than that of O_2. It also binds with the heme group in myoglobin (which is the "red" in red muscle). Its affinity for myoglobin is about 60 times that of O_2. Myoglobin stores and transports O_2 in muscle tissue, particularly in cardiac muscle. There is no measurable diffusion from an external atmosphere rich in CO into the blood or tissues of a dead body.

The stability of the COHb complex reduces the O_2-carrying capacity of the blood. Without O_2 and water, ATP (adenosine triphosphate) cannot be produced in the mitochondria of the cell and the cell dies (Feld 2002). Carbon monoxide also attaches to the heme group in Cya_3 oxidase, an enzyme that catalyzes production of ATP in the cell. CO also inhibits the transport of O_2 by myoglobin (particularly in cardiac muscle). Under hypoxic conditions, CO shifts from the blood into the muscle with a higher affinity for cardiac than for striated muscle (Myers and Cowley 1979). This may explain why low concentrations of COHb are sometimes found in fatalities.

Goldbaum, Orellano, and Dergari (1976) reported that merely reducing the hematocrit (the blood-carrying capacity) of dogs by as much as 75 percent did not result in death. Even replacing blood with blood containing 60 percent COHb by transfusion or infusion of CO through the peritoneal cavity did not result in death. Only when the CO was inhaled did deaths occur. This suggests that respiration of CO plays a critical role in causing death (not just its presence) (Goldbaum, Orellano, and Dergari 1976).

The mere presence of CO in the blood is not a sign of breathing fire gases. The normal body has COHb saturations of 0.5–1 percent as a result of degradation of heme in the blood. Higher concentrations (up to 3 percent) may be found in nonfire victims with anemia or other blood disorders (Penney 2000). Smokers can have levels of 4–10 percent since tobacco smoke contains a high concentration of CO. People in confined spaces with emergency generators, pumps, and compressors can have elevated, sometimes dangerous, COHb concentrations.

When a victim is removed from a CO-rich environment to fresh air, the CO is gradually eliminated. The higher the partial pressure of O_2, the faster the elimination.

In fresh air the initial concentration will be reduced by 50 percent in 250–320 min (approx. 4–5 hr). In 100 percent O_2 via mask, a 50 percent reduction can be achieved in 65–85 min (approx. 1 to 1.5 hr). In O_2 at hyperbaric pressures (3–4 atm), there is a 50 percent reduction in 20 min. (Penney 2000)

The time at which a blood sample is drawn from a subject must be noted, as well as the nature of any medical treatment (such as administration of O_2). The COHb saturation of blood in a dead body is very stable, even after decomposition has begun. CO poisoning kills many fire victims before they are ever exposed to fire. It can kill victims even some distance from a fire when they are not exposed to any heat or flames, but it is not the only factor in many fire deaths.

Levels of 4–5 percent CO_2 in air cause the adult respiratory rate to double. Levels of 10 percent cause it to quadruple. This increases the rate at which CO and other toxic gases are inhaled.

HYDROGEN CYANIDE

Hydrogen cyanide (HCN) is readily soluble in water (of blood plasma, cells, and organs, forming the CN radical. The CN radical combines with Cya_3 oxidase, inhibiting its action in the cells. The inhibition of Cya_3 oxidase prevents the formation of water and ATP, the basic route of respiration in the cell (Feld 2002).

Table 7.3 lists the tenability limits for incapacitation or death from exposure to CO, HCN, low O_2, and CO_2. The periods reflect 5 and 30 min, which are common benchmarks for narcotic products of combustion.

PREDICTING TIME TO INCAPACITATION BY CARBON MONOXIDE

The estimation of dosage levels for predicting times to incapacitation is an important concept. Under **Haber's rule** the dosage of toxic gases assimilated by an individual is assumed to be equivalent to the concentration. For example, a 1-hr exposure to a toxic gas at one concentration would be equivalent to a 2-hr exposure to one-half that concentration.

The Coburn–Forster–Kane (CFK) Equation. In certain cases, Haber's rule does not hold exactly true for exposure to CO. The relationship between concentration and uptake is linear only for high CO concentrations and not valid at extremely high concentrations. For lower concentrations, the time to incapacitation is an exponential

TABLE 7.3 ◆ Tenability Limits for Incapacitation or Death from Exposures to Common Toxic Products of Combustion

	5 min		30 min	
	Incapacitation	*Death*	*Incapacitation*	*Death*
CO (ppm)	6000–8000	12,000–16,000	1400–1700	2500–4000
HCN (ppm)	150–200	250–400	90–120	170–230
Low O_2 (%)	10–13	<5	<12	6–7
CO_2 (%)	7–8	>10	6–7	>9

Source: SFPE (1995, Table 2-8B[a]).

relationship and is described by the CFK equation. The CFK equation predicts that the half-time elimination of CO is a hyperbolic function of the ventilation rate (Peterson and Stewart 1975).

$$\frac{A[\text{HbCO}]_t - BV_{\text{CO}} - PI_{\text{CO}}}{A[\text{HbCO}]_0 - BV_{\text{CO}} - PI_{\text{CO}}} = e^{-tAV_bB},\qquad(7.11)$$

where

$[\text{HbCO}]_t$ = concentration of CO per blood at time t (ml/ml),
$[\text{HbCO}]_0$ = concentration of CO per blood at beginning of exposure (ml/ml),
PI_{CO} = partial pressure of CO in inhaled air (mm Hg),
V_{CO} = rate of CO production (ml/min),
A = derived constant,
B = derived constant, and
V_b = derived constant.

An obvious disadvantage of using the CFK equation is the number of variables needed. The CFK equation is appropriately used for exposure to CO concentrations less than 2000 ppm (0.2 percent), exposure durations greater than 1 hr, or estimation of time to death where COHb is 50 percent (Purser, in SFPE 2002a, 2-160).

The Stewart Equation. When dealing with predictions of time to incapacitation where CO concentrations are higher than 2000 ppm (0.2 percent) and COHb is less than 50 percent, a simpler equation is known as the Stewart equation, where

$$\%\text{COHb} = (3.317 \times 10^{-5})(\text{ppm CO})^{1.036}(\text{RMV})(t),\qquad(7.12)$$

where

CO = CO concentration (ppm),
RMV = respirations per minute of volume of air breathed (liters/min), and
t = exposure time (min).

Solving the Stewart equation for exposure time,

$$t = \frac{(3.015 \times 10^4)(\%\text{COHb})}{(\text{ppm CO})^{1.036}(\text{RMV})}.\qquad(7.13)$$

The standard inhalation values (RMV) are listed in Table 7.4, providing typical data for a man, woman, child, infant, and newborn (Health Canada 1995).

TABLE 7.4 ◆ Standard Inhalation Values (RMV; Liters per Minute)

Activity	Man	Woman	Child	Infant	Newborn
Resting	7.5	6.0	4.8	1.5	0.5
Light activity	20.0	19.0	13.0	4.2	1.5

Source: Derived from Health Canada (1995).

The standard approach for evaluating incapacitation from CO is to calculate the fraction of the CO per minute over 1 hr. During moderate activity inhalation, the human respirations per minute of volume of air breathed (RMV) is approximately 25 liters/min, and loss of consciousness occurs at 30 percent COHb (SFPE 2002a, 2-160). The formula for the fractional incapacitating dose valid for up to 1 hr is

$$F_{I_{CO}} = \frac{K(\text{ppm CO}^{1.036})(t)}{D},$$ (7.14)

where

$F_{I_{CO}}$ = fractional incapacitating dose,

t = exposure time (min),

K = 8.2925×10^{-4} for 25 liters/min RMV during moderate activity, $D = 30$ percent COHb, and

K = 2.8195×10^{-4} for 8.5 liters/min RMV at rest, $D = 40$ percent COHb.

The calculation for the fractional incapacitation dose for light activity is found by substituting these variables into equation (7.14), where

$$F_{I_{CO}} = \frac{(8.2925 \times 10^{-4})(\text{ppm CO}^{1.036})}{30}.$$ (7.15)

EXAMPLE 7.2   CO Incapacitation

An adult female is found unconscious in her bed by her rescuers at the scene of a house fire. Rough estimates suggest that she was exposed to a CO concentration of approximately 5000 ppm. Calculate the time to incapacitation and fractional incapacitation dose, assuming that the victim was at rest. Use Table 7.4 for RMV data.

Volume of air breathed   RMV = 6.0 liters/min (resting)
Loss of consciousness   COHb = 40 percent (resting)
CO concentration   CO   = 5000 ppm

Time to incapacitation $t = [(3.015*10^4)(40)]/[(5000)^{1.036}(6.0)] = 30$ min

Fractional incapacitation dose $F_{I_{CO}} = \left[\frac{(8.2925*10^{-4})(5000^{1.036})}{30}\right] = 0.188 \text{ min}^{-1}$

PREDICTING THE TIME TO INCAPACITATION BY LOW OXYGEN LEVELS

Anoxia (absence of oxygen) or *hypoxia* (low concentration of oxygen) is the condition of inadequate oxygen to support life. This can occur when air is displaced by another inert gas, such as nitrogen or carbon dioxide, by a fuel gas such as methane, or even by benign products of combustion such as CO_2 and water vapor. Normal air contains 20.9 percent O_2. At concentrations down to 15 percent O_2, there are no readily observable effects. At concentrations between 10 and 15 percent, disorientation (similar to intoxication) occurs and judgment is affected. At levels below 10 percent, unconsciousness and death may occur. Anoxia is aggravated by high levels of CO_2, which accelerate breathing rates. The effects of low-oxygen hypoxia include problems with memory and mental concentration, loss of consciousness, and death (Table 7.5).

TABLE 7.5 ◆ The Impact of Exposure to Low Oxygen Levels	
Oxygen Percentage	*Reported Effects*
14.14–20.9	No significant effects, slight loss of exercise tolerance
11.18–14.14	Slight effects on memory and mental task performance, reduced exercise tolerance
9.6–11.8	Severe incapacitation, lethargy, euphoria, loss of consciousness
7.8–9.6	Loss of consciousness, death

Source: Derived from SFPE (2002a, 2-161).

PREDICTING THE TIME TO INCAPACITATION BY HYDROGEN CYANIDE

Hydrogen cyanide (HCN) is another toxic gas in fires that incapacitates through biochemical asphyxia. As with CO, the time to incapacitation depends on the uptake rate and dosage (Purser, in SFPE 2002a).

The formula for time to incapacitation for 80- to 180-ppm HCN concentrations is

$$t_{I_{CN}}(\text{min}) = \frac{185 - \text{ppm HCN}}{4.4}, \qquad (7.16)$$

while the formula for time to incapacitation for HCN concentrations above 180 ppm is

$$t_{I_{CN}}(\text{min}) = \exp\left[5.396 - (0.023)(\text{ppm HCN})\right], \qquad (7.17)$$

and the fractional incapacitation dose per minute is

$$F'_{I_{CN}} = \left(\frac{1}{\exp\left[5.396 - (0.023)(\text{ppm HCN})\right]}\right). \qquad (7.18)$$

Note also that the term "exp" in equations (7.16) and (7.17) denotes the exponential form. As shown in Table 7.3, the HCN dosages required for incapacitation are much lower than the CO dosages.

EXAMPLE 7.3   HCN Incapacitation

An adult male is found unconscious in the waiting room after being exposed to toxic by-products from the fire in Example 7.1. A burning polyurethane plastic mattress cushion had produced an HCN concentration of approximately 200 ppm. Calculate the time to incapacitation and fractional incapacitating dose per minute.

HCN concentration	$\text{HCN} = 200\,\text{ppm}$
Time to incapacitation	$t = \exp\left[5.396 - (0.023)(200)\right] = 2.2\,\text{min}$
Fractional incapacitation dose	$F'_{I_{CN}} = \left(\dfrac{1}{\exp\left[5.396 - (0.023)(200)\right]}\right) = 0.45\,\text{min}^{-1}$

PREDICTING THE TIME TO INCAPACITATION BY CARBON DIOXIDE

Exposure to carbon dioxide has a wide range of effects, ranging from respiratory distress to loss of consciousness (see Table 7.6).

TABLE 7.6 ◆ The Impact of Exposure to Carbon Dioxide

Carbon Dioxide Percentage	Reported Effect(s)
7–10	Loss of consciousness
6–7	Severe respiratory distress, dizziness, possible loss of consciousness
3–6	Respiratory distress increasing with concentration

Source: Derived from SFPE (2002a, 2-161).

Along with being an asphyxiant displacing oxygen, carbon dioxide also increases the RMV, which in turn causes the individual to increase the uptake of the other toxic gases (Purser, in SFPE 2002a). This formula for the multiplication factor VCO_2 is

$$VCO_2 = \exp\left(\frac{CO_2}{5}\right), \tag{7.19}$$

$$VCO_2 = \left(\frac{\exp\left[(1.903)(\%CO_2) + 2.0004\right]}{7.1}\right). \tag{7.20}$$

The formula for the time to unconsciousness by carbon dioxide is

$$t_{I_{CO_2}} = \exp[6.1623 - (0.5189)(\%CO_2)] \tag{7.21}$$

and the fractional incapacitating dose per minute is

$$F'_{I_{CO_2}} = \left(\frac{1}{\exp\left[6.1623 - (0.5189)(\%CO_2)\right]}\right). \tag{7.22}$$

◆ 7.5 HEAT

The human body is capable of surviving exposure to external heat as long as it can moderate its temperature by radiant cooling of the blood through the skin and, more importantly, by evaporative cooling. This occurs internally via evaporation of water from the mucosal linings of the mouth, nose, throat, and lungs and externally via evaporation of sweat from the skin. If the core body temperature exceeds 43°C (109°F), death is likely to occur.

Prolonged exposure to high external temperatures, 80–120°C (175–250°F), with low humidity can trigger fatal hyperthermia. Exposure to lower temperatures accompanied by high humidity (which reduces the cooling evaporation rate of the water from the skin or mucosa) can also be lethal. Fire victims can die of exposure to heat alone even if they are protected from CO, smoke and flames. This may result in victims with minimal postmortem changes, although skin blistering and sloughing will occur after death.

PREDICTING THE TIME TO INCAPACITATION BY HEAT

For exposure to convected heat in a fire environment, the fractional incapacitating dose per minute is

$$F'_{I_h} = \left(\frac{1}{\exp \left[5.1849 - (0.0273)(T \, [°C]) \right]} \right). \qquad (7.23)$$

Assimilation of various data has formed the basis for the *Toxic and Physical Hazard Assessment Model* (Purser, in SFPE 2002a, 2-159). Data such as species concentration levels generated by fire models can serve as input to this hazard assessment model. According to this model, the normally accepted threshold for tolerance of radiant heat is 2.5 kW/m² for only a few minutes. Beside the burns to the skin, thermal damage to the upper respiratory tract can also occur.

Information needed to evaluate this hazard model come from two sets of information: concentration and time profiles of major toxic products and time, concentration, and toxicity relationships. The estimated toxic products within the victim's breathing zone include concentrations of carbon monoxide, hydrogen cyanide, carbon dioxide, oxygen, radiant heat flux, air temperature, and optical smoke density. Some of these values can be calculated by sophisticated computer models discussed previously.

INHALATION OF HOT GASES

Inhalation of very hot gases causes edema (swelling and inflammation) of mucosal tissues. This edema can be severe enough to cause blockage of the trachea and physical asphyxia. Inhalation of hot gases may also trigger *laryngospasm*, where the larynx involuntarily closes up to prevent entry of foreign material or *vagal inhibition*, where the breathing stops and the heart rate drops.

Rapid cooling of the inhaled hot gases occurs as the water evaporates from mucosal tissues so thermal damage usually does not extend below the larynx if gases are dry. If the hot gases include steam or are otherwise water saturated, evaporative cooling is minimized and burns/edema can extend to the major bronchi. If inhaled gases are hot enough to damage the trachea, they will usually be hot enough to burn the facial skin and mouth and singe the facial or nasal hair.

EFFECTS OF HEAT AND FLAME

The human body is a complex target when being affected by heat. Skin consists of two basic layers. The thin layer of *epidermis* (dead, keratinized skin cells) overlays a thicker dermal layer of actively growing cells in which are embedded the nerve endings, hair follicles, and blood capillaries that supply nutrients to the growing skin. Beneath the dermal layer is a layer of tough elastic connective tissue, subcutaneous fat, and, finally, muscle and bone.

Each of these components is affected differently by heat and flames. Application of heat can cause the epidermis to separate from the underlying *dermis* and form blisters in much the way paint or wallpaper blisters away from the wood or plaster beneath when heated. This occurs when the tissue reaches temperatures in excess of 54°C (130°F). The raised epidermal layer is very thin and is more easily affected by continuing heat, sometimes causing it to char. The epidermis can also separate in larger areas and form generalized skin slippage. Exposure of the denuded dermal layer can cause pain when its temperatures exceeds 43–44°C (110–112°F) (Purser, in SFPE 2002a, 2-159).

FIGURE 7.1 ◆ Effects of heat and flame shrink skin, eliminating wrinkles and changing facial contours. If it continues to shrink, it can split, leaving jagged, irregular, torn surfaces, as opposed to the sharply defined surfaces of knife cuts. This photo illustrates postfire recognition of knife cuts to the chest versus splitting of the skin on the arm. *Courtesy of University of Tennessee Regional Forensic Center, by permission.*

More prolonged exposure can destroy the proteins of the dermal layer and cause desiccation and discoloration. Higher heat fluxes can cause higher temperatures that cook and even char the tissues. As it desiccates, the skin shrinks, eliminating wrinkles and changing facial contours (making visual identifications of victims very risky). If it continues to shrink, it can split, leaving jagged, irregular, torn surfaces (as opposed to the sharply defined surfaces of knife cuts), as shown in Figure 7.1 (Smith and Pope 2003). Heat-split skin often demonstrates subcutaneous bridging of underlying tissue where cut skin does not.

If a victim survives for some time, this shrinkage can constrict blood vessels, so incisions are made (called escharotomies) through the dermal layer to relieve the pressure and maintain circulation. The investigator must recognize the effects of escharotomy or skin graft harvesting to distinguish them from fire effects if a fire victim has survived for some time after the fire.

Due to its small individual dimensions and low thermal mass, hair is affected very quickly by heat. Colors will change (typically changing to darker or redder colors). The hair shaft will bubble, shrink, and fracture as it singes. This shrinkage causes the curling seen as singeing. Its microscopic appearance is very distinctive (as opposed to cut or broken hair shafts). If hair is burned in large masses, it can form a black puffy mass.

FIGURE 7.2 ◆ Shrinkage of muscle and tendons can cause the joints to flex, causing what is called pugilistic posturing. *Courtesy of University of Tennessee Regional Forensic Center, by permission.*

If heating to the body is continued, the shrinkage can affect muscles. When the skin and muscles of the neck shrink, they can force the tongue out of the mouth. Shrinkage of muscle and tendons can cause the joints to flex, causing what is called ***pugilistic posturing***, as shown in Figure 7.2. This can cause bodies to move during fire exposure. If the body is on an irregular or unstable surface, this movement can cause the body to fall from a bed or chair and, possibly, change the direction of heat application, eliminating or obscuring previously protected areas. During legal cremations, pugilistic posturing has been observed after 10 min of exposure to flames at temperatures of 670–870°C (1240–1600°F) (Bohnert, Rost, and Pollak 1998).

Direct flame impingement, with its high-temperature gases of 500–900°C (930–1650°F) and high heat fluxes (55 kW/m^2), produces effects very quickly. Blisters will be formed in about 5 s, with charring induced some seconds later. Skin will be charred away in 5–10 min of direct flame contact, especially where stretched over the bone (joints, nose, forehead, skull) (Pope and Smith 2003). Very short but intense (flash) fire exposure can cause blistering of the epidermal layer without the sensation of pain (because the pain sensors are in the dermal layer beneath) and the heat takes longer to penetrate deeply.

Even in the absence of fire, prolonged exposure to higher temperatures causes desiccation and shrinkage of muscle tissue that causes flexion of joints (pugilistic posturing). Exposure to flames can cause combustion of muscle and fracturing of major limb bones where the bone is exposed.

Extreme heat causes bones to twist and fracture, with continued flame exposure (30 min or longer in observed cremations) causing calcination (where the charred organic matter is burned away). As the bones calcine, they become very fragile and may disintegrate of their own accord. The thin bones of the skull may delaminate, with the inner and outer layers failing separately. This delamination has given rise to thermally caused holes in the skull that have beveled edges similar to those produced by gunshots (Pope and Smith 2003). The lower-density major bones from elderly victims of osteoporosis have been seen to disintegrate under fire exposure more quickly and completely than bones of normal density (Christensen 2002). Tests have revealed that fire exposure can cause leakage of fluid through fissures in the exposed skull, but not explosion of the skull. Organs shrink as they desiccate and char, this requiring 30–40 min of cremation (Bohnert, Rost, and Pollak 1998). Bones can also shrink in length (giving rise to potential errors in ante-mortem height estimation).

Heat exposure can cause blood and fluids to accumulate, forming a hematoma in the ***extradural*** or ***epidural*** space between the skull and the tough bag of tissue (the *dura mater*) that surrounds the brain. With further heating, these fluids desiccate and char, producing a rigid, foamy, blackened mass. Physical trauma can cause subdural hematomas, where the blood collects between the dura and the brain. Fractures of the base of the skull have not been observed to be produced by fire exposure (Bohnert, Rost, and Pollak 1998). Fire exposure chars and seals exposed tissue, so as a rule, bodies exposed to an enveloping fire do not bleed. When the body is moved, the fragile layer of char can be broken, allowing body fluids to seep out.

Within certain considerations, the amount of fire damage on a body can be related to the duration of exposure if the intensity of the fire can be estimated from the fuel, ventilation, and distribution factors described elsewhere in this text. Factors such as combustion rates, thermal inertia, and relative combustibility of body components have been explored in recent studies (DeHaan 1999b).

FLAMES (INCINERATION)

When heat is applied to a surface, the rate at which it penetrates that surface is determined by the thermal inertia of the material (the numerical product of thermal capacity, density, and thermal conductivity). The thermal inertia of skin is not much different from that of a block of wood or polyethylene plastic (see Table 2.4). The pain sensors for human skin are in the dermis, about 2 mm (0.1 in.) below the surface. If heat is applied very briefly, there may not be any sensation of discomfort or pain. The longer the heat is applied, the deeper it will penetrate. The higher the intensity of the heat applied, the faster it will penetrate. Pain is triggered when skin cells reach a temperature of about 48°C (120°F) and cells are damaged if their temperature exceeds 54°C (130°F) (Stoll and Greene 1959).

Exposing skin to 2–4 kW/m² radiant heat for 30 s will cause pain. Cellular damage triggers blisters and skin slippage. Exposing skin to 4–6 kW/m² radiant heat for 8 s produces blisters (second-degree burns). Exposing skin to 10 kW/m² radiant heat for 5 s causes deeper partial thickness injuries. Exposing skin to 50–60 kW/m² radiant heat for 5 s produces third-degree burns destruction of the dermis (Stoll and Greene 1959).

BURNS

Even in the absence of fire or flames, prolonged exposure of body parts to heat that raises their temperature over 54°C (130°F) will cause desiccation, sloughing, and blistering, which can also be caused by exposure to caustic chemicals. It is often very difficult, if not impossible, to distinguish between ***burns*** suffered near the time of death (perimortem) and those inflicted after death (postmortem).

When gasoline or a similar volatile, low-viscosity, low-surface tension liquid fuel is poured on bare skin, some will be absorbed into the epidermis, but the bulk will run off, leaving a very thin film of liquid, particularly on vertical surfaces. This thin film will burn off very quickly (less than 10 s). The skin beneath may be spared completely or reddened (first-degree burns), except where folds of the skin or clothing have retained enough fuel to sustain longer burning. This will induce severe blistering and even charring in extreme cases. On horizontal skin surfaces, a deeper "pool" may be retained long enough to produce a "halo" or ring of blisters (second-degree burns) around the circumference of the pool (DeHaan 2002).

BLUNT FORCE TRAUMA

Blunt force trauma can also cause or contribute to the death of fire victims. Structural collapse or explosions can produce direct impact of solid materials onto victims. Falls or impacts with stationary surfaces (furniture or door frames) during escape attempts can induce blunt trauma that only careful examination can distinguish from an assault. Wound patterns, blood stains, or even trace evidence can be used to interpret blunt trauma injuries and establish whether they resulted from assault or the fire.

◆ 7.6 TIME INTERVALS

One of the problems outlined earlier is the time interval between exposure to a fire and its fatal aftermath. Death can occur nearly instantaneously or minutes or hours later. Under these conditions it is not difficult to connect the death to its actual cause. When a person dies weeks or even months after a fire, the cause can still be the fire, but the linkage can be obscured by the extensive medical events in between.

There is a range of effects that can determine the time interval between the fire and death. With ***instantaneous death***, vagal inhibition or laryngospasm occurs upon inhalation of flames and hot gases, causing cessation of breathing followed very quickly by death. Explosion trauma and incineration occurring by exposure to a fully developed fire (often as a result of structural collapse) will result in nearly instantaneous death.

The duration of ***seconds to minutes*** can occur with ***hyperthermia***, exposure to very hot but nonlethal gases or steam, or anoxia, the lack of oxygen. Toxic gases such as hydrogen cyanide or other pyrolysis products, asphyxia via the inhalation of carbon monoxide or blockage of airways by soot, exposure to flames, physical trauma, with loss of blood, internal injuries, and brain injuries can cause death within minutes.

In ***hours***, death from carbon monoxide, edema from inhalation of hot gases, burns (shock), and brain or other internal injuries can occur. Death due to the dehydration and shock from burns can occur ***days*** after the fire. Even ***weeks or months*** after the fire, deaths due to infections or organ failure triggered by fire injuries can occur.

Note that the ***cause of death*** is defined as the injury or disease that triggers the sequence of events leading to death. In a fire, the cause may be inhalation of hot gases, CO, or other toxic gases, heat, burns, anoxia, asphyxia, structural collapse, or blunt trauma. The ***mechanism of death*** is the biological or biochemical derangement incompatible with life. Mechanisms can be respiratory failure, exsanguination, and cardiac arrest. The ***manner of death*** is an assessment of the circumstances in which the cause was brought about. In the United States, these are most often homicide, suicide, accident, natural, or undetermined.

As one can appreciate, the longer the interval between the cause (the fire) and the onset of the mechanism of death (organ failure, septicemia, etc.), the more likely it is that the connection will be lost. This is especially true when the victim has been moved from trauma care hospitals to long-term care facilities, sometimes in other geographical areas. The investigator must be diligent to ensure that the cause of death is not listed on the final death certificate as some generic term such as respiratory failure, cardiac arrest, or septicemia.

SCENE INVESTIGATION

The reconstruction of the activities of a fire victim may depend on finding and documenting blood stains (such as from impact with a wall or door jamb or handprints on a wall), mechanical (blunt trauma) injuries, or artifacts found with the body. The nature of dress (street clothes, robe, nightgown) or objects (dog leash, jewelry, flashlight, fire extinguisher, house keys, phone, keepsakes, etc.) can provide clues as to what the person was doing prior to collapse.

The position (face up or face down) is not usually significant, as people known to have died during a fire have been found in all positions. Due to pugilistic posturing, the attitude of the body bears little reliable relation to antemortem actions. Young children tend to hide under beds or in closets, but finding them in other locations is not proof that they did not have the time or physical capacity to seek safety.

POSTMORTEM DESTRUCTION

A body exposed to fire can support combustion, the rate and thoroughness of which depend on the nature and condition of exposure of the body to the flames. The skin, muscles, and connective tissues will shrink as they dehydrate and char. If exposed to enough flame, they will burn and yield some heat of combustion, although quite reluctantly. The relatively water-logged tissues of the internal organs must be dried by heat exposure before they can combust, and that dehydration step increases their fire resistance and delays their consumption.

Bones have moisture and a high fat content, especially in the marrow, so they will shrink, crack, and split and contribute fuel to an external fire. The subcutaneous fat of the human body provides the best fuel, having an effective heat of combustion of the order of 32–36 kJ/g (DeHaan 1999b). Like candle wax, however, it will not self-ignite or smolder and will not normally support flaming combustion unless the rendered fat is absorbed into a suitable wick. This "wick" can be provided by charred clothing, bedding, carpet, upholstery, or wood in the vicinity (as long as it forms a porous, rigid mass). The size of the fire that can be supported by such a process is controlled by the size (surface area) of the wick.

Depending on the position of the body and its available "wick" area, fires supported by the combustion of a body will be of the order of 20–120 kW, similar to a small wastebasket fire. Such a fire will impact only items close to it and fire damage will often be very confined.

The flames produced by combustion of body fat will be 800–900°C (1475–1650°F) in temperature and if they impinge on the body surface, they can aid the destruction of the body. The process, if unaided by an external fire, is quite slow, with a fuel consumption rate of the order of 3.6–10.8 kg/hr (7–25 lb/hr). It is possible, given a long enough time (5–10 hr), that a great deal of the body can be reduced to bone fragments (DeHaan 1999b, 2001).

If a body is exposed to a fully developed room or vehicle fire, however, the rate of destruction will be much closer to that observed in commercial crematoria. In those cases, flames of 700–900°C (1300–1650°F) and a heat intensity of 100 kW/m^2 envelop the body and can reduce it to ash and fragments of the larger bones in 1.5 to 3 hr (Bohnert, Rost, and Pollak 1998).

SUMMARY OF POSTMORTEM TESTS DESIRABLE IN FIRE DEATH CASES

While not all of the following may be needed in all cases, it can be appreciated that once the body is released for burial or cremation, it will be too late for forensic examination.

The complexity of many fire death cases may necessitate finding the answers to questions that were not apparent earlier. Having comprehensive samples and data is the best route to a successful and accurate investigation.

- **Blood** (taken from a major blood vessel or chamber of the heart, not from the body cavity). Tested for COHb saturation, hydrogen cyanide, drugs (therapeutic and abuse), alcohol, and volatile hydrocarbons.
- **Tissue (brain, kidney, liver, lung).** Tested for drugs, poisons, volatile hydrocarbons, combustion by-products, CO (as a backup for insufficient blood).
- **Tissue (skin near burns).** Tested for vital chemical or cellular response to burns.
- **Stomach contents.** Tested for establishment of activities before death and possible time of death.
- **Airways.** Full longitudinal transection of airways from mouth to lungs to examine and document extent and distribution of edema, scorching/dehydration, and soot.
- **Internal body temperature.** Should be measured at the scene. It may be elevated due to hyperthermia antemortem.
- **X Rays.** Full-body (including associated debris in body bag) and details of teeth and any unusual features discovered (fractures, implants).
- **Clothing.** Remove and preserve all clothing remnants and associated artifacts.
- **Photographs.** General (overall), with close-ups of any burns or wounds, in color, with scale.
- **Postmortem weight.** Determine the postmortem weight of the body, *not* including the fire debris or body bag.

It is very useful for the fire investigator to be present when the postmortem is conducted, not only to ensure that all appropriate observations are made but also to be on hand to answer any questions that arise during the examination. Few pathologists have extensive knowledge of fire chemistry or fire dynamics, so the investigator is in a good position to advise the pathologist as to the fire conditions in the vicinity of the body.

◆ 7.7 CASE EXAMPLE 1—DEATH FOLLOWED BY FIRE: CAUSE AND DURATION

At about 9 A.M., Mr. John Doe (not his real name) discovered the severely burned bodies of his sister-in-law and his 3-year-old nephew in the dining room of their home. He had been called minutes earlier by his brother, Mr. Jim Doe, a salesman, who reported that he had left home that morning and had not been able to reach his wife (who was 8 months pregnant) by telephone. Concerned, he called his brother and asked that he stop by and check on her.

Upon arrival, John Doe found the house full of smoke but only small flames showing in the vicinity of the bodies that were sprawled amidst an area of burned carpet in the dining room. He reportedly batted out the small flames with a hand towel and then withdrew to call emergency 9–1–1. The local fire department responded by 9:10 A.M. and, forewarned of the conditions within, limited its response to one firefighter with a pressurized water tank who entered the building, sprayed a very small amount of water on or near the body to extinguish the small flames still present, and then withdrew along the same path. Except for opening windows to vent the accumulated smoke, that was the extent of fire suppression.

Jim Doe returned to the house while the investigation was beginning and claimed that he had left the house at about 6:00 that morning and that everything had been in order. He had stopped to put gasoline in his car and then at a restaurant

FIGURE 7.3 ◆ The fire damage was limited to an oval area of carpet about 2 m² in size, centered on the two bodies. The mother was sprawled roughly face-down and the child was spread-eagled on his back alongside her. *Courtesy of Elk Grove, Fire Department, Elk Grove, CA, by permission.*

while making his sales calls for the morning, establishing his presence elsewhere after 6:00 A.M.

The fire damage was limited to an oval area of carpet about 2 m² in size, centered on the two bodies. The mother was sprawled roughly face-down and the child was spread-eagled on his back alongside her as shown in Figure 7.3. There were areas of scorching on the walls and baseboards of the dining room near the entry to the kitchen and two areas of damage to the vinyl floor of the kitchen immediately adjacent to the mother's lower right leg. There were areas of scorching on her legs adjacent to these floor patterns. The fire investigators quickly determined through laboratory analysis that automotive gasoline had been poured on the carpet around the bodies and in nearby areas of the living room dining room, kitchen, and hallway (Figure 7.4).

Both victims had zero COHb levels and no soot detected in the airways. Both bodies bore signs of physical (blunt force) trauma, and the child's cause of death was ascribed to severe multiple head wounds. Blood stain patterns indicated that the mother had been beaten or kicked in the vicinity of where the bodies were found but fire damage to her face, neck, and upper torso precluded the establishment of an exact cause of death (Moran 2001).

The major fire-related issues pertained to the duration of the fire: what had fueled the fire, how long a body would burn in such circumstances, and what the mass loss rate for a body is when it represents the main fuel package for a fire (as opposed to being involved in a massive enveloping fire fueled by furnishings and other external fuel load).

Carpet and pad from the scene were recovered in large quantities to serve as comparison samples and test targets for fire testing. The carpet was 95 percent polypropylene and 5 percent nylon face yarns bonded to a polypropylene mesh backing. The pad

FIGURE 7.4 ◆ Fire investigators quickly determined through laboratory analysis that automotive gasoline had been poured on the carpet around the bodies and in nearby areas of the living room, kitchen, and hallway. *Courtesy of Elk Grove, Fire Department, Elk Grove, CA, by permission.*

was low-density polyurethane foam. Testing revealed that this carpet/pad combination could be ignited to support a self-sustaining combustion by the ignition of a small quantity of gasoline.

The burning carpet was observed in tests to produce small flames, 5–8 cm (2–3 in.) high that advanced to consume approximately 0.5 m^2 of carpet per hour. Based on the limited amount of carpet burned very deeply (i.e., in direct contact with the gasoline pool fire), it was estimated that less than 2 liters of gasoline was used. Such a quantity would be expected to produce a fire of approximately 500 kW if ignited on carpet, with a duration of 2 to 3 min. Testing confirmed these estimates.

As can be seen in the floor plan in Figure 7.5, the living room, dining room, kitchen, and family room formed one large room, with a high vaulted ceiling extending throughout those areas, while a partial-height wall separated the living/dining portion from the family room/kitchen. The vaulted ceiling over the location of the bodies was an estimated 3.6 m (12 ft) from the floor and bore no heat damage. A flame plume from a 500-kW fire approximately 1 m (3.3 ft) in diameter would be (per Heskestad) about 1.8 m (6 ft) high and would not be expected to produce any ceiling damage.

The wall nearest the main plume would have been the south dining room wall and would have been more than a meter away. All walls were painted gypsum wallboard and none bore any heat damage. There was no heat damage to any of the wood furniture in the dining room. There was a narrow, arc-shaped burn mark on the carpet identified as a gasoline trail between the entry door and the bodies but there was no damage to any furnishings except to the carpet.

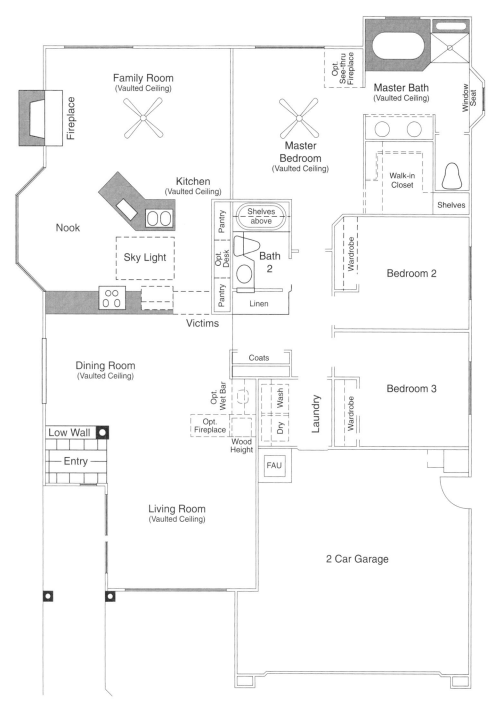

FIGURE 7.5 ◆ Floor plan showing the living room, dining room, kitchen, and family room, which formed one large area with a high vaulted ceiling. *Courtesy of Bruce Moran, Sacramento County District Attorney's Laboratory of Forensic Services, Sacramento, CA, by permission.*

Several areas of the carpet bore scorched and melted areas whose appearance could be reproduced by the burning of a splattering of small droplets of gasoline on the surface of such synthetic carpet pile. Smoke had penetrated throughout the house and soot had lightly coated all horizontal surfaces and condensed out on vertical surfaces (showing the outlines of wall studs, ceiling joists, and other concealed structural elements).

Blood spatters on surfaces and coatings of soot on floors showed that objects such as vases, boxes, and cutlery that were strewn about to simulate a burglary had been placed in those positions after the blood was splattered but before the fire. The areas of scorching and sooting on the vinyl flooring of the kitchen were identified as being produced by gasoline or a similar ignitable liquid of relatively low boiling point by comparison to tests of such fluids on similar floor coverings.

The most significant fire damage was to the bodies of the two victims. No clothing could be identified on the child's body and the remains of a knit top and panties on Mrs. Doe would not have constituted a significant fuel load.

A reasonably small quantity of gasoline had apparently been used and so attention was focused on the combustion of the bodies as a major factor in establishing the time frame for the fire. Because Mrs. Doe had been receiving prenatal care, there was good documentation of her antemortem body weight. Compared to the postmortem weight, it was estimated that some 12 kg of body weight had been lost. No such data of comparable quality were available for the child victim, so it was decided that the mass lost from Mrs. Doe's body would be the most reliable measure of establishing duration of the fire.

Testing was conducted at the California Bureau of Home Furnishings (BHF) room calorimeter to establish the heat release rates and mass loss rates of body fat. Pork fat was selected, for its similarity to human subcutaneous fat and availability. Tests of various small (1- to 2-kg) quantities of pork fat (with skin) wrapped in cotton cloth established that small fires, 20 to 70 kW, could be maintained by combustion of the rendered fat.

The size of the fire was controlled largely by the surface area of the wick (the charred cloth), while the duration of the fire was controlled by the supply of fuel. Mass loss rates appeared to be of the order of 1–2 g/s (3.6–7.2 kg/hr). Testing was also conducted in the cone calorimeter at BHF using a 10×10-cm tray, with paraffin wax as a control. It was found that at 35 to 50 kW/m^2 incident radiant heating, pork fat would melt and smoke but could not be ignited to support a continuous flame unless a cotton cloth wick were placed across the surface of the fat (DeHaan 1999b). The radiant heat would then char the cotton cloth melting the fat beneath. The melted fat wicking up via the charred cloth could be readily ignited. Testing established that the effective heat of combustion for both pork and human fat was approximately 34 kJ/g and that their rate of heat release (per unit surface area of fuel) were roughly the same, 200–250 kW/m^2. The observed properties of melting behavior and ignitability for human and pork fat were also indistinguishable, confirming that test results using pork fat or pig carcasses would yield results applicable to human cadavers (DeHaan 1999b).

A final series of tests in which pig carcasses clothed in knit garments and ignited by means of a 1-liter pour of gasoline confirmed that the charred garments and charred carpet and pad served as a wick and would support combustion of the fat being rendered out of the carcass. This combustion produced a smoky fire of approximately 50 kW, with a mass loss rate of approximately 1.5 g/s (5.4 kg/hr) for times exceeding 1 hr. Further, the gasoline pour would burn off in approximately 3 min while producing a fire of 250–400 kW. This brief fire would initiate fire in the surrounding carpet that would continue to burn for various times, consuming carpet at a rate of about 0.5 m^2/hr.

Of interest in this case was the observation that the gasoline-fed fire was not of sufficient duration to cause the skin to shrink and split and allow the rendered fat to escape to support external flames. Only after the clothing and carpet had burned for 10 min or longer could the rendering process be seen to be supporting the fire.

The fire behavior supported by tests and data were considered when reaching the conclusion that the fire had been burning for 2 to 4 hr at the time of its extinguishment. Since the fire was still burning at approximately 9 A.M., it was concluded that it was ignited using gasoline sometime between 5 and 7 A.M. This time frame, combined with conflicting statements made by the suspect, led to his prosecution. He was convicted of three counts of second-degree murder and one count of arson. He is serving consecutive sentences for a total of 60 years to life.

This case is courtesy of Bruce Moran, Sacramento County District Attorney's Laboratory, Sacramento, California.

◆ 7.8 CASE EXAMPLE 2—EXECUTION BY FIRE

The burned body of a young woman was found in a remote agricultural area. She bore burns over most of her body and was dressed in the burned remains of cotton denim pants, a long-sleeved cotton shirt, and a bra, all identified by brand name tags that survived the fire (Figure 7.6). Her ankles had been bound together with a heavy leather belt wrapped over her cotton socks, prior to the fire (Figure 7.7).

FIGURE 7.6 ◆ Victim of murder by burning as found at scene. Repeated scuff marks with limited scorching in soil adjacent to torso and limbs show movement during the fire. Impressions of creased fabric of shirt on soil next to torso were produced as victim rolled on ground. Charred residue of denim jeans remains on left leg and waist. Charred cloth fragments away from body also show movement after and while burning. Shoe print in lower left corner is partially overlaid by "leg" fabric impression showing sequence.

FIGURE 7.7 ◆ Scuff marks in soil near toes and feet show repeated movement. Flexion of toes is due to heat shrinkage of tendons. Feet of victim bear only first and second degree burns from cotton socks burning off. Sock material trapped beneath leather belt around ankles burned longer (supported by gasoline wicking out) inducing charring of skin beneath.

The victim was found lying on her back with her arms and legs partially flexed. There was a burned pile of clothing some 8 ft 8 in. (2.6 m) away from her head (probably in a cloth duffel bag). The clothing and other contents indicated that this bag belonged to the victim.

There were a number of small areas of scorching or charring visible on the dry soil nearby, isolated from both the clothing and the body. While these could have been the result of isolated burning splashes of flammable liquid, they were more consistent in appearance with having been the result of contact between the burning clothing and the soil.

Examination of the scene revealed that there was a faint pattern of wrinkled fabric (as from garments like shirt and pants) impressed into the dry sandy soil between the body and the pile/bag of clothing. There was an additional pattern of wrinkled fabric in the soil to the right of the body adjacent to the torso and upper right leg.

The soil at the fire scene revealed additional information. There were deep scuff marks in the soil, of the type made by bare or stocking feet, under and beside the toes of the victim. Scorched areas of soil to the right side of the body were displaced and disturbed by patterns of movement of the soil, probably by movement of the right arm and right leg of the victim. There were a number of shoe prints and tire prints in the vicinity of the body, some of which overlaid the scuff marks, but in most cases the scuff marks obscured the shoe impressions, confirming sequencing and timing.

Laboratory tests revealed the presence of gasoline in the clothing of the victim, in the burned clothing pile, and in the soil in several areas. There was no gasoline container found in the vicinity. No estimate could be made of the quantity of gasoline that had been used. No other fuels were present except for the clothing.

There was extensive scorching and blistering of the skin across nearly all exposed areas of the victim's body. The back was more burned than the front of the torso, and most of the clothing was burned away, particularly on the back and sides. There were signs of vital reaction to many burned areas of skin. The burns were only partial thickness (second degree) over nearly all of the body, with the only exception being deep charring of the ankles and lower legs immediately adjacent to the cotton socks held in place by the leather belt. Nearly all of the scalp hair was burned away, reduced to black masses. There were charred remains of cloth clutched in the fingers of the right hand.

Postmortem examination revealed internal heat/flame damage to the mucosa of the throat and larynx as a result of inhalation of extremely hot gases. Toxicological analysis reported a COHb level of approximately 14 percent. The fire issues here were whether gasoline had been poured on the victim while she was conscious and what the result of such an event would be.

Re-creations in the laboratory using a dressmaker's mannequin approximately the same height as the victim demonstrated that two full revolutions of a body that size as it rolled on the ground resulted in the same displacement as seen at the scene between the body and the clothing pile. Due to the taper of the body from the shoulders to the feet, the path taken by a body rolling in this manner is in the form of an arc rather than a straight line.

A gasoline fire is known to produce heat transfer sufficient that direct flame contact induces blistering (partial-thickness burns) in about 5 s at 50 kW/m^2. Based on the distribution of burn damage to the body and the clothing, it was concluded that the fire damage could not have been the result of gasoline having been poured over an inert, unconscious body. The absence of protected areas and the extensive damage to the back and buttocks could have resulted only from the victim's consciously moving. The charred cloth clutched in the hand away from other fire damage indicated a conscious attempt to grasp the burning clothing during the fire.

While flexion of the major joints as a result of fire exposure can result in some movement of an unconscious or even a deceased body, the pattern of cloth/fabric impressions in the soil was the result of a rolling movement of the body. The scuff marks in the soil to the right side of the body were most likely the result of movement of the right arm and right leg of the victim while she was lying on her right side. The body apparently fell or rolled onto its back, where it assumed its final resting position. Movement of the feet and toes would have produced the deep scuffs found there. Fire damage to the clothing and the isolated scorch marks on the soil were also consistent with conscious movement of the body.

The duration of a gasoline fire would be of the order of 2 min or less (assuming that less than a gallon was used). The duration would be shorter if the gasoline was spread on bare skin or thin cotton fabric. Such a short-duration fire would produce the limited-depth burns found here, with the only exception being the gasoline-soaked cotton socks, which acted as a wick to sustain flames long enough to induce deep burns to adjacent tissue. The presence of shoe prints overlying some of the scuff marks in the soil was an indication that the victim was set afire in the vicinity of the clothing bag and that her assailants were present during (at least) the initial stages of the fire as she rolled on the ground in an attempt to extinguish the flames. It was concluded that she was conscious for many seconds during this fire.

The three people who carried out this execution were acquaintances of the victim who decided that it would be interesting to dispose of her in this manner and transported her in a truck while they purchased the can and the gasoline. All three have pled guilty to murder charges.

◆ 7.9 CASE EXAMPLE 3—FATAL VAN FIRE

At about 5:30 on a cold November morning a farmer heard shouts of alarm coming from across the river. He looked out to see a camping-style van fully engulfed in flames. He phoned the fire department, but due to the remote location of the site, it was 15 min or more before fire crews arrived and another 10 min before the fire was extinguished.

Inside the van were the remains of an adult female victim who had reportedly been asleep in the back of the van with her boyfriend at the time the fire was discovered. The boyfriend, after awakening and reportedly discovering the smoke and limited flames within the van, had allegedly tried to rouse the victim, who was nonresponsive, and then escaped the van by butting his head against a rear window. He attempted to rescue his friend from outside but discovered that all but one of the doors to the van were locked and he could not reach her through the one unlocked door due to a blanket hung between the cargo compartment and the driver's compartment. He claimed that they had been camping on the river just for a day. Both were "recreational" drug users, and he had a record for breaking-and-entering and minor drug charges.

Investigators called to the scene discovered that all the windows to the "Sportsman-style" van had been broken but could not tell if any had been broken by mechanical force (the side and rear windows being tempered glass, which shatters into similar small pieces whether broken by thermal shock or mechanical impact) but all appeared to have been soot covered. The van had been gutted, with most damage to the cargo and driver's compartments (Figure 7.8) and there were signs that the fire had penetrated into the engine from inside the van. The sheet metal of the roof had been badly distorted by the fire. There were no signs that the vehicle had been in operation at the time of the fire. The vehicle's fuel tank was still half-full and the fuel system was normal.

Fire damage outside the vehicle was limited to the grass immediately around and under it; there were no indications of an external grass fire growing to involve the vehicle. The van was heavily loaded with bags and boxes of clothing, food, camping supplies, tools, and blankets. There were a propane camping lantern and camp stove with two 1-lb propane bottles, but they were not in use. All exposed fuels in the van were fire damaged.

A friend who had visited the man and woman the night before the fire said that they were eating snacks (with no cooking) and were using small votive candles for light. Luckily the fire chief in this jurisdiction always kept a point-and-shoot camera in his vehicle and was able to get several photos of the still-smoldering interior of the vehicle before the contents were removed to permit overhaul. The upholstery of the seats and dash and paneling of the cargo compartment were completely consumed (Figure 7.9). The fire damage clearly demonstrated that the fire within this compartment had gone to flashover, with ventilation supplied by the windows (closed at the start of the fire but failing as the fire grew).

The decedent was found in a partial sitting position on the right side of the cargo compartment (Figure 7.10) facing the rear of the van. The body had suffered massive

FIGURE 7.8 ◆ Exterior damage to the gutted van, with most damage to the cargo and driver's compartments. *Courtesy of Solano County Sheriff's Dept., Fairfield, Calif., by permission.*

FIGURE 7.9 ◆ Interior of the vehicle immediately after extinguishment and before overhaul. The victim is sitting facing the camera, with her back resting against the back of the passenger seat. *Courtesy of Solano County Sheriff's Dept., Fairfield, Calif., by permission.*

FIGURE 7.10 ◆ Fire victim as found in the vehicle. Note the extensive destruction by flames, with exposed ribs, spinal column, and internal organs. The skull has calcination and fracture damage resulting from prolonged fire exposure. *Courtesy of Solano County Sheriff's Dept., Fairfield, Calif., by permission.*

fire damage; nearly the entire right half of the body had burned away (exposing the spinal column), with major bones reduced to fragments. The victim had a 41 percent COHb level, with no alcohol and very low levels of amphetamine metabolites. She was young (23 years old), healthy, and physically unimpaired. Extensive fire exposure had destroyed the soft tissues of the head and neck and had caused the exposed skull to calcine and fracture into a number of pieces. There were no underlying hemorrhages indicating antemortem trauma. The fire had compromised the lungs and airways such that the presence of combustion products could not be determined.

The COHb level left no doubt that the victim had been alive and breathing for some time during the fire and had died as a result of exposure to smoke and flames. The extensive damage to the body was ascribed by the pathologist to incineration by a flammable liquid fire, despite later laboratory findings that revealed no petroleum distillates in the small segment of carpet recovered from beneath her.

No flammable liquid residues were detected on the exterior of the van or beneath it. The extensive damage to the van, the absence of accidental ignition sources, and the nonoperational status of the van led investigators to conclude that the fire had been deliberately ignited with a flammable liquid on or around the decedent, which prevented her escape. The boyfriend was charged with murder.

The surviving victim (boyfriend) was found outside the van dressed in T-shirt and undershorts (he was the person shouting for help that the farmer heard). He had a coating of soot on his face, arms, and hands but no burns. His facial hair (including moustache) was unburned but hair at the top of his head was singed. He had fresh abrasions on his knuckles, one on the top of his head, and scratches and abrasions on his lower legs. Tiny fragments of glass were found in his scalp hair. His blood was not sampled immediately but was taken later and was found to bear low CO concentrations, low levels of amphetamine metabolites, and no alcohol. He claimed that he and the decedent had gone to sleep side by side wrapped in blankets on the floor of the cargo compartment and awakened to discover the fire. He denied any conflict with the decedent.

A review of the evidence (by one of the authors) at the request of the local district attorney's investigator resulted in the conclusion that the physical, medical, pathological, and toxicological evidence was more consistent with an accidental fire occurring as described by the accused than with a murder with flammable liquid.

A fire-related human behavior analysis revealed that if the survivor had been sleeping on his back in a smoke-filled van, his face and arms would be expected to be coated with soot but not exposed to the hot gas layer until he sat up and attempted escape. At that time, his scalp hair would be exposed to the highest temperatures. Beating at the window glass or fumbling in the dark for the door locks could have produced the abrasions on his right hand. Butting his head against the rear window until it broke could have produced the abrasion on the top of his head and the glass fragments in his hair. Sliding out the high rear window would be expected to induce cuts and scratches to the lower legs.

The 41 percent COHb level in the decedent is similar to the levels encountered in victims of accidental house fires. In the opposing hypothesis, throwing gasoline on someone in sufficient quantities to prevent his/her escape would be likely to leave unburned residues after the fire in protected areas under the body and would also be likely to produce a flash fire of sufficient size to induce singed facial hair on someone close by (i.e., the thrower). The ignition of a quantity of gasoline vapor in the vicinity of a victim's face can induce rapid cessation of breathing so that CO levels in the blood would be very low. The first few seconds of a gasoline vapor/air fire are also very low in CO, so the gases that are inhaled are unlikely to produce an elevated COHb level. Such a scenario would not produce the glass fragments, singed scalp hair, or leg injuries. A gasoline-fueled fire would also be unlikely to induce such extreme damage to the decedent's body unless it involved massive quantities of fuel.

The fire issues involved were the following.

- What were the contributions of the fuel load in the van?
- Did the size of the van (compartment) and the breakage of windows play a role?
- Could a fire in a closed van produce a fire that would be escapable and produce the physical and clinical evidence seen on the survivor?
- Could such a fire progress to flashover?
- If the fire did progress to flashover, could postflashover fire produce the damage to the body and to the van without the presence of any ignitable liquids?

A long-wheelbase Sportsman van similar to the one in the incident was obtained by the District Attorney's office. Based upon calculations of the window area and height, it was determined that when all the windows failed, enough air would be admitted to the van to support a fire of 3.2 MW. The internal volume of the van was calculated to be 9.8 m^3, so there was considerably more air inside even a closed van than there would be in the average passenger vehicle.

The van obtained (Figure 7.11) had nearly the same arrangement of windows but had one solid side panel instead of a large window so a vent panel was cut through the metal and left attached by a hinge so it could be opened during the fire test to simulate failure of the window by thermal shock. The interior finish was duplicated, with paneling, carpeting, and seats similar to the original.

FIGURE 7.11 ◆ The prototype van obtained had nearly the same arrangement of windows but had one solid side panel instead of a large window so a vent panel was cut through the metal and left attached by a hinge so it could be opened during the fire test to simulate failure of the window by thermal shock. *Courtesy of Solano County Sheriff's Dept., Fairfield, Calif., by permission.*

FIGURE 7.12 ◆ Wooden tool boxes and a wool blanket curtain were fitted to simulate the original, and some 120 kg of clothing, blankets, plastic crates, cardboard, and miscellaneous combustibles was added, estimated to be about one-half of the added fuel mass in the original van. *Courtesy of Solano County Sheriff's Dept., Fairfield, Calif., by permission.*

Wooden tool boxes and a wool blanket curtain were fitted to simulate the original and some 120 kg of clothing, blankets, plastic crates, cardboard, and miscellaneous combustibles was added, estimated to be about one-half of the added fuel mass in the original van, as shown in Figure 7.12.

Temperatures were monitored via three thermocouples mounted on a tree in the center of the van near the position of the decedent and one thermocouple near the rear door where the survivor allegedly made his escape. Ignition was produced with an open flame in ordinary combustibles and the van closed up. A misplaced votive candle was probably the ignition source for the actual fire. Given the nature of the fire and the fuels available, it was very unlikely to have been a smoldering source.

In the test, within 3 min heavy smoke had formed throughout the van and temperatures in the smoke layer ranged from 80 to 100°C (176 to 212°F), as shown in Figure 7.13. Soot was observed condensing on all windows. The rear window was broken from outside by mechanical impact at 3 min 45 s. Gas layer temperatures exceeded 600°C (1112°F) by 7 min, and flashover occurred as windows began to fail from thermal shock. The side vent was opened as all the windows failed between 6 and 8 min into the fire. By 8 min, temperatures in the hot gas layer exceeded 1000°C (1832°F). Temperatures of over 600°C (1112°F) were maintained for over 12 min as the fire consumed the fuel load and went into decay. Total burn time was 30 min. Damage to the exterior and interior of the van duplicated that in the original, as shown in Figures 7.14 and 7.15.

Temperatures and heat fluxes in excess of those found in commercial crematoria were measured in the center of the van. The victim's body would have been exposed to these intense fire exposures from her right side, near the center of the compartment, where the postflashover burning would have been most intense. The fuel load available

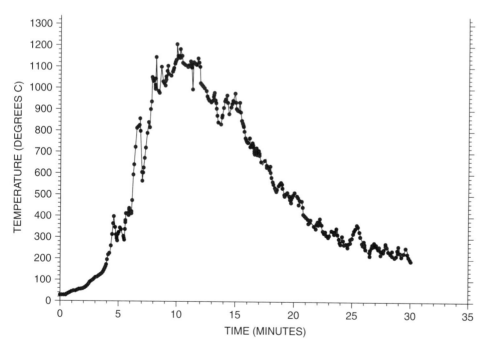

FIGURE 7.13 ◆ Test results for fire in a similar vehicle documenting the interior temperature versus time at a thermocouple located near the victim's location. *Courtesy of Fred Fisher, Fisher Research & Development, Inc., Vacaville, Calif., by permission.*

FIGURE 7.14 ◆ Thermal damage to exterior of prototype test van. *Courtesy of Solano County Sheriff's Dept., Fairfield, Calif., by permission.*

FIGURE 7.15 ◆ Thermal damage to interior of prototype test van. *Courtesy of Solano County Sheriff's Dept., Fairfield, Calif., by permission.*

in this test limited the duration of the fire, but not its intensity. The intensity was determined by the surface area of fuel involved in the postflashover burning and the ventilation available to support the combustion.

While no heat release data were captured during this test, the test team, with their extensive experience, estimated that the heat release rate was in excess of 3 MW during the peak of the fire (at 10 to 12 min). Since the ventilation inside the test van was the same as in the fatal van, the size and intensity of the fires would have been equivalent. With the more substantial fuel load in the scene van, it would have been capable of supporting a full postflashover fire for much longer than the test van could.

Witness observation by the farmer confirmed that the van was fully engulfed when he first saw it and it burned that way until extinguished some 25–30 min later. Based on the results of this test, the district attorney agreed that an accidental fire could not be ruled out and, in fact, was more consistent with the physical evidence. All charges were dropped.

Test data in this case is courtesy of Fred Fisher, P.E., of Vacaville, California. The authors thank Professor R. Brady Williamson of University of California, Berkeley, Dr. Charles Fleischman of University of Canterbury, Christchurch, New Zealand, and members of the Solano County Fire Investigation Team for their contributions and for making this test possible.

◆ **7.10 CONCLUSIONS**

In a fatal fire, the cause of death and the cause of the fire are independent but linked by circumstances. Each must be established, and only then can the link between them be determined.

Accidental fires can accompany deaths by accident, suicide, homicide, or even natural causes. Incendiary fires can be associated with homicide (as a direct cause of death or simply as part of the crime event) but also with accidental or natural-cause deaths. For a fire death investigation to be successful (i.e., accurate and defensible), the following guidelines must be observed.

- **Treat as a crime scene.** Every fire with a death or major injury should be treated as a potential crime scene and not prejudged as accidental. The scene should be secured, preserved, documented, and searched by qualified personnel acting as a cooperative team.
- **Documentation is essential.** This includes accurate floor plans, with dimensions and major fuel packages included, and comprehensive photographic coverage. Photos must include presearch survey photos, photos during search and layering, and photos of all views of the body prior to removal, during removal, and during postmortem examination. This documentation is essential to proper reconstruction (as described in Chapter 4).
- **Avoid moving the body.** The body must not be moved until it has been properly examined by the fire investigator and the pathologist or coroner's representative and thoroughly documented by photos and diagrams. The debris under and within 3 ft of the body should be carefully layered and sifted. All clothing or fragments should be preserved. Artifacts (jewelry, weapons, etc.) must be documented and collected.
- **Assess the fuels.** The fire investigator has to assess the fuels already at the scene (structure as well as furnishings), the role fuels may or may not have played in ignition, flame spread, heat release rates, time of development, and creation of flashover conditions.

- **Perform full forensic exams.** Every fire death deserves a full forensic postmortem including toxicology and X-rays. Toxicology samples should be tested for alcohol and drugs, as well as CO, and should include both blood and tissue. The clothing should remain with the body and be documented and evaluated in situ before removal, if possible, then properly preserved. The internal (liver) body temperature should be taken as soon as possible (preferably at the scene).
- **Examine pets.** Deceased pets should be X-rayed and necropsied. Injuries to living pets should be noted and documented. Blood from deceased pets should also be tested for COHb saturation.
- **Examine living victims.** "Nonfatal" burn victims should be photographed and blood samples taken for analysis later if needed. External clothing (pants, shoes, shirt) should be saved and properly preserved.
- **Fully appreciate the fire environment.** Pathologists and homicide detectives must appreciate the fire environment—temperatures, heat and its transfer, flames, and smoke—and the distribution of fire products and the variables of human response to those conditions. In best practice, the pathologist visits the scene and sees the body in situ to appreciate its conditions of exposure (to flame, heat, and smoke), the nature of debris, and its location and position.
- **Carry out forensic reconstructions.** A full reconstruction may involve criminalistics evidence such as blood spatter or transfers, fingerprints, tool marks, shoe prints, and trace evidence. The criminalist should be part of the scene team along with the homicide detective, fire investigator, and pathologist.

As we have seen, a death involving fire is not a simple exposure to a static set of conditions at a single moment in time. Fire is a complex event, and a fire death investigation is even more complex and challenging. A coalition of talents and knowledge working together as a team is the only way to get the right answers to the big questions: What killed the victim? Was the fire accidental or deliberate? and How did those two events interact?

■ ■

Problems

7.1. Discuss the common problems and pitfalls associated with fatal fire investigations. Compare them with a recent ongoing case in the national media. What parallels can you find?

7.2. Referring to Example 7.1, determine the visibility of the same fire in an enclosed hallway measuring 3 × 15 m with a ceiling height of 2.5 m.

7.3. Referring to Example 7.2, calculate the fractional incapacitation dose for a male, a child, and an infant in the same scenario.

■ ■

Suggested Reading

Brogan, R. 2000. The Speed Street fire—Was it murder by—or prior to—the fire? *Fire and Arson Investigator* 50(July):17–21.

DeHaan, J. D. 2002. *Kirk's fire investigation,* 5th ed., chap. 15. Upper Saddle River, NJ: Prentice Hall.

DeHaan J. D., and F. L. Fisher. 2003. Reconstruction of a fatal fire in a parked motor vehicle. *Fire and Arson Investigator* 53(2): 42–46.

Levin, B. C., et al. 1990. Analysis of carboxyhemoglobin and hydrogen cyanide in blood from victims of the Dupont Plaza Hotel fire—Puerto Rico. *Journal of Forensic Science* 35(1)l:151–168.

Purser, D. A. 1995. Toxicity assessment of combustion products. In *SFPE handbook of fire protection engineering,* 2nd ed., chap. 2–8. Quincy, MA: SFPE/NFPA.

Society of Fire Protection Engineers. 2000. *Engineering guide to predicting 1st and 2nd degree skin burns*. Bethesda, MD: SFPE.

CHAPTER **8** ◆ # Fire Testing

Testing for fire reconstruction purposes can span the range from simple field tests requiring no equipment to extensively instrumented, full-scale test burns. Fire tests can help confirm or reject hypotheses about the fire's ignition or spread, test and validate the predictions of computational models, or those of experienced investigators, or demonstrate the roles of various factors in the fire and its effects on occupants. This chapter explores, in summary form, many of the tests useful in fire investigation.

◆ 8.1 ASTM TESTS

A number of test standards and methods applicable to fire investigation are listed in Table 1.3 (see Chapter 1). Several of the ASTM tests can be used to assess the ignitability of materials or the nature and speed of flame propagation, and they are discussed briefly here. However, as you will see, there are limitations to the application of results of these tests to an actual case. Many limitations have to do mostly with the geometry of the sample.

A flame will spread much more quickly upward than it will downward or outward in the same fuel. A fire started at the top edge of a sofa back will spread much more slowly than if the same fabric were ignited at the base of the sofa. A fire ignited at the top of hanging draperies will be much more likely to cause failure of the hooks and drapery rod with collapse of the draperies, possibly resulting in drop-down ignition of materials beneath, where draperies ignited at the bottom are often nearly completely consumed before they cause failure of their supports.

Fire resistance properties today are sometimes built into a combination of layers of fabrics and backings that result in acceptable ignition resistance. Separating and testing the layers apart from one another may yield very misleading results. The residual moisture in test samples can affect their ignitability and so specimens are often conditioned at 24–48 hr at specific conditions prior to testing.

FIGURE 8.1 ◆ Testing for fire reconstruction purposes can span the range from simple field tests requiring no equipment to extensively instrumented full-scale test burns. *Courtesy of Michael Dalton, Knox County Sheriff's Office, by permission.*

Most ASTM standard methods include an advisory comment regarding use of the method to compare materials or their performance under controlled conditions to assess the possible contributions a material might make in a real fire and not to use results to predict what a product *will* do when exposed to a real fire. Some of the tests reference other entities such as the National Fire Protection Association (NFPA) and the U.S. Consumer Product Safety Commission (CPSC).

- ◆ **Flammability of Clothing Textiles (Title 16 CFR 1610-U.S.).** This test requires that a piece of fabric that is placed in a holder at a 45° angle and exposed to a flame for 1 s not ignite and spread flame up the length of the sample in less than 3.5 s for smooth fabrics or 4.0 s for napped fabrics.
- ◆ **ASTM D1230, Standard Test Method for Flammability of Apparel Textiles (not identical to 16 CFR 1610).** CPSC requires fabrics that are introduced into commerce to meet requirements of 16 CFR 1610. This test is suitable for textile fabrics as they reach the consumer or for apparel other than children's sleepwear or special protective garments. A sample of fabric 2 × 6 in. (50 × 150 mm) is held at a 45° angle in a metal holder and a controlled flame is applied to the bottom end for 1 s. The time required for the flame to proceed up the fabric a distance of 5 in. (127 mm) is recorded. This test is reportedly less expensive and time-consuming than 16 CFR 1610.
- ◆ **Flammability of Vinyl Plastic Film (Title 16 CFR 1611-U.S.).** This test requires that vinyl plastic film (for wearing apparel) placed in a holder at a 45° angle and ignited not, to burn at a rate exceeding 1.2 in. per second.

- **Flammability of Carpets and Rugs, Title 16 CFR 1630 (Large Carpets) and 16 CFR 1631 (Small Carpets).** A specimen of carpet is placed under a steel plate with an 8-in. (20-cm)-diameter circular hole. A methenamine tablet is placed in the center of the hole and ignited. The duration and heat release rate of the tablet duplicate those of a typical dropped match. If the specimen chars in any direction more than 3 in., it is considered a "fail." Note that this test is the standard one for "approval" of carpets for normal occupancy use and that the test is not a severe one. The ambient radiant heat flux is minimal because the sample is tested at room temperature. The ignition source is a modest 50- to 80-W flame of brief duration. Testing has shown that carpets that pass this test will be readily ignited and spread flames if the radiant heat flux is larger as a result of the ignition source being larger and more prolonged. The carpet is tested in the flat, horizontal position so the same carpet mounted vertically may behave very differently.

- **ASTM D2859, Standard Test Method for Ignition Characteristics of Finished Textile Floor Covering Materials.** This method uses a steel plate 9 in. square and 0.25 in. thick with an 8-in.-diameter circular hole. A methenamine tablet is placed in the center and ignited in a draft-free enclosure. Samples are thoroughly oven-dried and cooled in a desiccator before being tested. The product fails if charred portion reaches within 1 in. (25 mm) of the edge of the hole in the steel plate. Eight specimens are tested. The Flammable Fabrics Act (FFA) regulations require that at least seven of the eight specimens pass this test.

- **Flammability of Mattresses and Pads (16 CFR 1632-U.S.).** A minimum of nine regular tobacco cigarettes is burned at various locations on the bare mattress—quilted and smooth portions, tape-edge tufted pockets, and so forth. The char length of the mattress surface must not be more than 2 in. (50 mm) from any cigarette in any direction. The test is repeated with the cigarettes placed between two sheets covering the mattress.

- **Flammability of Children's Sleepwear (16 CFR 1615 and 1616-U.S.).** Each of five 35 × 10-in. specimens is suspended vertically in a holder in a cabinet and exposed to a small gas flame along its bottom edge for 3 s. The specimens cannot have an average char length of more than 7 in., no single specimen can have a char length of 10 in. (full burn), and no single sample can have flaming material on the bottom of the cabinet 10 s after the ignition source is removed. This is required for finished items (as produced or after one washing and drying) and after the items have been washed and dried 50 times.

- **ASTM E1352, Standard Test Method for Cigarette Ignition Resistance of Mock-Up Upholstered Furniture Assemblies.** This test uses reduced-scale mock-ups (18 × 22 in. [0.45 × 0.57 m]) of plywood, upholstered with simulations of seat, backrest, and armrests to test ignitability of furniture to dropped, smoldering cigarettes. The test is used for furniture to be used in public occupancies, nursing homes, hospitals, and the like. The cigarettes are positioned on each of the various surfaces and along crevices between seats, armrest, and back. The distance of char extension from each cigarette or ignition by open flame is recorded for comparison.

- **ASTM E1353, Standard Test Method for Cigarette Ignition Resistance of Components of Upholstered Furniture.** This test uses mock-ups to test individual components—cover fabrics, welt cords, interior fabrics, and filling or batting materials in the form, geometry, and combination in which they are used in real furniture. This test uses single cigarettes but each is covered with a single layer of cotton sheeting that retains more heat and makes it a more severe test than open-air placement.

- **ASTM E648, Standard Test Method for Critical Radiant Flux of Floor-Covering Systems Using a Radiant Heat Energy Source.** This test tests a horizontally mounted floor covering specimen (8 × 39 in. [0.2 × 1.0 m]) with a gas-fired radiant heater mounted at a 30° angle above it. This produces a radiant heat flux of 1–10 kW/m^2. This test uses a gas pilot burner. The distance charred along the sample indicates the minimum radiant heat flux for ignition and propagation.

The ASTM's (1999a) *Fire Test Standards,* 5th ed., lists methods for testing other materials per the ASTM Committee E-5 on Fire Standards.

- **ASTM D1929, Standard Test Method for Determining Ignition Temperature of Plastics (ISO 871).** This test uses a cylindrical hot-air furnace to heat the test sample in a pan. A thermocouple monitors the temperature of the sample while the temperature is adjusted so that pyrolysis gases venting through the top of the chamber can be ignited by a small pilot flame held near it. This establishes the flash ignition temperature (FIT). The same apparatus can be used to establish the spontaneous ignition temperature (SIT) by eliminating the pilot flame and observing the sample visually for flaming or glowing combustion (or a rapid rise in the sample temperature). SITs for common plastics are 20–50°C higher than FITs by this technique. The test requires 3 g of material for each test in the form of pellets, powder, or cut-up solids or films.

- **ASTM E659, Standard Test Method for Autoignition Temperature of Liquid Chemicals.** A small sample (100 μl) of liquid is injected into the mouth of a glass flask heated to a predetermined temperature and the flask is observed for the presence of a flash flame inside the flask and a sudden rise in internal temperature (as monitored by an internal thermocouple). If no ignition is observed, the temperature is increased and the test repeated until the material ignites reliably. This method establishes the "hot flame" autoignition temperature (AIT) in air. Ignition delay times may also be recorded. The temperature at which small sharp rises occur in the internal temperature alone is the "cool flame" autoignition temperature.

  The method can also be used for solid fuels that melt and vaporize or sublime completely at the test temperatures, leaving no solid residues. The test conditions are controlled by the heat transfer between the glass flask and the fuel introduced and the confined geometry of the spherical flask. In real-world ignitions, the nature of the surface and the manner of contact will determine the convective heat transfer coefficient and may modify times or temperatures. Any geometry (tube or enclosure) that keeps the fuel in contact with the hot surface is going to produce ignitions at lower temperatures than an open, flat surface where buoyancy of the heated vapors can take the fuel away from the heated surface.

- **ASTM D3675, Standard Test Method for Surface Flammability of Flexible Cellular Materials Using a Radiant Heat Energy Source.** This method uses a gas-fired radiant heat panel (12 × 18 in. [0.3 × 0.46 m]) in front of an inclined 6 × 18-in. (0.15 × (0.46-m) specimen of material, oriented so that ignition (with a pilot flame) first occurs at the upper edge of the sample and the flame front moves downward. The rate at which the flame moves downward is observed and a flame spread index is calculated. The test is suitable for any material that may be exposed to fire. At least four specimens must be tested.

- **ASTM D3659, Standard Test Method for Flammability of Apparel Fabrics by Semi-Restraint Method.** This test simulates the burning characteristics of a garment hanging vertically from the shoulders of a wearer. Test specimens are 6 × 15 in. (0.15 × 0.38 m) (five specimens required) and are oven-dried and weighed prior to testing. Hung vertically from a cross bar, the flame of a small gas burner is positioned against the bottom edge of the fabric for 3 s and then removed. The weight (percentage area) destroyed by the flame and the time required are the criteria for comparison. (Cross-reference to FF 3-71, Flammability of Children's Sleepwear, Sizes 0–6X.)

- **ASTM E84, Standard Test Method for Surface Burning Characteristics of Building Materials.** This is the "Steiner Tunnel" test, which mounts specimens on the underside of the top of an insulated tunnel 25 ft (7.6 m) long, 12 in. (0.15 m) high, and 17.75 in. (0.45 m) wide. A gas burner at one end ignites the sample and the rate of flame spread along the length of the specimen is observed and recorded. The test specimen shall be at least 20 in. (0.51 m) × 24 ft (7.27 m). An optical density measurement system (source and photocell) is mounted in the vent pipe of the apparatus. Samples of products of combustion can be taken downstream of the photometer. The "upside-down" configuration of this test apparatus is not often found in real-world situations except for combustible ceiling coverings.

- **ASTM E1321, Standard Test Method for Determining Material Ignition and Flame Spread Properties.** This method tests for the ignition and flame spread properties of a vertically oriented fuel surface when exposed to an external radiant heat flux. The results can be used to calculate the minimum radiant heat flux and temperature needed for ignition and flame spread. It uses a large, vertically mounted specimen and a gas-fired radiant panel heater mounted at an angle to it. A gas pilot flame is used, and the rate and distance of flame spread along the length of the specimen are recorded.

- **ASTM E800, Standard Guide for Measurement of Gases Present or Generated during Fires.** This guide describes methods for properly collecting and preserving combustion gas samples during fire tests and analytical methods for O_2, CO, CO_2, N_2, HCl, HCN, and oxides of nitrogen and sulfur, using gas chromatography, infrared, or wet chemical methods. As we have seen, some fuels produce high concentrations of toxic or irritant gases under various fire conditions. Testing materials known to be present in a fatal fire may yield clues as to what role their combustion gases played in the fatality.

◆ 8.3 FLASH AND FIRE POINTS OF LIQUIDS

Tests for flash and fire (flame) points of ignitable liquids can be as simple as placing a few drops of the fuel in a petri dish or watch glass at room temperature and waving a lighted match near the surface of the liquid.

Ignition will confirm that the material is a Class I A or B Flammable Liquid with a flash point below room temperature 23°C (75°F). More reliable testing requires more controlled evaporation and confinement of the vapors with a more reproducible ignition source. These tests are described briefly below.

Flash point testing in general involves taking a quantity of liquid fuel and placing it in the cup of a water bath whose temperature is gradually raised. The cup may be open to the atmosphere or closed with a small shutter. A very small flame is introduced periodically and the cup is observed for the presence of a brief flash of flame. The fire point of many liquids may be established by the open-cup method by increasing the temperature until a self-sustaining flame is established rather than just a "flash" of flame.

Closed-cup testers retain some of the vapor produced and make it easier to ignite. Closed-cup flash points are typically a few degrees lower than open-cup determinations of the same fuel.

- **ASTM D56, Standard Test Method for Flash Point by Tag Closed Tester.** Applicable to low-viscosity liquids with flash points below 93°C (200°F). Requires 50 ml of liquid for each test. (Sometimes abbreviated TCC.)

- **ASTM D92, Standard Test Method for Flash and Fire Points by Cleveland Open Corp.** Applicable to all petroleum products with flash points above 79°C (175°F) and below

400°C (752°F) except fuel oils. Requires at least 70 ml for each test. (Sometimes abbreviated COC.)

- ◆ **ASTM D93, Standard Test Methods for Flash Point by Pensky–Martens Closed-Cup Tester.** Applicable for petroleum products with a flash point in the range of 40–360°C (104–680°F) including fuel oils, lubricating oils, suspensions, and higher-viscosity liquids. Requires at least 75 ml of fuel for each test.

- ◆ **ASTM D1310, Standard Test Method for Flash Point and Fire Point of Liquids by Tag Open-Cup (TOC) Apparatus.** Applicable for liquids with flash points between 0 and 325°F (-18 to 165°C) and fire points up to 325°F (165°C) (note: will work on subambient temperatures). Fire point criterion: when fuel ignites and burns for at least 5 s. Requires 75 ml of sample for each test.

- ◆ **ASTM D3278, Standard Test Method for Flash Points of Liquids by Small-Scale, Closed-Cup Apparatus.** Suitable for flash point determination of fuels with a flash point between 0 and 110°C (32 to 230°F) in small quantities (2 ml required for each test). Employs a microscale apparatus sold as a Setaflash tester. Test methods similar to those for ISO 3679 and 3680. Note: will work on subambient determinations.

- ◆ **ASTM D3828, Standard Test Method for Flash Point by Small-Scale Closed Tester.** Similar to D3278. Used to establish whether a product will flash at a given temperature, using 2–4 ml of sample for each test.

Due to the multiplicity of flash point test methods, the ASTM (2000) offers a *Standard Test Method for Selection and Use of ASTM Standards for the Determination of Flash Points of Chemicals by Closed-Cup Methods* (ASTM E502-84).

◆ **8.4 CALORIMETRY**

Classical ***bomb calorimetry*** is used to measure the total heat of combustion of fuels (as in ASTM D2382). The term *bomb* refers to a sealed vessel that can withstand internal pressures. This involves burning a weighed specimen in a sealed container with an excess of oxygen until it is completely oxidized. The heat generated by the combustion is measured by measuring the increase in temperature of the sealed "bomb" and using its specific heat and mass, calculating the heat released.

This test is obviously not practical for the large fuel packages involved in most fires and yields the total heat of combustion, not the effective heat of combustion (which allows for thermal losses within the material from pyrolysis and gasification).

Because almost all common fuels yield nearly the same amount of heat per weight of oxygen consumed (13 kJ/g), if one were to measure the amount of oxygen consumed in the combustion process, the heat generated could be easily calculated. This is called ***oxygen consumption*** (or depletion) ***calorimetry***. If a specimen is burned in air and the waste products are all drawn into a system of ductwork, the oxygen, CO, and CO_2 concentrations can be measured. If the ventilation rate into the test chamber and through the exhaust duct is measured, the amount of oxygen consumption can be measured. Optical sensors in the ductwork can also monitor the optical density and obscuration of smoke products generated. Note that both the oxygen consumption and the heat release rate are calculated.

This approach has been refined into three general categories of size and application: cone, furniture, and room calorimetry. In the cone calorimeter, developed by Babrauskas at NIST and described in ASTM E1354, a 4 × 4-in. (10 × 10-cm) specimen in a metal

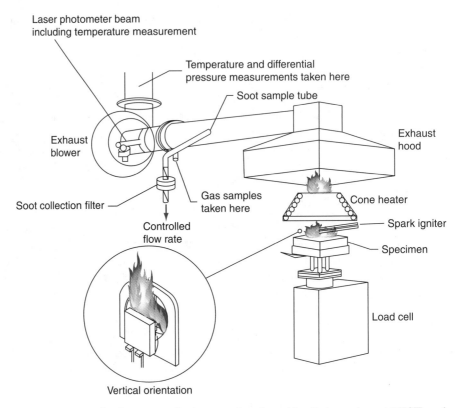

FIGURE 8.2 ◆ In the cone calorimeter, developed by Babrauskas at NIST and described in ASTM E1354, a 4 × 4-in. (10 × 10-cm) specimen in a metal tray is exposed to a uniform incident radiant heat flux from a cone-shaped electric heater mounted above it. *Courtesy of Dr. Vytenis Babrauskas, Fire Science & Technology, Issaquah, Washington, by permission.*

tray is exposed to a uniform incident radiant heat flux from a cone-shaped electric heater mounted above it (as in Figure 8.2).

The radiant flux striking the sample can be controlled by adjusting the temperature of the heater element. A small electric arc source is introduced to ignite the plume of smoke generated by the heating and then removed. The flame and smoke vent upward through the center of the cone heater and the gases are ducted to where the flow rate, oxygen, CO, and CO_2 concentrations and smoke obscuration are measured. The entire specimen tray is mounted on a sensitive electronic balance, so the mass loss rate is continually calculated. The analysis calculates the heat release rate (total and per unit surface area of fuel), mass loss rate, and effective heat of combustion. Even though the samples tested are reduced scale, testing has shown that the results have a high degree of correlation to real-world fire performance (Babrauskas 1997). The system will work with liquid or solid fuels (even low-melting point thermoplastic materials) in the horizontal configuration. Wood and other rigid fuels can also be tested in the vertical configuration.

Furniture calorimeters were developed to test the heat release rate of single items of real furniture in full scale by the same method. Here, the furniture is mounted on a load cell and burned in the open, with all combustion products drawn into a vent hood.

The measurement of flow rate, oxygen, CO, and CO_2 concentrations is carried out in the ductwork. The calculated heat release rate can then be combined with the mass loss data from the load cell to calculate the effective heat of combustion.

When multiple pieces of furniture, along with carpet, wall linings, and other fuels, are to be evaluated as a fuel package, the ***room calorimeter*** is used. In this case a room is built (often to a standard size of $10 \times 12 \times 8$ ft [$3 \times 3.9 \times 2.4$ m]) and the entire room becomes the "collector" for the exhaust vent. The combustion products vent out through the door opening and are drawn up into the exhaust hood and measured as before. This arrangement allows for the measurement of heat release rate constantly as fire spreads from one item to another, even culminating in near-flashover conditions.

◆ 8.5 FURNISHINGS

Tests for ignitability of upholstered furniture vary widely, from cigarette ignition to small flame to larger gas-fired burners of known heat output (typically 17–40 kW). They may be applicable to standardized, small-scale mock-ups of furnishings to the actual items taken from the production line. Pass/fail criteria may be based on observable flame spread, penetration, percentage mass loss, peak heat release rate, quantity of smoke produced, or toxicity of fire gases.

TEST METHODS

Various test methods in use today include those described under the U.S. Code of Federal Regulations (CFR), British Standards (BS), International Standards Organization (ISO), ASTM, and California Bureau of Home Furnishings Technical Bulletins (TB). Krasny, Parker, and Babrauskas (2001) have assembled a comprehensive description and discussion of the current techniques, summarized in Table 8.1. See also *Kirk's Fire Investigation,* 5th ed., for additional information (DeHaan 2002).

As with the bench-scale tests described earlier, there are limitations and cautions regarding applying the results of these tests to the reconstruction of "real" fires. Cautions include requirements for conditioning samples for 24–48 hr at a particular temperature and humidity. The investigator should be aware of what influence such variables might have on the final result. The geometry of the test compartment is also important. The same piece of furniture may yield different heat outputs or fire patterns if tested in a corner or against a wall as opposed to its behavior if tested in the center of a large room that minimizes radiant feedback or ventilation effects.

The location, manner, means, and duration of ignition can play a significant role in the fire performance of furnishings. A vinyl-covered chair seat, for instance, provides a long-lasting barrier if an ignition source is placed on the seat. (PVC upholstery acts like an elastomer and tends to form a thick char.) But even a small flaming source placed under the seat where the flame can contact the polyurethane padding will readily ignite the chair.

CIGARETTE IGNITION OF UPHOLSTERED FURNITURE

Cigarette ignition of upholstered furniture was a major fire cause for decades due to the predominant use of cotton or linen upholstery fabrics over cellulosic padding of cotton, coir, kapok, or similar materials, all of which were easily ignited by a glowing cigarette.

TABLE 8.1 ◆ Summary of Furniture Item Flammability Tests

Test	Cigarette Ignition	Flame Ignition	HRR	Smoke	Toxicity
ISO 8191	X	X			
BS 5852	X	X			
ATSM E1352	X				
ASTM E1353	X				
NFPA 261	X				
NFPA 260	X				
Calif. TB116, 117	X				
Upholstered Furniture Act Council	X				
BIFMA	X	X			
CFR 1632	X				
BS 6807	X	X			
Calif. TB 133		X	X	X	X
Calif. TB 129		X	X	X	X
Calif. TB 121		X		X	X
ASTM E1537		X		X	X
ASTM E1590		X		X	X
ASTM E162		X			
ASTM D3675		X			
ISO 5660			X	X	X
ASTM E1474			X	X	X
ASTM E1354			X	X	X
NFPA 264A			X	X	X
ASTM F1550M				X	
ISO TR 5924				X	
ASTM E662				X	
ASTM E906			X	X	
ISO TR 9122					X
ISO TR 6543					X
ASTM E1678					X

Source: Derived from Krasny, Parker, and Babrauskas (2001).

Holleyhead (1999) has published an extensive review of the literature regarding cigarette ignition of furniture. A great deal of misinformation once existed, however, about the time required for such ignition to result in flaming combustion. Investigators erroneously concluded that such ignitions always required 1–2 hr, inferring that any flaming ignition in a shorter time must have been the result of deliberate ignition by open flame.

Extensive testing by the California Bureau of Home Furnishings and NIST (then NBS), first reported in the early 1980s, demonstrated that the process was variable in time requirements. Transition to flaming combustion was reported in as little as 22 min from the placement of a glowing cigarette between the cushions or between the cushions and an arm or backrest. Others required 1 to 3 hr, while some never transitioned to open flames and smoldered for hours before self-extinguishing. Even identically built mock-ups demonstrated a wide variety of behaviors.

Tests by the authors have demonstrated similar time and ignition results. If the fabric is torn and the cigarette drops into direct contact with a non-flame retardant-treated cotton padding, transition to flame has been observed in as little as 18 min in still air. The presence of drafts can influence the progress and increase the heat release rate of smoldering cellulosics. One informal test by the author with a cigarette wrapped loosely in a cotton towel and left outside in a light, variable wind resulted in flames about 15 min after placement. Hand-rolled cigarettes present a lower ignition risk because they tend to self-extinguish much more than do commercially made tobacco cigarettes. The mechanism for transition from smoldering to flame has so far defied modeling and accurate predictions.

Latex (natural) rubber foam has also been ignited by prolonged contact with a glowing cigarette. It should be noted that the cigarette itself will be consumed in a maximum of 20–25 min after being lit. Ignition times longer than that obviously result from a self-sustaining smolder in the surrounding fuel being initiated by the cigarette. This is confirmed by the significant quantities of white smoke being produced for some minutes prior to flame (far more than can arise from a single cigarette).

◆ 8.6 SCALABLE TESTS

A great deal of useful information can be obtained from building scaled-down replicas of rooms or furnishings for testing, at reduced cost and complexity compared to full-scale re-creations. These are particularly valuable for testing ignition and incipient fire hypotheses (where the interaction of room surfaces or conditions does not play a major role).

Where radiant heat or ventilation factors are involved, the test must take into account that all factors in a fire do not scale down in the same linear fashion as the room dimensions. For instance, ventilation through an opening is controlled by the area of the opening and the square root of its height. It is possible to calculate corrections for such factors, but it is not as simple as building a doll house, with all doors and windows in the same proportion as in full scale. Radiant heat flux falls off with the inverse square $(1/r^2)$ of the separation distance so scale models must keep those relationships in mind as well.

Floor or wall materials reduced in thickness to scale proportions may respond to thermal insult as "thermally thin" targets if their thickness is less than 1 mm and may respond differently than "thick" sections of the same material. Speed of smoke movement and time to flashover will also not be "to scale."

FLUID TANKS

A great deal of the information used to create and test the models of smoke and hot gas movement discussed in previous chapters was developed in *fluid tanks* with no fire or smoke at all. Since movement of smoke in a building is driven by the buoyancy of the hot gases versus the density of normal room air, the same process can be simulated by using two liquids of different densities (such as fresh water and salt water). With one fluid dyed, its movement and mixing can be easily observed and recorded.

Often a scale (one-eighth to one-fourth) model of the room(s) is built of clear Plexiglas and immersed and inverted in a large tank of the lighter (less dense) liquid. The heavier liquid is then introduced into the model in a manner to simulate the generation of the gas from a fire in the room. Scaling factors are important and the relative densities and viscosities of the liquids have to be calibrated to simulate the mixing of gases in a buoyant flow, but the technique can provide answers to some investigative problems and confirm or reject computational models and hypotheses (Fleischmann, Pagni, and Williamson 1994).

FIELD TESTS

Identification of the first fuel ignited is critical to the reconstruction process. If the first fuel is susceptible to open flame ignition and not to glowing or hot surface ignition (such as natural gas, gasoline vapors, or polyethylene plastic), that helps focus the scene search on possible ignition sources of the appropriate type. Cellulosic fuels or other natural materials such as wool are much more easily ignited by hot surface sources like discarded cigarettes or glowing electrical connections. Most such materials can be reliably identified by visual inspection, but information as to their behavior when ignited is much more "on point" for the fire investigator.

A simple *flame* or *ignition susceptibility test* using a match or lighter applied to one corner while the small test sample is held in still air will reveal how readily it will ignite and whether it will support flaming combustion when the ignition source is removed as described in *NFPA 705* (NFPA 1997). Under these conditions, cellulosic materials will support a yellow flame with a gray smoke. When blown out, they tend to support glowing combustion. The ash left behind will be gray to black in color and powdery or crumbly in texture. There will be no hard-melted droplet of residue.

Thermoplastic synthetic materials will melt as they burn, so melted, burning droplets will result. They also tend to shrink and shrivel as they melt and often burn with a blue-based flame, with smoke ranging from negligible to white (polyethylene) to heavy and inky black (polystyrene). They do not support smoldering combustion.

Thermosetting resins are usually more difficult to ignite than other fuels. They tend to smolder when flame is removed, produce an aggressive smoke, and leave a hard, semiporous residue. Elastomers (rubbers) can be either natural (latex) or synthetic and can behave like either. Some will burn very readily; others like silicone rubber burn much more reluctantly. Silicone rubber leaves a brilliant white, powdery ash where others leave a hard, dark porous mass. Samples of any unknown material suspected of being the first fuel ignited should be taken if there is any doubt of their contribution to the ignition or growth of the fire.

Caution should be exercised in interpretation of such informal ignition tests. The presence of fire retardants may significantly affect ignition and flame spread. It should also be noted that materials like carpet are much more easily ignited from a corner or edge than if the same ignition source were applied to the center of the specimen. Ignition and flame spread are enhanced by vertical sample orientation and may be retarded by draft or moisture in the sample, so the results should be noted but not taken as proof of identity (NFPA 1997).

FULL-SCALE FIRE TESTS

Considerable knowledge has been gathered from fire tests conducted in real buildings slated for demolition. It is rare that the dimensions, ventilation, and construction materials of a test building will match those conditions of a specific fire that is being studied, but a great deal of reliable knowledge about fire behavior has been gleaned from these tests. Aside from physical suitability, there are often issues of fire exposure to surrounding properties, environmental concerns, logistics of access, and limitation of fire protection services. If such buildings are used, provisions should be made for multiple videocamera viewports and the inclusion of thermocouple and radiometers as described below.

One author found a house nearly identical to one in which six fire fatalities occurred. The test house had been scheduled for demolition. It was repaired and refurnished to duplicate the initial fire scene. Thermocouples, gas analyzers, and videocameras were used to capture fire tests with accelerated and nonaccelerated ignitions. The test data revealed that a nonaccelerated fire could have been responsible (DeHaan 1992).

Because of a number of training-fire deaths in the United States, fire departments have been considerably more reluctant to expose their personnel to the risks of large fires in original buildings. The majority of fire personnel recognize the value of training fire-fighters in such structures and will plan "training burns" in accordance with the safety provisions outlined in *NFPA 1403—Standard on Live Fire Training Evolutions* (NFPA 2002). The investigator should be aware that those provisions include prefire venting of ceilings and roofs, removal of most doors and all glass, and limited fuel loads (no ignitable liquids or gases). The resulting fire behavior and, most importantly, fire patterns therefore do not generally conform to those features of fires in buildings with normal furnishings, doors, glass windows, and intact ceilings and roofs. It is suggested that fire investigators not attempt to use such firefighting exercises to gather behavior/pattern data.

Full-scale rooms that duplicate particular conditions and control other variables can be purposely built for fire tests. The criteria for room fire experiments are detailed in ASTM E603-01, *Standard Guide for Room Fire Experiments* (ASTM 2003). This guide addresses assembling tests that will be used to evaluate the fire response of materials, assemblies, or room contents in real fire situations that cannot be evaluated in small-scale tests. Provisions for measuring the optical density of smoke, temperatures, and heat fluxes in the compartment are described. The documentation and controls necessary are also described.

◆ 8.7 CASE EXAMPLE 1—CUBICLE CONSTRUCTION

Effective room mock-up cubicles can be assembled at low cost for full-scale tests. Based on a design by Mark Wallace, these are basically four 8×8-ft (2.43×2.43-m) wood-framed panels (typically wood studs on 2-ft centers) with gypsum wallboard with a similar panel resting on top to form a ceiling (see Figure 8.3).

A larger unit can be constructed easily, with the plans shown in Figure 8.4. The design provides for a "knock-down" wall that can be removed easily after the test to facilitate documentation.

Doors and windows can be cut as needed. Be sure that a header is left above each door (10–18 in. deep). Cubicles are easily erected on four wood pallets covered with 0.5-in. sheet plywood or on 2×4- or 4×4-in. joists. Since these are intended for short-duration (<30-min) tests, lack of insulation will not usually affect the results of interior fires. Electrical outlets can be added easily.

Viewports are easily added by cutting holes of approximately 10×12 in. in one or more walls near floor level. An unframed piece of ordinary window glass is then glued to the *interior* surface of the wall using silicone sealant (allowing 24 h to harden before the fire). The absence of a frame means that the entire sheet of glass is exposed to the same heat flux and will not develop the same stresses that cause failure before flashover.

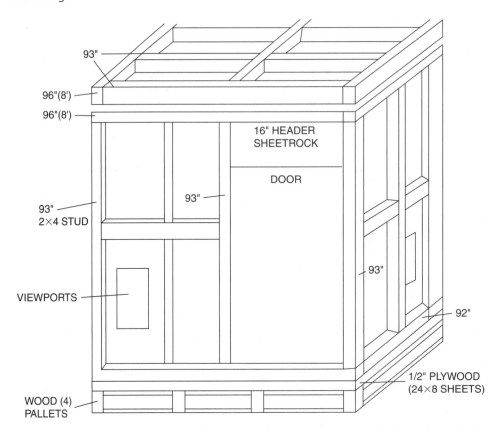

FIGURE 8.3 ◆ Effective room mock-up "cubicles" can be assembled at a low cost for full-scale tests. *Courtesy of John DeHaan.*

Experience reveals that such windows will usually last until flashover. Heat-resistant glass such as that used in fireplace screens can ensure integrity postflashover if that is desired.

Thermocouples can be easily added through small holes drilled through the gypsum where desired. A minimum of three thermocouples on one wall is suggested: one near the ceiling, one midlevel, and one approximately 6 in. above the floor at a point well away from the door or other vents. A second set can be installed on the opposite side of the room or additional thermocouples can be placed inside the door header (to record flameover) or above target fuel packages. Data are best captured on a datalogger (such as an Omega 550), although they can also be captured on a videocamera monitoring digital outputs of several individual pyrometers at once for later manual transcription.

Such cubicles allow creation of roomlike environments that closely simulate real rooms (additional 4 × 8-ft segments can be added for larger rooms) while keeping safety risks to a minimum (no risks of toxics, asbestos, structural collapse, etc.). One entry wall can be framed separately so it can be easily opened and laid flat on the ground after the fire for ease of examination and photography. Video recordings should be made of all tests using multiple cameras and synchronized time clocks where possible. A visual and audio cue should accompany ignition so that it marks $t = 0$ on all recordings simultaneously. It should be remembered that the small size (2.4 × 2.4 m) will produce flashover in less time (approximately 30 percent less) than a typical 3 × 4-m room given the same fuel load.

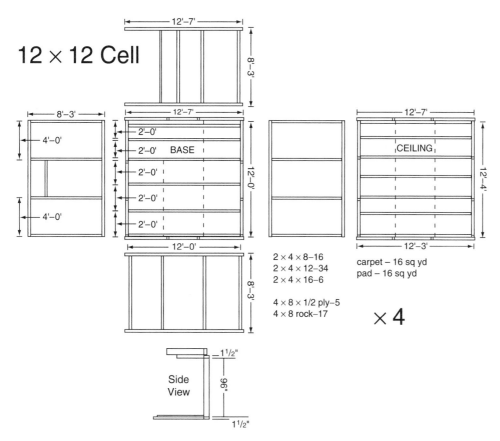

FIGURE 8.4 ◆ Example plans for a larger test cubicle unit that can be constructed easily. The design provides for a "knock-down" wall that can be removed easily after the test to facilitate documentation. *Courtesy of Mike O'Brien, Chief, Suisun City Fire Department, Suisun City, California, by permission.*

Scientifically valid, small-scale tests can be conducted to test flame spread ignitability without special equipment as long as possible roles of the test variables are taken into account. For instance, carpet or furniture can be tested in the open, with the realization that it would burn more quickly (producing greater heat release rates) if tested in a compartment. This is due to the radiant heat reflected back from walls and ceiling.

Carpets need to be secured to the floor in the same manner as in the actual structure. Synthetic carpets shrink and curl as they burn. If this lifts the edges, the combustion of the carpet will be seriously enhanced and distort the heat release rate and fire patterns. Carpets must be backed with the same pad as during the fire being duplicated, since interaction between the pad and the carpet can cause self-sustaining combustion where neither material might sustain flame spread by itself.

The ignition source used should duplicate the one known to be involved or various kinds should be tested. For instance, polypropylene carpet will resist ignition by a single dropped match at room temperature, as it will pass the required methenamine tablet test; however, a slightly larger ignition source such as a burning crumpled sheet of newspaper or any additional external heat flux and the carpet will self-sustain combustion. Once ignited, such carpet will sustain combustion at rates up to approximately 0.5–1 m^2/hr with small flames of about 2–3 in. (5–7 cm) tall. (Figures 8.5(a), (b), (c))

(a)

FIGURE 8.5 ◆ (a) Interior of 2.4 × 2.4-m cubicle after 20-min fire. Bed on right was ignited by open flame to clothing on top and required nearly 16 min to be completely involved. The room went to flashover less than 1 min later, igniting carpet and chair in left-rear corner (with pig haunch and legs). Note the extensive combustion of the corner of the bed nearest the door (ventilation effect). Glass viewport in rear corner failed prior to flashover. *Courtesy of J. D. DeHaan.*

(b)

FIGURE 8.5 ◆   (b) View from doorway of same compartment
after removal of bed, chair, and remaining carpet from left side of
compartment. Note exposed areas of carpet and 0.5-in. plywood
floor consumed in less than 4 min of postflashover fire and top-
down burning of 2×4 wood floor joists. No accelerants were
used. *Courtesy of J. D. DeHaan.*

If a fire test is conducted in a compartment, care must be taken that the major fuel
packages are located in the same position in each test (or in the same position as in
the scene being replicated). Packages in corners will burn differently than packages
against the wall or in the middle of the room. A large fire located away from an open
door will burn differently than the same fire set near the door (Figure 8.5).

(c)

FIGURE 8.5 ◆ (c) Results of postflashover fire in furnished 2.4 × 2.4-m office cubicle. Note extensive destruction of synthetic carpet and combustion of drywall paper on right. Glass viewport in right-rear corner failed prior to flashover. Note clean-burn ventilation pattern extending up the rear wall and extensive damage to carpet beneath the opening. There was only carpet and pad in the right-rear corner. *Courtesy of John DeHaan. Tests courtesy of California Association of Criminalists and Huntington Beach Fire Department.*

Examples of the valuable data that can be obtained from cubicle tests were obtained in the Iowa cubicle tests, conducted September 25, 2002, in Waterloo. In these tests, the Iowa Chapter of the International Association of Arson Investigators (IAAI) and the Bureau of Alcohol, Tobacco and Firearms(ATF) built two cubicles of nearly identical fuel loads.

The cubicles each contained a synthetic upholstered chair, a wood dresser, curtains, a kitchen chair, and synthetic carpet with polyurethane padding. The first cubicle was ignited by an open flame to the skirting of one chair. This fire required 10.5 min to grow to flashover. The second cubicle was ignited with gasoline across the middle of the floor, with flashover occurring in approximately 70 s. Both cubicles experienced the same maximum temperatures and same damage to the carpet and wood floor. The tests also served as a good example of physical reconstruction and data gathering.

Figures 8.6–8.16 detail the Iowa cubicle tests.

These photos are courtesy of Special Agent Mike Marquardt, CFI-ATF; the tests, courtesy of the Iowa Chapter of the IAAI. Data are courtesy of David Sheppard, a Fire Protection Engineer at ATF's Fire Research Laboratory, Ammendale, Md.

FIGURE 8.6 ◆ Prefire view of the furnished 2.4 × 2.4-m cubicle with drywall walls and cellings. The fire was ignited on the skirt of the chair on the left. *Courtesy of S/A Mike Marquardt, CFI, ATF, Grand Rapids, Mich., by permission.*

FIGURE 8.7 ◆ Postfire view after 13 min, which was postflashover for 5 min. *Courtesy of S/A Mike Marquardt, CFI, ATF, Grand Rapids, Mich., by permission.*

FIGURE 8.8 ◆ Pattern on floor showing intense charring of plywood and irregular destruction of carpet and pad, most extensive at the door. *Courtesy of S/A Mike Marquardt, CFI, ATF, Grand Rapids, Mich., by permission.*

FIGURE 8.9 ◆ Same cubicle with furniture replaced in its prefire positions. Note the floor-to-ceiling combustion direction patterns on second chair and dresser. *Courtesy of S/A Mike Marquardt, CFI, ATF, Grand Rapids, Mich., by permission.*

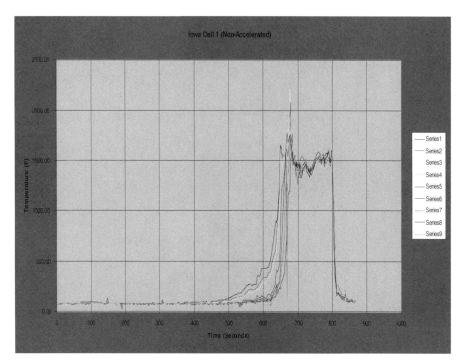

FIGURE 8.10 ◆ Temperature plot for the thermocouple tree in the center of the compartment at 0.3-m (1-ft) intervals. Note postflashover temperatures of approximately 816°C (1500°F). *Courtesy of S/A Mike Marquardt, CFI, ATF, Grand Rapids, Mich., by permission.*

FIGURE 8.11 ◆ Prefire views of cubicle furnished in the same fashion as previously. The thermocouple tree is in the center of the cubicle. *Courtesy of S/A Mike Marquardt, CFI, ATF, Grand Rapids, Mich., by permission.*

FIGURE 8.12 ◆ Cubicle fire approaches flashover temperature at 60 s. *Courtesy of S/A Mike Marquardt, CFI, ATF, Grand Rapids, Mich., by permission.*

FIGURE 8.13 ◆ Postfire view after 4 min of gasoline-accelerated fire. The ceiling and front drywall collapsed from fire-fighting action. Note protected areas on walls behind both large chairs. *Courtesy of S/A Mike Marquardt, CFI, ATF, Grand Rapids, Mich., by permission.*

FIGURE 8.14 ◆ Floor patterns indistinguishable from that of nonaccelerated fire. Note destruction of carpet and padding and charring of floor in areas toward door where no accelerant was poured. *Courtesy of S/A Mike Marquardt, CFI, ATF, Grand Rapids, Mich., by permission.*

FIGURE 8.15 ◆ Reconstruction of furniture placement. Note fire patterns on chairs outward from the rear center of the floor, where the accelerant was poured. *Courtesy of S/A Mike Marquardt, CFI, ATF, Grand Rapids, Mich., by permission.*

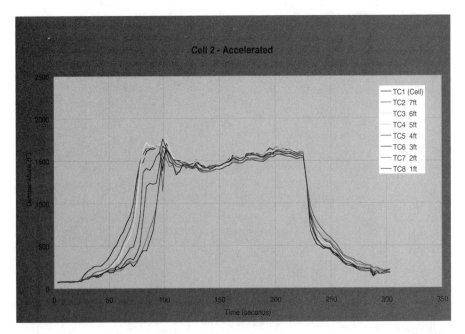

FIGURE 8.16 ◆ Temperature plot for thermocouple tree. Note ultrafast fire growth to flashover at 75–90 s. Postflashover period is approximately 130 s. *Courtesy of S/A Mike Marquardt, CFI, ATF, Grand Rapids, Mich., by permission.*

The most sophisticated (and most expensive) form of full-scale testing is the re-creation of an entire compartment in a laboratory with calorimetry, radiometry, continuous gas sampling, and video monitoring. An excellent example of this is the re-creation of the entire seating bay of the Stardust Disco, conducted at the Fire Research Station (*Anatomy of a Fire*—Video, [FRS 1982]).

The single-component or furniture mockup tests described previously are useful for gauging ignitability and fuel properties, but only a full-scale reconstruction (such as the Stardust test) can reveal the complex interactions between the growing fire and the various fuels that, in that case, resulted in rapid growth to flashover of a small initial fire.

There are very few facilities in the world that can provide the resources necessary. NIST, Factory Mutual Engineering, Underwriters Laboratories, California Bureau of Home Furnishings (Department of Consumer Affairs), Aberdeen Proving Grounds, and Southwest Research Institute also have large burn laboratory and testing facilities. These agencies have provided invaluable testing for many fire investigators over the years, often at no cost to public agencies.

The ATF Fire Research Laboratory (FRL), Annendale, Maryland, is a scientific fire research laboratory dedicated to supporting many research needs of the fire investigation community. It also houses ATF's National Forensic Laboratory. The FRL was designed by a team of fire scientists, engineers, and fire investigation specialists from NIST, Factory Mutual Engineering, Underwriters Laboratories, Hughes Associates, and the University of Maryland.

An example of a large-scale test, simplified for the purposes of a single demonstration, is shown in Figures 8.17, 8.18, 8.19, and 8.20, corresponding to 3, 8, 11, and

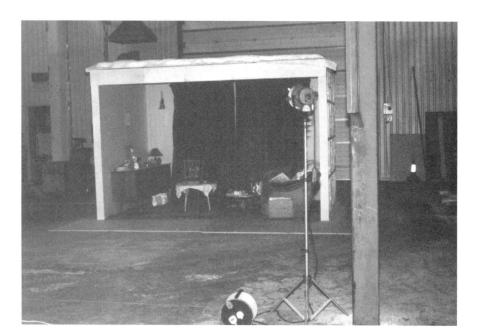

FIGURE 8.17 ◆ Full-scale living room fire test at 3 min after ignition, with only wastebasket and newspapers alight (incipient fire stage). *Courtesy of John DeHaan. Test courtesy of FRS, Building Research Establishment, Garston, U.K., by permission.*

FIGURE 8.18 ◆ Full-scale living room fire test at 8 min after ignition. Draperies have burned almost completely; drop-down ignites chair and table in far-left corner (growth phase). *Courtesy of John DeHaan. Test courtesy of FRS, Building Research Establishment, Garston, U.K., by permission.*

FIGURE 8.19 ◆ Full-scale living room fire test at 11 min after ignition. Fire is postflashover, with maximum heat release rate of 5.2 MW observed at 10.7 min. Carpet fully involved. *Courtesy of John DeHaan. Test courtesy of FRS, Building Research Establishment, Garston, U.K., by permission.*

FIGURE 8.20 ◆ Full-scale living room fire test at 14 min after ignition. Wood cabinet at left is alight. Filled with the remaining draperies, it will sustain a large fire that grows to 2 MW, then decays. Sofa and chairs are almost completely consumed (decay phase). *Courtesy of John DeHaan. Test courtesy of FRS, Building Research Establishment, Garston, U.K., by permission.*

14 min after initiation of the fire, respectively. The FRS staff assembled an 8.2 × 12.4 × 8-ft (2.5 × 3.75 × 2.4-m) re-creation of a living room (lounge) under the 9 × 9-m calorimeter in the Large Burn Hall at the FRS.

The structure was wood frame with ceramic fire insulation board liner (to permit reuse of the basic structure). The interior was lined with gypsum plasterboard, and the floor was covered with vinyl-backed carpet tiles (over a layer of sand to protect the concrete floor of the burn hall). There was a shallow (10-in. [0.25-m]) header across one side, which was otherwise left entirely open to simulate a "patio room" enclosure. This very large opening would provide enough air to ensure that the fire would not be ventilation limited even if it went to flashover. Using the relationship for maximum mass flow of air through an opening of area A_0 and height h_0,

$$\dot{m}_{\text{air}} = 0.5A_0\sqrt{h_0} = (0.5)(8.06)(1.46) = 5.88 \text{ kg/s}, \qquad (8.1)$$

and since 1 kg of air will support 3 MJ of heat release,

$$\dot{Q}_{\text{max}} = (3000 \text{ kJ/m}^3)(\dot{m}) = 17{,}640 \text{ kW}. \qquad (8.2)$$

The ventilation limit for this room would be of the order of 15–17 MW. This calculation assumes 100 percent efficiency. At 50 percent, $\dot{Q}_{\text{max}}$ would be 8825 kW.

The furnishings included non-flame-retardant draperies (no window), wood tables, chairs, and cabinets, a polyvinyl chloride bean bag chair, miscellaneous newspapers and books, and a three-seat sofa of recent production. This sofa was made using flame-retardant fabrics that delayed the growth of the fire substantially.

Three Chromel/Alumel thermocouples were placed in the room to monitor gas temperatures at ceiling and nose height. Data from those thermocouples were logged every second during the test. The fire gases were collected in the calorimeter hood above (not visible in Figure 8.17).

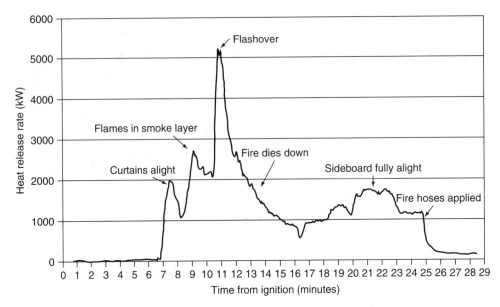

FIGURE 8.21 ◆ Heat release rate recorded in full-scale fire test. *Courtesy of FRS, Building Research Establishment, Garston, U.K., by permission.*

Heat release rate and CO, CO_2, and O_2 gas concentrations were monitored continuously in the calorimeter. Continuous video recordings were made and still photos were taken by observers at 30-s to 1-min intervals. Ignition was of the newspapers in the wastebasket at the far end of the sofa (near the bean bag chair), by open flame. The heat release rate of this test is shown in Figure 8.21, and the temperatures are shown in Figure 8.22. They can be compared to the development stages of the fire shown in Figures 8.17–8.20.

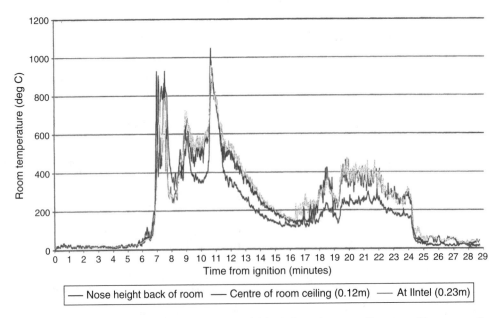

FIGURE 8.22 ◆ Recorded temperatures within full-scale room fire test. *Courtesy of FRS, Building Research Establishment, Garston, U.K., by permission.*

This fire was allowed to decay to 25 min and the remaining flames were then extinguished by water spray. Due to the sustained postflashover burning in the room, there were no reliable indicators of duration or direction of propagation remaining on visual inspection.

This demonstration showed that when a single major fuel package such as a living room sofa is flame retardant, the growth to flashover can be delayed considerably (since a 1990 vintage sofa would have promoted flashover in a similar room in less than 3 min if ignited by open flame in a similar fashion). Data from such tests can be used to confirm the accuracy of time/condition predictions of models, corroborate witness statements, demonstrate what stage of a fire might have incapacitated or killed a victim trapped in it, or demonstrate the role furnishings or fire protection systems play in preventing future fires.

Problems

8.1. Find an example of a legitimate fire testing program in your state. Visit or call this organization and determine what type of testing it performs.

8.2. Obtain a collection of at least 10 tests involving the use of a calorimeter.

8.3. Research the methods of conducting field fire testing.

Suggested Reading

DeHaan, J. D. 2001. Full-scale compartment fire tests. *CAC News* Second Quarter: 14–21.

DeHaan, J. D. 2002. *Kirk's fire investigation,* 5th ed., chap. 10. Upper Saddle River, N.J.: Prentice Hall.

Holleyhead, R. 1999. Ignition of solid materials and furniture by lighted cigarettes: A review. *Science and Justice* 39(2):75–102.

Krasny, J. F., W. J. Parker, and V. Babrauskas. 2001. *Fire behavior of upholstered furniture and mattresses.* New York: William Andrews.

National Fire Protection Association. 1997. *NFPA 705—Recommended practice for a field flame test for textiles and films.* Quincy, MA: NFPA.

National Fire Protection Association. 2000. *NFPA 555—Guide on methods for evaluating potential for room flashover.* Quincy, MA: NFPA.

Urbas, J. 1997. Use of modern test methods in fire engineering. *Fire and Arson Investigator* 47(December):12–15.

CHAPTER **9** **Case Studies**

Mr. Mac, the most practical thing that you ever did in your life would be to shut yourself up for three months and read twelve hours a day at the annals of crime. Everything comes in circles. . . . The old wheel turns, and the same spoke comes up. It's all been done before, and will be done again.

Sir Arthur Conan Doyle,
"The Valley of Fear"

The purpose of this chapter is to introduce the concept of detailed case histories as a learning tool. As with any science, particularly that of forensic fire scene reconstruction, documented case histories play a large role in introducing underlying concepts, expanding the knowledge of fire science, and suggesting new areas of research to explain the phenomena documented. Examples are offered where the data collected in controlled tests can be of value in formulating or testing hypotheses.

The cases presented here reflect many of the concepts raised previously in this textbook. While many principles are discussed, the case analyses are not all conclusive, leaving the reader to apply additional concepts in reviewing, discussing, or expanding upon the case histories.

◆ 9.1 CASE EXAMPLE 1—UNUSUAL FIRE DEATH

At approximately 2:30 P.M., the fire department responded to a "smoke showing" call from an apartment complex just 200 yards from their station. Discovering smoke coming from the closed door of one apartment and there being no response to a knock, the firefighters forced the door open. The movement of the door was limited by an object lying on the tile entry area behind it. A limited flaming fire was showing on this object.

The fire was quickly extinguished with a mist hose stream. No further fire damage was found to the rooms or their furnishings, as shown in Figure 9.1. The burning object turned out to be the body of the apartment's occupant, a middle-aged female (Figure 9.2).

The victim's body had suffered considerable fire damage to the upper-left torso, shoulder, and upper arm, with penetration through skin and subcutaneous body fat to underlying connective tissue in the affected areas. The body had apparently been lodged against the base of the wood entry door and an inverted "V" pattern of char

FIGURE 9.1 ◆ This fire was quickly extinguished with a mist hose stream. Soot deposits are visible on the walls and ceiling. No further fire damage was found to the rooms or their furnishings, except for limited charring to the inside of the door. *Courtesy of Santa Ana Fire Department, Santa Ana, Calif., by permission.*

FIGURE 9.2 ◆ The victim's body suffered considerable fire damage and was found lodged against the base of the wood entry door. An inverted "V" pattern of char extended from near the floor to some 3.3 ft (1 m) up the door. Smoke staining is noticeable surrounding the charred area. *Courtesy of Santa Ana Fire Department, Santa Ana, Calif., by permission.*

extended from near the floor to some 3.3 ft (1 m) up the door. Smoke staining surrounded the charred area. Smoke had permeated through the entire one-bedroom apartment and soot had settled on all horizontal surfaces and condensed on walls, ceilings, and cabinets, thus revealing their internal structures by differential condensation.

At postmortem exam, the victim was found to have been a nonsmoker in good health, with a COHb saturation of <3 percent but a blood alcohol concentration of 0.31 percent weight per volume. She was known to be an alcoholic and had been seen at 10:30 A.M. under the influence. She was wearing the remains of slacks and a loose top. Fragments of a cloth placemat were found under the body.

Due to the minimal firefighting efforts and absence of fire damage, a trail of burned cloth fragments and melted plastic was easily tracked from her location (the only entrance/exit) back to the kitchen's electric range. One element of the range was still energized and was partially covered with a heat-damaged sheet of aluminum foil (Figure 9.3).

FIGURE 9.3 ◆ One element of the range was still energized and was partially covered with a heat-damaged sheet of aluminum foil. Note the frozen-food dish and drink glass to the left of the range controls. *Courtesy of Santa Ana Fire Department, Santa Ana, Calif., by permission.*

Wrappers found in the kitchen showed that microwave meals in plastic dishes were commonly prepared, but there was no microwave oven. Apparently, she was using the electric range as an impromptu hotplate when the plastic dish ignited. She apparently used a placemat (one missing from the table) and rushed for the front door with the flaming dish. The flames ignited her loose top garment and she was overcome by the inhalation of flames before any substantial CO was produced or inhaled. She collapsed behind the door and the fire, fueled by rendering body fat, with the placemat and her clothing acting as a wick, burned until she was discovered.

For analysis of the plume against the door using the NFPA 921 flame height calculation from Chapter 3 and a wall effect factor of $k = 2$,

$$H_f = 0.174 \, (k\dot{Q})^{2/5}, \tag{9.1}$$

$$\dot{Q} = \frac{79.18 \, H_f^{5/2}}{k}, \tag{9.2}$$

where

H_f = flame height (m),

$\dot{Q}$ = heat release rate (kW),

k = wall effect factor, with

$\quad$ 1 = no nearby walls,

$\quad$ 2 = fuel package at wall, and

$\quad$ 4 = fuel package in corner, and

$$\dot{Q} = \frac{(79.18) \, (1.0)^{5/2}}{2} = 39.6 \, \text{kW.} \tag{9.3}$$

This heat release rate is in the range of values observed during previously published tests using pig carcasses (DeHaan 1999b).

Calculations of the heat release rate needed to produce plume damage on the rear of the door and calculations of the approximate body mass loss (5–10 kg) all indicated a fire of some 30–50 kW had burned for about 2 hr (based on DeHaan 1999b). In the behavioral analysis, note the cocktail glass to the left of the range controls and the elevated blood alcohol level, which also suggest alcohol impairment of the victim's judgment in this case.

This case is courtesy of Robert Eggleston, Santa Ana (California) Fire Department (retired).

◆ 9.2 CASE EXAMPLE 2—HEAT RELEASE RATE HAZARD OF CHRISTMAS TREES

Each year, Christmas trees are involved in a number of household and commercial fires. The National Fire Protection Association (NFPA) estimated 500 tree fires per year in the United States. In 1998, Christmas trees were the first item ignited in nearly 300 home fires, causing 18 deaths, 108 injuries, and nearly $21 million in property loss.

To assess this hazard and address some of the misinformation about ignition, heat release rates, tree varieties, and flame retardants, Dr. Vytenis (Vyto) Babrauskas and his colleagues conducted an extensive study of Christmas trees and their contributions to fires. The results of one major part of this study are presented in this section (Babrauskas 2002).

INTRODUCTION

Burning items that produce a high heat release rate are the ones most likely to lead to fire injuries or fatalities. This is true not only because of possible thermal injuries, but also because the production of toxic gases is highly correlated with heat release rates (Babrauskas 1997, 1998; Babrauskas and Peacock 1992). In most households, upholstered furniture and mattresses are the combustible items that are likely to produce the highest heat release rate values. A mattress may produce 500 to 1500 kW (kilowatts), and an upholstered chair about the same, while a large sofa may produce 2000 to 3000 kW (Babrauskas and Krasny 1985; Babrauskas et al. 1997). Other occupant goods, such as bookcases, television sets, and dressers, normally show heat release rate values only a fraction of the above. Exceptions to consider are Christmas trees.

Several studies examining the heat release rates of Christmas trees have appeared in the literature. Ahonen and co-workers in Finland tested a fresh-cut tree and two that represented a week's indoor storage. When ignited with a small source, the fresh tree showed heat release rate of about 80 kW, while the others showed a rate of about 600 kW (Ahonen, Kokkala, and Weckman 1984).

The California Bureau of Home Furnishings tested nine Christmas trees ignited with a match (Damant and Nurbakhsh 1994). Their results also showed that moisture was important, with very dry trees giving 900–1700 kW. A fresh-cut tree, however, could not be ignited by small ignition sources, such as matches.

Stroup and others (1999) at NIST conducted fire tests of Scotch pine Christmas trees using a large calorimeter. Radiative and total fluxes were measured at a location 2 m from the tree centerline. Seven of the trees were allowed to dry for approximately 3 weeks and burned easily when ignited with an electric match. An eighth tree, which could not be ignited, was freshly cut and kept in water. These trees ranged in height from 23–3.1 m and in weight from 9.5–20 kg. Heat release rates were measured from 1620–5170 kW.

TEST PROCEDURES

Babrauskas and Colleagues tested Christmas trees in a heat release rate calorimeter capable of measuring over 10,000 kW. The basic features of the calorimeter are described in ASTM E1354-02d, the *Standard Test Method for Heat and Visible Smoke Release Rates for Materials and Products Using an Oxygen Consumption Calorimeter* (ASTM 2002b). The tests were run in a large laboratory hall where air supply is effectively unrestricted. Christmas trees are commonly placed in a corner of a room, so to simulate the small amount of radiant heat feedback due to walls, an "open room" was built. Using a floor plan of 2.4 × 3.6 m (8 × 12 ft), two walls 3.0 m (10 ft) high were erected, as was a ceiling. The remaining two walls were absent, apart from short skirts below the ceiling used to channel the exhaust gases to the collection hood. The walls were made of noncombustible fiber-reinforced cement board. The center of each Christmas tree was located 0.9 m (3 ft) from each wall, which left a slight clearance in front of the walls. The test Christmas tree was placed on a load platform, which allowed a continuous record of mass loss to be obtained. In the heat release rate calorimeter, smoke production and CO and CO_2 production were measured along with the heat release rate. In addition, four radiometers and a large number of thermocouples were installed at various locations.

The trees tested were all Douglas fir (*Pseudotsuga menziesii*) and were approximately 2 m (6.5 ft) tall. Various watering programs were used, ranging from displaying trees in stands that were always filled with water to the other extreme of never watering the tree. To examine systematically the effect of watering programs, a majority of the trees were burned following a 10-day display period. These data were supplemented with data from tests with additional trees where the length of the display period and care were varied.

Before each test, the trees were weighed. The weight of the trees ranged from 6.4 to 22.4 kg, which was largely due to differences in the moisture content of the trees. Apart from the 22.4-kg specimen, which had a very high moisture content and did not show a high heat release rate, the next-heaviest specimen was 16 kg. The average weight was 9 kg.

In the formulation of the test program, it was hypothesized that the moisture content at the time of burning would be the main controlling variable (in contrast, for example, to the length of time the tree was displayed). Thus, in the study, the moisture content of the foliage was accurately determined. In previous research programs, not enough trees were tested, nor were accurate moisture measurements available in order to estimate the role of moisture. Moisture content was determined by sampling branches, weighing them, then oven-drying them at slightly above 100°C (212°F) and weighing them again. Moisture content (MC) is defined as

$$MC = \frac{Mass_{wet} - Mass_{dry}}{Mass_{dry}}. \tag{9.4}$$

Note that since the definition uses the dry mass in the denominator, the moisture content can easily exceed 100 percent. When possible, water potential measurements were also taken using a pressure chamber.

In most U.S. households, there will be a significant amount of additional combustibles associated with Christmas trees—decorations of various sorts, wrappings at the base, and gift packages. In the tests described here, no tree decorations were used. Located at the base, however, was a representative collection of gift boxes. The boxes were cardboard and had customary Christmas wrappings and ribbons. Most of the boxes contained a small amount of tissue paper and no other combustibles. Two toys, a plastic truck and a stuffed animal (peak heat release rate = 6 kW) were the only nonpackaging fuels included. The packages were arranged in an identical manner around the base for each test.

Three tests were run without packages and two other tests were run without the simulated walls to determine the effect, if any, these items had on the fire development. To start the fire, the procedure specified by the California State Fire Marshal for fire testing of Christmas trees (a Bunsen burner with a 38-mm-high flame; time of application = 12 s) was first applied to a branch about halfway up the tree. The burner simulates a small flame source such as a cigarette lighter. If the tree would not ignite from this source, then the same flame was applied to a gift package below the tree or to a stuffed toy located in the tree.

RESULTS

Photographs of the test are shown in Figures 9.4 through 9.10. The first quantitative aspect that was determined in these tests was the moisture content needed for the California test burner, applied directly to the tree, to cause ignition of the tree. The results indicated that MC = 50 percent was a suitable demarcation between trees that ignited from this source and trees that did not. Of 48 trees tested, only 3 failed to obey this rule, with 1 test being inconclusive. All the trees that did not ignite from the small flame ignited from the burning gift-package assembly or stuffed toy.

FIGURE 9.4 ◆ An example tree ready for testing. All test trees included gift packages under the tree and a single stuffed toy animal in the middle of the tree. Ignition was by a small flame to a branch, by a small flame to the stuffed toy, or by a small flame to gift packages. No tree failed to burn from at least one of these ignition locations. *Courtesy of Dr. Vytenis Babrauskas, Fire Science & Technology, Issaquah, Wash., by permission.*

FIGURE 9.5 ◆ Fierce burning from a dry (MC = 11 percent) tree. Peak heat release rate = 2700 kW. *Courtesy of Dr. Vytenis Babrauskas, Fire Science & Technology, Issaquah, Wash., by permission.*

296

FIGURE 9.6 ◆ Postfire condition of tree shown burning in Figure 9.5. Only the main trunk remains. *Courtesy of Dr. Vytenis Babrauskas, Fire Science & Technology, Issaquah, Wash., by permission.*

The gift-package assembly, when tested by itself, showed peak heat release rate values of 98–118 kW, with four replicates being tested. Thus, its own heat release rate was insignificant compared to that of a tree. Visual observations indicated the process of ignition in the tests where the gift packages ignited the tree. The packages burn at a moderate but steady rate, and the fire effectively circles around the base of the tree. The hot combustion products go up and this has the effect of progressively drying out the tree branches. By the time the packages have stopped burning, the branches are dry enough that they catch fire and burn.

Figure 9.11 shows the heat release rate curves for three tests. The MC = 20 percent and MC = 38 percent specimens were ignited by the small flame burner placed next to a branch. It can be seen that about 50 s elapses before any detectable heat release rate occurs, but after that point the burning is exceedingly fierce. In about 1 min the tree is essentially burned up, with only minor burning going on after that point.

The trunks and the large branches never burned in these tests, although in a room fire, other items would likely contribute to the fire and this would generally cause even these members to be consumed. In the tests reported here, about one-fourth to one-third of the mass of each tree remained unburned.

The specimen of MC = 70 percent showed a distinctly different behavior. It took about 80 s before there was significant heat release rate. Again, about two-thirds to three-fourths of the mass was consumed for the package-ignited trees, but their time course was much slower. They typically showed a curve with two or more "humps" and the time required for the bulk of the tree to be burned was about 300–500 s, instead of 80 s.

FIGURE 9.7 ◆ A rapidly burning tree will ignite combustibles nearby. Here nearby fabric targets are shown being ignited. *Courtesy of Dr. Vytenis Babrauskas, Fire Science & Technology, Issaquah, Wash., by permission.*

FIGURE 9.8 ◆ This Christmas tree would not burn when a small flame was applied to a tree branch. Applying the same flame to a small stuffed animal hung on the tree ignited the toy and the tree. Peak heat release rate = 525 kW. *Courtesy of Dr. Vytenis Babrauskas, Fire Science & Technology, Issaquah, Wash., by permission.*

FIGURE 9.9 ◆ A fire-retardant (FR) tree burning. Some FR specimens generated a great deal of smoke and mist. *Courtesy of Dr. Vytenis Babrauskas, Fire Science & Technology, Issaquah, Wash., by permission.*

The moistest package-ignited tree tested had a MC = 154 percent. This tree showed a peak heat release rate of 444 kW, which is still a sizable fire. Even this tree lost 60 percent of its mass during the test. Because MC data were available for each tree, a relation between MC and peak heat release rate could be found (Figure 9.12).

However, two trends are very clear: (1) An increasing MC decreases the peak heat release rate; and (2) none of the trees are "nonburning," or even of negligible fire hazard if ignited with a large ignition source. The equation derived from the dotted curve in Figure 9.12, which can be used for predictive purposes of heat release rate based upon the moisture content of average-size, non-FR trees, is

$$\dot{Q} = 4050 - 734 \ln(\text{MC}), \tag{9.5}$$

where ln denotes natural logarithm.

CONCLUSIONS

Christmas tree fires are well documented in fire-incident investigations and have the potential for highly catastrophic results. For example, there is a detailed report from FEMA (David 1991) concerning a Christmas tree fire that killed seven people. Some experimental data are available from previous laboratory studies, but these were limited and did not explore a wide range of treatments, nor did they quantify the effect of moisture content. The present study shows that trees with a MC below 15 percent will give heat release rate values in the range of 2000 to 3000 kW. These values pertain to a common tree height of 2 m. Taller or shorter trees can be expected to give proportionately different results.

The test results confirmed that MC (apart from tree species and tree height, which were not tested) was the main variable governing the burning characteristics. The

FIGURE 9.10 ◆ This FR-treated tree burned vigorously and essentially burned up completely. *Courtesy of Dr. Vytenis Babrauskas, Fire Science & Technology, Issaquah, Wash., by permission.*

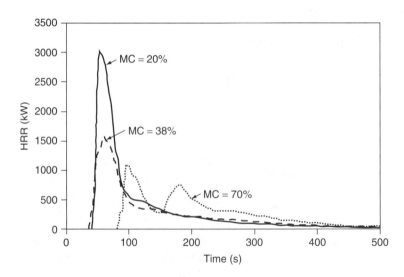

FIGURE 9.11 ◆ Example heat release rate curves at several moisture levels. *Courtesy of Dr. Vytenis Babrauskas, Fire Science & Technology, Issaquah, Wash., by permission.*

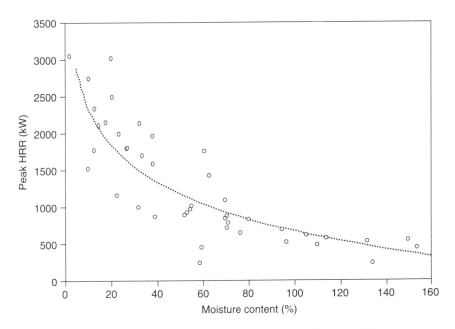

FIGURE 9.12 ◆ The relation between MC and HRR for non-FR trees.
All trees Douglas fir, ≈2 m high. *Courtesy of Dr. Vytenis Babrauskas, Fire
Science & Technology, Issaquah, Wash., by permission.*

length of the display period and the quality of the watering regimen both serve to determine the MC.

Douglas firs with an MC of more than 50 percent were found to be unlikely to be directly ignited by small flames. However, such small ignition sources can readily ignite gift-wrapped packages, and this secondary ignition can then result in a serious tree fire. Defective lights or electrical cords are the most commonly reported ignition sources associated with Christmas tree fires. The arcs that occur in the failure of such devices are physically quite small, but an electrical fault may preheat combustibles in the vicinity of the fault prior to overt arcing occurring. This ignition mechanism has not been studied in our work but deserves to be investigated in some detail.

It is normally estimated that a small to moderate-size room can be driven to flashover by a fire of about 1500 kW if ventilation is through a single open doorway (Babrauskas 1984). This type of calculation is easier to perform for a bedroom than for a living room, where complex, interconnected spaces are often involved. Nonetheless, heat release rate values of the kind documented here represent a highly significant hazard, since in actual houses, Christmas trees are unlikely to be adjacent to cement board surfaces. Instead, combustible curtains, chairs, sofas, and other items will likely be close by, and these can be expected to ignite readily from the kinds of fires that have been documented here.

Good and prudent advice is frequently given to consumers to keep their Christmas trees well watered. This is essential since, if tree basins are allowed to run dry, moisture levels drop. None of the trees that were continually displayed in water during these tests burned when exposed to the small flame ignition source. However, it is not uncommon for trees to be displayed for much longer periods of time than were used in these tests. Trees are commonly bought a few days after Thanksgiving and may not be taken down until Epiphany, a time span of about 6 weeks. Based on previous postharvest studies, Douglas firs would not be expected to be able to maintain a high moisture level for this length of time even if they were displayed in water. Other species of Christmas trees such as Noble fir or Fraser fir that are able to maintain high moisture levels for extended periods of time would be much better suited for use under these conditions. This probably explains why these species are now the most commonly purchased Christmas trees on the market.

This case and testing are courtesy of Dr. Vytenis Babrauskas, Fire Science and Technology, Inc., Issaquah, Washington.

◆ 9.3 CASE EXAMPLE 3—VEHICLE FIRES AND EXPLODING GASOLINE CANS

Several cases have been documented where a person is burned to death as a result of a vehicle fire, and a witness claims that a plastic container of gasoline in the passenger compartment had leaked or spilled and that a fire had (somehow) been ignited in the vapors created. The fire then "must have" traveled back to the container and ignited its vapors, causing it to explode in a massive fireball that prevented rescue of the occupant. Such scenarios raise several issues that are testable by the means described in this text.

IGNITABILITY OF VAPORS

Assuming that there is some portion of the passenger compartment that contains vapors within the flammable range (1.4–7.5 percent in air), what is required to ignite them? In his extensive review, Holleyhead (1996) demonstrated that ignition of gasoline vapors by a hot surface, in particular, glowing cigarettes, is nearly impossible.

Referring to the literature shows that the autoignition temperature (AIT) of automotive gasoline is of the order of 400–500°C and the temperature of the glowing coal of a cigarette can be as high as 800–900°C. Before concluding that the higher source temperature means that there would be ignition, these two numbers have to be evaluated in the context of the contact and heat transfer between the source and the fuel.

If a cigarette is not being puffed, the hot coal has a layer of ash surrounding it. This ash has a very low thermal conductivity that minimizes heat transfer. Gasoline has a minimum ignition energy (MIE) of 0.24 mJ at ideal mixture so at least this amount of energy has to be transferred. The laminar flow around the coal also minimizes contact between the hot surface and the fuel molecules. As a result of these effects, there is an insufficient quantity of heat transferred to cause ignition.

If the cigarette is being puffed, the ash layer is largely absent and the fuel might be drawn into intimate contact with the porous glowing coal. When this occurs, however, the gases are being drawn through the coal so quickly that there is insufficient contact time for enough heat to be transferred, and the oxygen content of the smoke stream is so low that a flame will not be able to propagate. The one exception is if there is an impurity in the tobacco or the paper that permits a brief flicker of flame to occur as the cigarette is being puffed. In that case, the flame, if in contact with an ignitable mixture, may ignite it. For this reason, hand-rolled cigarettes pose a slight risk, whereas tests of commercially made tobacco cigarettes have never demonstrated an ignition of gasoline vapors (Holleyhead 1996).

Gases with a lower minimum ignition energy such as hydrogen and acetylene (MIEs of 0.01–0.02 mJ) and carbon disulfide vapor (with its very low AIT) have been shown to be ignitable by a glowing cigarette. Hot surface ignition, such as from the electric element of the vehicle's cigarette lighter, is not possible due to its relatively low temperature and the flat surface contact area. An electric arc caused by switching off electrical components of the vehicle or the flame of a lighter being used will cause ignition of the accumulated vapors if they are present at the correct concentrations and in contact with the ignition source. Since most such connections are well concealed in the instrument panel of a modern vehicle, such contact would be very difficult to achieve.

VAPOR CONCENTRATIONS AND HUMAN TENABILITY

That raises the second issue—vapor concentrations and human tenability. The concentration of gasoline vapor in air must exceed 1.4 percent (14,000 ppm) to be ignitable under normal ambient conditions. The normal human nose is sensitive to gasoline vapors at concentrations below 20 ppm. Gasoline vapors are an irritant to the eyes and to the mucosal membranes of the nose and throat at concentrations above 200 ppm and will be intolerable at concentrations above 2000 ppm, even in the absence of a sense of smell. This means that gasoline vapors will be detectable and a gasoline-vapor environment will be untenable at concentrations far below their lower flammable limit (LFL).

While gasoline vapors are two to three times heavier than air, they will tend to sink in still air. The movement of air in a vehicle caused by vehicle movement, fan, passenger movement, or even thermal currents from warm bodies will circulate, mix, and dilute the vapors. It would be very difficult, if not impossible, for an ignitable concentration of over 14,000 ppm in one part of the passenger compartment to be tolerated by conscious passengers.

SOURCE OF VAPORS

The third issue is the source of the vapors. Liquid fuel must evaporate before it can be ignited. As we have seen previously, the rate of evaporation is controlled by the surface area of the liquid pool, temperature, and nature of the surface. An open fuel container will produce vapors only from the exposed fuel surface; that is, one with a cap loosely in place will leak vapors, but at a very slow rate.

If the liquid is spilled on the seat or floor of the vehicle during transport, it will evaporate much faster. If the surface area of the supposed spill can be estimated, one can estimate the maximum evaporation rate based on published data (DeHaan 1995). Only the fuel that is in the vapor state at the time of ignition will support that ignition (with ongoing flame dependent on a continuous source of vapors).

The problem, then, is putting a suitable ignition source (flame or electrical arc) into contact with a suitable accumulation of vapor. If a floor-level spill is hypothesized, the ignitable vapors may be found only at or near floor level (due to the dilution and mixing in air). Laboratory testing of the floor mats should reveal the residues of significant amounts of liquid. Careful examination of the vehicle may reveal the presence of a suitable ignition source (dropped matchstick, broken wire to seat motor, etc.). Except under the most unusual vehicle conditions, the ignition source will be found in the same "compartment" as the fuel vapor accumulation (i.e., trunk, engine, passenger). Interviewing a witness may reveal the circumstances of ignition.

THE CONTAINER

Assuming that there was a spill (ignored by passengers), enough vapor, a suitable source, and ignition of the accumulated vapors, one would expect a deflagrating combustion of the vapors in the vehicle, but would there be an explosion of the vapors in the can?

The answer is in the vapor pressure of the gasoline. At normal ambient temperatures above 0°C (32°F), the combined vapor pressure of the most volatile components in automotive gasoline (butane, isobutane, pentane, hexane) is sufficient (>200 mm Hg) that (if there is any significant quantity of liquid fuel in the container) the vapor concentration in the container is well above the upper 7.5 percent flammability limit of the vapors. This means that only the vapors that have escaped the container and mixed adequately with the surrounding air to be within their flammable range will support combustion.

The result is not an explosion in the container but a modest flame supported at the mouth of the container by the vapors coming from within. As shown in the test in Figure 9.13, a small flame will be supported at the loose cap. If it is a plastic container, this flame will eventually melt (in 1–10 min) the neck and top of the container, exposing a larger surface area of fuel. In most cases the container will burn from top down, melting at the fuel level inside as it burns. If there is enough boilover of excess fuel, the sides will collapse and release a large liquid pool. This process nearly always leaves liquid fuel trapped in the melted folds of the base of the container that is readily detected by careful visual inspection of the remnant in the lab. If it is a metal can, the soldered seams may eventually melt and the heat conducted through the metal causes faster and faster evaporation of the remaining fuel but no explosion. Only when the container is nearly empty (or if it was so at the start of the fire), there may

FIGURE 9.13 ◆ Fire test sequence (top to bottom) for evaluating a 20-liter plastic can filled with gasoline, with a loose cap. After 8 min, a small fire at the cap is supported by vapors leaking from the cap. After 19 min, the top has melted away, producing an equivalent pool fire of 400 kW, and after 24 min, spillover of boiling gasoline occurs. *Courtesy of New Zealand Police, Christchurch, New Zealand, by permission.*

be a stage where there is not enough fuel vapor to sustain a saturation (equilibrium) vapor pressure inside, and the mixture inside the can may sustain a deflagration as it boils dry.

This was the mechanism thought to be responsible for the fuel tank explosion on TWA Flight 800, where a small quantity of low-volatility fuel in a large fuel tank reached ignition conditions from being heated externally prior to takeoff (Fulghum 1997).

THE EXTERNAL FIRE

When there is a spill onto the floor mat of the vehicle or the entire top of the container is melted away, the heat release rate of the resulting fire can be estimated from existing data on pool fires. What is needed to determine this heat release rate is the estimated surface area of the pool.

When melted away, the exposed top of the plastic container, for instance, is approximately 16 × 16 in. (0.4 × 0.4 m). This produces an effective "pool" 0.16 m^2 in area.

In order to estimate the heat release rate of fire emitted from the exposed top of the burning container, we refer to Chapter 2, equation (2.5), when the horizontal burning area of a material is known along with the burning rate per unit area (mass flux):

$$\dot{Q} = \dot{m}'' \, \Delta h_{eff} \, A, \tag{9.6}$$

where

$\dot{Q}$ = total heat release rate (kW),

$\dot{m}''$ = effective mass burning rate per unit area (g/m^2-s),

Δh_{eff} = heat of combustion (kJ/g), and

A = burning area (m^2).

Table 2.3 lists the typical range of values of mass flux and heat of combustion for gasoline (see Chapter 2). Substituting them into the equation for heat release rate, the maximum value for $\dot{Q}$ will be

$$\dot{Q} = (53 \text{ g/m}^2 \cdot \text{s}) \, (43.7 \text{ kJ/g}) \, (0.16 \text{ m}^2) = 370 \text{ kW}. \tag{9.7}$$

Similar calculations can be carried out for the area of the footwell or floor mat assuming maximum area and saturation. This fire will be sustained if there is adequate ventilation as the fuel level regresses at 1–4 mm/min until the container fails or all the fuel is consumed.

◆ 9.4 CASE EXAMPLE 4—ASSESSING CONSTRUCTION AND WALL COVERINGS

This case study emphasizes the importance of assessing building features such as construction and wall coverings to evaluate their possible contributions to fire spread and production of unusual burn patterns, as well as the importance of developing a valid hypothesis by comparing evidence from the scene to witness-reported information and features of the uninvolved portion of the structure.

A fire in a three-story, wood-frame university dormitory building claimed five lives and injured eight, all residents of the building. The 47 × 124-ft (14.3 × 37.8-m) building contained 46 studio apartments, with each floor having a central hallway served by enclosed stairwells at each end. The walls and ceilings of the halls and stairwells were plasterboard. The stairs themselves were wood, and the hallway floors were plywood covered with vinyl floor tiles. The walls of the first (ground) and third floor halls and the stairwells were covered with thin (approximately $\frac{3}{16}$-in. [3- to 4-mm]) plywood paneling nailed to the walls and held in place with wooden baseboards and wainscoting.

The stairwell doors were fire-rated gypsum core doors, fitted with spring closers. However, they also had flip-up door stops and interviews with residents indicated that they were frequently blocked open. The doors to individual rooms were hollow-core wood interior doors. There were heat detectors in the stairwells and halls directly connected to the fire department's headquarters alarm panel, but no fire suppression system.

The fire extensively damaged the east end of the second floor and the stairwell to the third floor. There was no fire damage to the first floor, but damage did extend about halfway down the east stairs toward the first floor. There was less damage on the third floor and significant portions of the wood paneling survived. It was the burn patterns on this paneling that initially suggested that multiple fires had been set in the third floor hallway using flammable liquids.

There were 20 or more areas where localized fire damage to the flooring had occurred and extended up the walls in burned areas 2–3 ft (0.61–0.9 m) high, with most being 6–12 in. (0.15–0.3 m) wide. The burned areas were on both sides of the third floor hallway at somewhat regular intervals (Figure 9.14) and ranged from the paneling being scorched or lightly charred to its being completely destroyed (along with the baseboard), exposing the plasterboard beneath. All these areas were found to be located where the paneling was not nailed tightly to the wall studs (i.e., centered between the nailings at 16- or 32-in. centers) (Figure 9.14), allowing the

FIGURE 9.14 ◆ Areas of localized fire damage to the third-floor flooring and walls at regular intervals along the length of the corridor. *Courtesy of Sgt. Paul Echols, Carbondale [Ill.] Police Department, by permission.*

paneling to bulge slightly away, thus allowing an air space to develop and producing a thermally thin fuel.

This thermally thin material was exposed to radiant heat from the hot gas layer that filled the corridor from the stairwells and from the burning paneling that had ignited in the hot gas layer, peeled away, and dropped to the floor. The wainscoting secured most of the lower half of the paneling in the third-floor hall, leaving these isolated V patterns coming from floor level.

Investigation revealed that ceiling penetrations in the stairwells allowed fire gases to penetrate the floor joist space between the second and the third floors. The joists were arranged longitudinally (along the length of the building), providing a channel for air movement into the third-floor hallway. The plywood flooring had been nailed to these longitudinal joists with no provision for cross-support where the plywood sheets butted against one another.

Floor tiles that overlapped these butt joints were seen to be cracked and broken. Floor tiles along the walls were loose or displaced from the flexing across the joints, and a few had been replaced. This allowed air from the joist space to escape through these joints and permit more intense burning of the adjacent baseboard and paneling as they were heated by radiant heat from the fire gases trapped in the hall. Where a butt joint coincided with a thermally thin area of paneling, a localized area of damage was created (Figure 9.15) from the radiant heat from the intense/deep hot gas layer (as evidenced by the burned areas of walls above the wainscoting). The "burn-throughs" of flooring/wall structure occurred at 4- or 8-ft (1.22- or 2.44-m) intervals where the plywood flooring joined.

One large area of such damage was originally thought to be the result of a large burning pool of ignitable liquid against a wall. However, interviews with survivors (supported by physical evidence) revealed that one rescuer had crawled through this

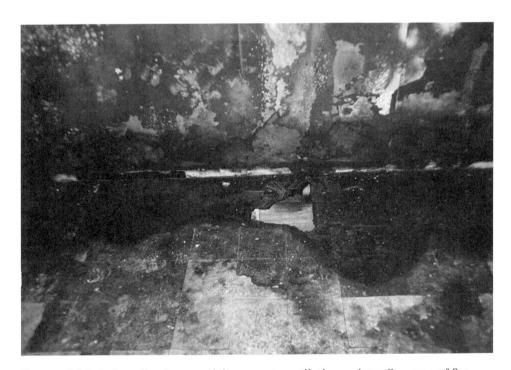

FIGURE 9.15 ◆ Localized areas of damage at ventilation points. *Courtesy of Sgt. Paul Echols, Carbondale [Ill.] Police Department, by permission.*

very area dragging a victim during the fire and had not been exposed to an ignitable liquid or localized fire at that time. This illustrates the value of thorough interviews and comparing information from the scene to that from witnesses to validate or exclude a hypothesis. There were also areas of localized scorching on floors and baseboards that were clearly consistent with the brief-duration fires fueled by portions of the upper paneling peeling off the wall and igniting in the hot gas layer. These would be classified as drop-down fires. Examination of the floor on the second floor and undamaged paneling on the first floor confirmed the features described.

This fire was ultimately identified as an arson, with two origins being identified in piles of clothing and trash outside the door of one apartment immediately next to the stairs and in the stairwell immediately adjacent. No flammable liquids were involved. The staircases were affected by fall-down, buoyant flow, and radiation processes. Figure 9.16 shows the extension of fire from the second floor down the stairs by radiant heat and fall-down of the paneling, with a firefighter standing in the open door in the second-floor hallway.

Figure 9.16 is of the stairs between the second and the third floors immediately above one of the origins of the fire. Much more extensive damage to risers (including penetrations) and to wall coverings and handrails was present in this stairwell than in the one below it. Of note, all of the fatalities and most of the eight injuries occurred in rooms on the third floor (not on the floor of origin).

FIGURE 9.16 ◆ Extension of the fire down the stairs by radiant heat and fall-down of the paneling. *Courtesy of Sgt. Paul Echols, Carbondale [Ill.] Police Department, by permission.*

Thin materials are usually thought to be draperies, loose garments, calendar pages, and the like. They can also be multilayered materials that delaminate when exposed to heat. This includes plywoods, veneers, and some wallpapers or paints that blister due to moisture or solvent trapped beneath an impermeable outer skin, and wall or floor coverings that can separate from a solid underlayment due to adhesive or fastener weaknesses. Skin can also display this effect under some conditions as a blister forms and the distended epidermal layer loses contact with the dermal layer beneath.

The generally accepted equations for thermally thin and thermally thick materials from Chapter 2 (equations [2.11] and [2.12]) are repeated here. In the case of thermally thin conditions, for $\delta \leq 1$ mm,

$$t_{ig} = \frac{T_{ig} - T_\infty}{\dot{q}''/\rho c_p \delta}. \tag{9.8}$$

Under thermally thick conditions, $\delta > 1$ mm,

$$t_{ig} = \frac{\pi}{4} k\rho c_p \left(\frac{T_{ig} - T_\infty}{\dot{q}''}\right)^2, \tag{9.9}$$

where

t_{ig} = time to ignition (s),

k = thermal conductivity (W/m-K),

ρ = density (kg/m^3),

c_p = specific heat capacity (kJ/kg-K),

δ = thickness of material,

T_{ig} = piloted fuel ignition temperature,

T_∞ = initial temperature, and

$\dot{q}''$ = radiant heat flux (kW/m^2).

From Table 2.5 (Chapter 2), the characteristics of yellow pine are a thermal conductivity, k, of 0.17 W/m-K, a density, ρ, of 640 kg/m^3, a specific heat capacity, c_p, of 2.85 kJ/kg-K, and a thermal inertia, $k\rho c_p$, of 0.255 kW^2-s/m^4-K^2. With a piloted ignition temperature, T_{ig}, of 450°C, an ambient initial temperature, T_∞, of 20°C, and a paneling surface thickness, δ, measuring 3 to 4 mm, the thermally thick time-to-ignition formula is used:

$$t_{ig} = \left(\frac{\pi}{4}\right)(0.255)\left(\frac{450 - 20}{\dot{q}''}\right)^2 \tag{9.10}$$

or, simplified,

$$t_{ig} = \frac{37012.0}{(\dot{q}'')^2}. \tag{9.11}$$

In evaluating a range of possible heat fluxes, 20, 30, 40, and 50 kW/m^2, the calculated times to ignition for the pinewood paneling are 93, 41, 23, and 15 s, respectively. Where this paneling was beginning to delaminate, $\delta = 1$ mm and thin-material ignition would be expected in approximately 26 s at 30 kW/m^2.

Therefore, for low heat fluxes (<50 kW/m^2) and short durations (<30 s), thin materials are much more likely to be ignited (or even melted) than are thick objects made of the same materials. These considerations have to be kept in mind when documenting indicators of heat damage.

◆ 9.5 CASE EXAMPLE 5—PLUME HEIGHT AND TIME OF BURNING BODY

Early one morning, at 8:20 A.M., employees noted a plume of smoke coming from brush at the edge of an industrial park. Responding to what was thought to be an unattended campfire of the homeless, the fire department found the fire to be consuming the body of a young woman.

The small flames were quickly doused and the scene preserved by 8:28 A.M. The body was found lying on its back, severely burned (approximately 50 percent of the body weight remaining, with the lower legs burned off). Remnants of a heavy cotton blanket or drapery were found with the body. Next to the body was a burned, carpet-covered board of the type found in camper shells and trailers, braced on some low-hanging branches, as shown in Figure 9.17.

The brush above the body bore an "inverted-V" pattern of damage, extending from ground level, involving charred branches 0.6–0.9 m (2–3 ft) from the ground, scorched leaves and branches at 1.5 m (5 ft), and withered and scorched leaves at 2.1 m (7 ft). As shown in Figure 9.18, there was no sustained combustion of the brush, since the body was located in a natural gap somewhat between the shrubs.

The body was not readily visible from the street due to a gentle slope downward. The victim was identified as a young local woman. Her COHb was less than 10 percent (negligible, due to her history as a smoker). She had a low blood alcohol concentration

FIGURE 9.17 ◆ Responding fire department personnel found the fire consuming the body of a young woman next to a burned board. *Courtesy of Sgt. Bruce M. Wiley, San Jose Police Department, San Jose, Calif.*

FIGURE 9.18 ◆ Photo of the body location shows an "in-verted-V" pattern of thermal damage to foliage of shrubs overhanging the position of the body. Scorching and shriv-eling extended to a peak about 2.1 m above the ground where the body burned. Char-ring extended some 0.6–0.9 m from the ground. *Courtesy of Sgt. Bruce M. Wiley, San Jose Police Department, San Jose, Calif.*

and tested negative for drugs. There were no signs of a struggle at the scene, no weapons, and no blood. The dry hardpan soil had been walked over repeatedly by fire personnel so no shoeprints were recovered. Based upon postmortem weight and an estimated antemortem weight from the victim's mother, the mass loss rate of the body was estimated as 30 kg.

The main issues were time of death, cause of death, and duration of burning. Due to massive destruction of the soft tissue, no wounds could be identified (an informant later said that she had been shot in the side of the neck around midnight) and time of death was not established (although there were no signs of decomposition).

As far as establishing the time of burning, the McCaffrey (1979) method was used for estimating the centerline maximum temperature rise, velocity, and mass flow rate of a plume as described in Chapter 3, equations (3.15), (3.16), and (3.17). The temper-ature profiles for the smoke plume above various size fires can be estimated as

$$T_0 - T_\infty = 21.6 \, \dot{Q}^{2/3} \, Z^{-5/2}, \tag{9.12}$$

where

T_0 = maximum ceiling temperature (°C),

T_∞ = ambient air temperature (°C),

$\dot{Q}$ = total heat release rate (kW), and

Z = measurement along the centerline (m).

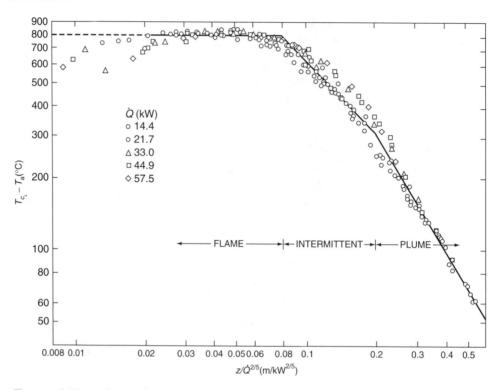

FIGURE 9.19 ◆ Centerline fire plume temperatures. *From Principles of Fire Behavior, 1st ed., by J. Quintiere © 1998. Reprinted with permission of Delmar Learning, a division of Thomson Learning: www.thomsonrights.com.*

These relationships correspond with the plot in Figure 9.19 (Quintiere 1998, 142).

A fire of about 100 kW at ground level would be calculated to have gas temperatures just about right ($T = 100$–$150°C$ at 2 m, $450°C$ at 1 m). This maximum probably occurred as the cotton blanket/curtain wrapped around the victim burned, but assuming that it was sustained during the burning of the body, the plume damage observed sets the maximum heat release rate of the body at 100 kW.

Based on data from DeHaan (1999b, 2001), a burning body by itself via the wick mechanism will demonstrate a mass loss rate of about 3 g/s (10.8 kg/hr). Allowing for fluid loss, time to ignition, and a total mass loss of about 30 kg, this yields a minimum time of about 4 hr. Anything less than 1 g/s (30 kW) would not sustain itself, and this fire was still burning, with flames showing in the crotch area of the body when it was found and extinguished at 8:28 A.M. That minimum rate is 3.6 kg/hr, which sets a maximum time of about 6 hr. The body had been placed face down on a wide board with carpet, and as the limbs flexed, the body apparently shifted and rolled off the board and onto its back. The leaf litter under the body was much less badly burned than the board (where nearly all the ashes and debris were found), and there was no protected area on the body. Nearly all the skin was burned away from all surfaces of the torso.

Based on this calculation and physical evidence (burned blanket, absence of flammable liquid residues, death before fire exposure), which corroborated the witness statements, the individuals responsible were tried and convicted of murder.

This case study is courtesy of Officer Robert Froese, San Jose Police Department, San Jose, California.

In 1981 a man died as the result of third-degree burns over most of his body. His boarding-house roommates reported that he had returned to the house a short time earlier, well under the influence of alcohol, and was last seen asleep on the living room sofa.

Some time later, the roommates, who were watching television in another room, heard shouts, saw a flash of flame, and were horrified to see the victim stagger out of the living room with his clothes on fire. He walked to an adjacent bedroom, where an attempt was made to smother the flames with a blanket, and then to a bathroom, where his clothes were stripped off and dropped in the bathtub.

A neighbor had seen the flames and notified the fire department. Fire crews arrived to find the victim in the bathroom with a trail of burned clothing and burned blankets in the adjacent bedroom. The fire in the living room was limited to the sofa base and its cushions and was extinguished with a minimum of water. Scene investigation revealed a glass jar with residues of gasoline on the floor of the dining room, burns to the cushions of the sofa (Figure 9.20), and numerous isolated surface burns

FIGURE 9.20 ◆ Scene investigation revealed burns to the seat cushions of the sofa and a glass jar with residues of gasoline on the floor of the dining room. *Courtesy of Norman Fehle, LAPD Arson (Retired), by permission.*

of the carpet in front of the sofa. Residues of gasoline were identified in the carpet, the victim's clothing, and the jar. The victim died some 18 hr later from external burns (98 percent second and third degree) and inhalation of hot gases.

A roommate confessed to having thrown gasoline on the victim as he allegedly attempted to rise from the sofa and assault him. The accused claimed that the victim had repeatedly assaulted him physically and sexually in recent days and that he had made plans to stop him if he tried it again. Roommates had seen him in the house with a glass jar of clear liquid similar to gasoline minutes before the fire.

Authorities reasoned that the victim had been asleep on the sofa and that the accused had poured the gasoline on him as he lay on his side. The front of the sofa had been consumed and there were limited burns to the loose seat and backrest cushions of the sofa; the scattering of burns across the carpet in front of the sofa was noted but not interpreted at the time. The accused was convicted of murder (with a torture enhancement!) and sentenced to life.

During the appeal process it was requested that the evidence of the burn patterns be reexamined. Unfortunately, most of the scene photos are of a quality too poor to permit reprinting but the sketch in Figure 9.21 demonstrates the critical features. The front face of the sofa (which appeared to be cotton padded) was burned nearly to floor level and fire had penetrated under the cushions.

The seat cushions appeared to be cotton covering over foam rubber and had been thrown on the floor during extinguishment. Of the surfaces visible in the photos, only one bore visible damage and it was limited to longitudinal burns parallel to the edges.

FIGURE 9.21 ◆ Plan view sketch of the couch, pillows, and remains of burned clothing. *Courtesy of John DeHaan.*

The two backrest cushions still in place bore only limited, isolated burn damage. There was a scattering of lightly burned areas on the carpet in front of the sofa extending some 3 ft away from it. These appeared to be the scorched or melted areas created when a small quantity of flammable liquid burns off the face of the carpet without penetrating the pile of the carpet.

Because the victim had walked about the house while on fire, all of his clothing had been burned, leaving only charred remnants in various rooms. The postmortem photos and report revealed generalized third-degree burns uniformly distributed over most of his body. As a result, no interpretation could be offered as to the area of clothing first ignited.

Since synthetic fabrics can burn with an intensity similar to that of gasoline, the damage they can produce cannot always be distinguished from that produced by gasoline. An attempt was made to reconstruct the apparent burn damage to the carpet and sofa. It was thought unusual that a fire ignited by pouring gasoline on a reclining body would damage the front of the sofa cushions, but not the tops, or the adjacent backrest cushions. Extensive spillage or splatters on the carpet some distance from the sofa would also not be an expected consequence of a pour on top of the cushions.

Using a sofa of approximately the same proportions with loose seat and backrest cushions, tests were conducted to establish the nature of splash and splatter patterns created by throwing a liquid toward the sofa. The sofa was partially covered with brown wrapping paper, a glass jar of the same capacity as that recovered at the scene was used, and isopropyl alcohol was used as the liquid. Isopropyl alcohol has viscosity and surface tension properties very similar to those of gasoline so it will darken paper surfaces where it contacts and yet will evaporate completely after a short period of time and is nontoxic.

A separate cushion propped against the front of the sofa simulated the presence of a body sitting on the edge of the sofa or standing immediately in front of it. The "thrower" was standing approximately 6 ft from the front of the vertical cushion. Figure 9.22 shows the splatter pattern created when the alcohol (about 300 ml) was thrown low against the target (about knee height). A very heavy concentration is seen in a single "pool" at the base of the cushion. Although it cannot be seen on the fabric, a quantity of liquid was detected on the front of the sofa cushions to both sides of the target. It was decided to cover these cushions with paper to detect this side splatter.

Figure 9.23 shows the fluid still in flight when thrown higher up on the cushion, while Figure 9.24 shows the wetted areas about 30 s later. Note the wetted areas on the front of the cushions and the more limited distribution of concentrated (pooled) liquid but more discrete splatters on the floor in front of it.

Figure 9.25 shows the effect of throwing the liquid against a textured fabric surface rather than a smooth paper one. Note the distribution of discrete spots on the floor and limited splatter on the front edge of the seat cushions. Figure 9.26 shows the distribution on a horizontal surface. The main splatter reaches about 70 in. from the near edge (where the forward motion of the jar was checked). The farthest splatter is about 100 in. from the near edge. This large concentrated pool does not match the distribution on the carpet at the scene.

Of the limited series of tests conducted it appeared that the photos in Figures 9.24 and 9.25 represented the pattern of distribution of gasoline noted at the scene. These tests showed that it was more likely that the gasoline was not poured on the recumbent victim as he slept but thrown at him from some distance as he sat on the edge of the sofa or attempted to rise from it. These tests supported the accused's contention that he acted in self defense and he was released on parole having served 18 years for the crime.

FIGURE 9.22 ◆ The splatter pattern created when alcohol (about 300 ml) was thrown low against the target, at about knee height. *Courtesy of John DeHaan.*

◆ 9.7 CASE EXAMPLE 7—ACCIDENTAL FIRE DEATH

A woman was found dead on the landing outside her bedroom door. She was covered in soot and her body was outlined in soot on the carpet around her. She was partially undressed as shown in Figure 9.27.

There was a large amount of soot and mucus in the victim's nostrils. Postmortem blood analysis revealed that she had a high COHb concentration, a significant blood alcohol level, and high levels of antidepressants. There was no evidence of violence.

FIGURE 9.23 ◆ The fluid still in flight when thrown higher up on the cushion. *Courtesy of John DeHaan.*

A very localized fire had taken place in/under a nightstand in her bedroom, as shown in Figure 9.28.

Due to the very limited fire there was little suppression activity necessary and care was taken to preserve the scene for the investigator. Note the minimal plume extension from beneath the nightstand and radiant heat-driven extension into the sisal carpet. Under the nightstand was a small, plastic-cased heater (plugged in). The remains of clothing were found charred into the remains of the heater, having apparently fallen from the chair. Note the shoes and throw rug in disarray. The bedclothes were disturbed, there was a TV remote control on the bed, and prescription medicines on the table and desk, along with a large glass of alcohol and a partial bottle of alcohol.

FIGURE 9.24 ◆ The wetted areas about 30 s later. Note the wetted areas on the front of the cushions and the more limited distribution of concentrated (pooled) liquid but more discrete splatters on the floor in front of it. *Courtesy of John DeHaan.*

The scenario and physical evidence were entirely consistent with the victim's partly undressing while consuming alcohol and medications, then reclining on the bed to watch television. She apparently fell asleep after partially undressing and dropping clothing haphazardly onto the chair. The clothing fell off the chair, onto the heater, and ignited into a small, but very smoky, fire. The deceased then awoke and attempted to escape but was overcome once she stood and started breathing in the accumulated smoke layer more deeply, reaching the landing before collapsing from the combined effects of CO, smoke, alcohol, and drugs.

FIGURE 9.25 ◆ The effect of throwing the liquid against a textured fabric surface rather than a smooth paper one (case example 6). *Courtesy of John DeHaan.*

This case is courtesy of Ross Brogan, New South Wales Fire Brigade, Greenacre, NSW, Australia.

◆ **9.8 CASE EXAMPLE 8—PIZZA SHOP EXPLOSION AND FIRE**

An early-morning explosion and fire did considerable damage to a pizza shop. A considerable quantity of gasoline had been poured inside the premises (Figure 9.29). The remains of a plastic oil can were found melted on the floor, with residues of gasoline still inside the container (Figure 9.30).

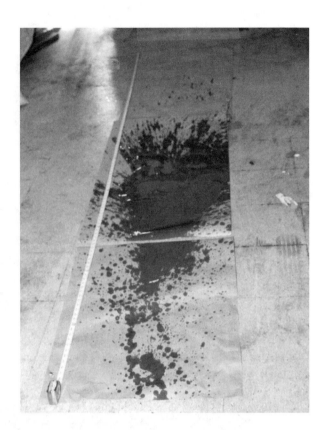

FIGURE 9.26 ◆ The effect of throwing the liquid against a horizontal smooth paper surface (case example 6). *Courtesy of John DeHaan.*

FIGURE 9.27 ◆ A partially undressed woman found dead on the landing outside her bedroom door, covered in soot (case example 7). *Courtesy of Ross Brogan, New South Wales Fire Brigade, Fire Investigation & Research Unit, Greenacre, NSW, Australia, by permission.*

FIGURE 9.28 ◆ A localized fire in and under the bedroom nightstand. Note the plume extension from beneath the nightstand, radiant heat-driven extension into the carpet, heater (plugged in), shoes and throw rug in disarray, prescription medicines on the table and desk, and glass and bottle of alcohol (case example 7). *Courtesy of Ross Brogan, New South Wales Fire Brigade, Fire Investigation & Research Unit, Greenacre, NSW, Australia, by permission.*

FIGURE 9.29 ◆ Interior of pizza shop showing fire damage from a large gasoline pour on the floor (case example 8). *Courtesy of Ross Brogan, New South Wales Fire Brigade, Fire Investigation & Research Unit, Greenacre, NSW, Australia, by permission.*

FIGURE 9.30 ◆ Melted and scorched plastic oil container (center) still contains gasoline. Note the direction indicators of melted plastic and scorched items on the shelf, which act as vectors pointing toward heat. *Courtesy of Ross Brogan, New South Wales Fire Brigade, Fire Investigation & Research Unit, Greenacre, NSW, Australia, by permission.*

FIGURE 9.31 ◆ Broken glass forced outward from the explosion within the shop. Note the toilet paper roll inside the door with paper extending up the stair to the right. *Courtesy of Ross Brogan, New South Wales Fire Brigade, Fire Investigation & Research Unit, Greenacre, NSW, Australia, by permission.*

FIGURE 9.32 ◆ A scorched and burned athletic shoe indicates the path of flight up the steps. *Courtesy of Ross Brogan, New South Wales Fire Brigade, Fire Investigation & Research Unit, Greenacre, NSW, Australia, by permission.*

The gas stove in the kitchen (with pilot light) was identified as the ignition source. The rear door appeared to have been broken by the explosion and then exposed to a sooty fire. The remains of a toilet paper roll trailer were found adjacent to the rear door (the roll bearing gasoline residues) (see Figure 9.31). Near the rear stairs was a badly scorched athletic shoe with surface scorching and melting consistent with brief exposure to flames, as shown in Figure 9.32.

Farther away were blood droplets and the remains of burned clothing in the alleyway leading to the apparent location of the "getaway car" (Figure 9.33). Two individuals arrived at a local hospital at about the same time as the fire report, each reporting a different address as the location where his burns and other injuries were obtained. The fire service responded to both addresses and found no fires at either. One suspect was apparently inside the structure pouring the gasoline while the other was outside preparing the trailer. Bloodstains from the step and parking area were matched to one of the suspects.

The individual inside succumbed to his burn injuries some 3 months after the fire. The surviving suspect eventually implicated the pizza shop owner in a "hired torch" scheme.

This case is courtesy of Ross Brogan, New South Wales Fire Brigade, Greenacre, NSW, Australia.

FIGURE 9.33 ◆ Burned clothing remnants between rear door and vehicle. Bloodstains on pavement from suspect injured by explosion-propelled glass. *Courtesy of Ross Brogan, New South Wales Fire Brigade, Fire Investigation & Research Unit, Greenacre, NSW, Australia, by permission.*

◆ 9.9 CASE EXAMPLE 9—STAIRWELL SMOKE FILLING

This fire occurred at approximately 7:00 P.M. in a two-story, two-bedroom apartment. The fire was reported by an occupant, who then left, closed the door to the apartment, and awaited fire crews outside. No mention was made for some time that another occupant, an elderly adult male, was in his bedroom at the end of the upstairs hallway, presumably asleep, as was his habit at that time.

The fire had involved a large quantity of newspapers, magazines, and food containers that had been strewn on the stairs from the midflight landing down to the foyer on the main floor, as shown in Figures 9.34 and 9.35. The main structure was reinforced concrete and masonry. The interior walls were painted plasterboard.

There was a vinyl wallpaper (woodgrain) on one wall of the stairwell. The hall floors and steps were carpeted. The entry foyer floor was vinyl tile. The flame plume had extended primarily along one wall, with areas of consumption of wall covering right to the gypsum at the base of the steps, with reduced blistering and charring of the paint up to a height of about 2.7 m (9 ft). The ceiling over the stairs was about 5.2 m (17 ft) above the floor of the foyer; the stairs were 0.8 m (34 in.) wide. The hot smoke layer was recorded by the discoloration and blistering of the paint on the walls of the hall and stairwell, and was at approximately 2.2 m (7 ft) from the foyer floor level. The

FIGURE 9.34 ◆ View from Foyer to midstain landing, with hall and the victim's bedroom door to the right. *Courtesy of Halton Regional Police Service, Oakville, Ontario, Canada, by permission.*

upper hallway was approximately 6.1 m (20 ft) in total length, 2.4 m (8 ft) in height, and 0.9 m (35 in) in width. A hallway door to a second bedroom was tightly closed (as was a furnace room door) at the time of the fire, as minimal smoke/heat penetration was noted in these rooms. The door to a 2.2 × 2.4 × 2.4-m (7 × 8 × 8-ft) bathroom at the top of the stairs was open and there was extensive heat and smoke damage to the room and its contents.

The responding firefighters found the closed entry door to be burning from within and fire extending out into the common hallway. The door was forced open and a substantial fire was discovered in the stairwell and in the entry foyer. The fire was extinguished with fog/water application. There was no immediate search for victims, as no one was reportedly present. The victim was discovered when firefighters opened windows in the upstairs bedrooms to vent smoke and steam.

FIGURE 9.35 ◆ View of the hallway from the landing. The victim's bedroom was at the end of the hallway. The smoke pattern confirmed that the door to the bedroom was partially open during the fire. Note the lines of demarcation of damaged and undamaged wall surfaces. *Courtesy of Halton Regional Police Service, Oakville, Ontario, Canada, by permission.*

The door at the end of the hallway was reportedly nearly closed when firefighters searching the apartment entered (Figure 9.35). The bedroom (approximately 13 × 12 ft [4 × 3.7 m]) was heavily charged with smoke. The windows were closed.

The victim was found on his bed by the sound of his labored breathing. He was removed from the room and resuscitated with oxygen but succumbed to inhalation injuries and extensive first- and second-degree burns on the front half of his torso and face the next day. The level of the hot smoke layer in the room was 1.5–1.8 m (5–6 ft) from the floor as evidenced by discoloration of the walls and the softening of a plastic clock face on one wall. The collapse of a plastic lamp shade on a table and light smoke deposits visible on various surfaces throughout the room as well as the observations of the firefighter indicated that smoke had largely filled the room. Upon admission to hospital the victim was found to have a COHb saturation of 23.3 percent (after some 40 min on oxygen). He was seen to have soot in the nostrils.

The fire damage was limited to the stairwell, and the issue became "What were the conditions in the bedroom" such that the victim could have suffered those injuries? He was observed to have no singed hair and he appeared to be dressed in a cotton T-shirt and underpants that did not appear to be thermally damaged (his outer clothing was found on a sofa in the room). It was conjectured that the fire gases produced by the stairwell fire had penetrated around the partially open door during the course of the fire and its suppression. The radiant and convected heat of the accumulated hot gases (supplemented in part by the steam generated by the suppression activity) induced first- and second-degree burns on the victim.

The heat release rate of a fire at the foot of the stairs can be estimated by applying the flame height calculation method from *NFPA 921* (NFPA 2001), described in Chapter 3, on fire pattern analysis:

$$H_f = 0.174 \, (k\dot{Q})^{2/5}, \tag{9.13}$$

$$\dot{Q} = \frac{79.18 \, H_f^{5/2}}{k}, \tag{9.14}$$

where

H_f = flame height (m),

$\dot{Q}$ = heat release rate (kW), and

k = wall effect factor, with

    1 = no nearby walls,

    2 = fuel package at wall, and

    4 = fuel package in corner.

The area of direct flame contact (both intermittent and continuous) is determined by establishing the area of most damage to the wall covering, which extended to a height, H_f, of approximately 2.7 m (8.85 ft) from the foyer floor as shown in Figure 9.36. Calculating for the heat release rate, $\dot{Q}$, with a wall effect factor of $k = 2$ (fuel package at wall),

$$\dot{Q} = \frac{(79.18) \, (2.7)^{5/2}}{2} = 474 \, \text{kW (or} \sim 500 \, \text{kW).} \tag{9.15}$$

Due to the position of the fire, we can assume that the smoke production of a 500-kW fire accumulated in the stairwell and hallway above, as shown in Figures 9.34 and 9.35. Using the Zukoski (1978) method for smoke production rates, the equation for estimating the mass flow rate of a plume above the flame at 20°C (68°F) is

$$\dot{m}_p = 0.065 \, \dot{Q}^{1/3} \, Y^{5/3}, \tag{9.16}$$

where

$\dot{m}_p$ = rate of smoke-filled gas production (kg/s),

$\dot{Q}$ = total heat release rate (kW), and

Y = distance from virtual point source to bottom of smoke layer (m).

FIGURE 9.36 ◆ View down the corridor from the bed-room toward the stairwell. Note the demarcation sur-face effect of smoke and heat patterns on the far wall. *Courtesy of Halton Regional Police Service, Oakville, Ontario, Canada, by permission.*

For a fire of $\dot{Q} = 500 \text{ kW}$ and $Y = 5$ m (16.4 ft) at a time close to its start, the estimated smoke production rate along the centerline is

$$\dot{m}_p = (0.065) \, (500)^{1/3} \, (5)^{5/3} = 7.51 \text{ kg/s}. \tag{9.17}$$

As the corridor fills, the height becomes $Y = 2.5$ m (8.2 ft),

$$\dot{m}_p = (0.065) \, (500)^{1/3} \, (2.5)^{5/3} = 2.37 \text{ kg/s}. \tag{9.18}$$

With a maximum density of 1.2 kg/m^3 these fill rates, at a minimum, are 6.1 and 1.94 m^3/s. With an average T of 200°C, the density would be reduced by a factor of 273/473 (T_a/T_g), or approximately 0.7 kg/m^3, and the filling rates would be, corre-spondingly, 10.6 and 3.3 m^3/s.

If the upper hall (including the staircase) is 6.1 m long × 2.4 m high × 0.9 m wide (20 × 7.87 × 2.95 ft), its volume is only 13.2 m^3. Even allowing for the filling of the bathroom, which measures 2.2 × 2.4 × 2.4 m (7.2 × 7.9 × 7.9 ft), with a volume of 13.7 m^3, the hall and bathroom (26.9 m^3) are going to be filled with smoke in 5 s or less after the fire reaches 500 kW. The buoyancy of the gases accumulating at the hall door would cause them to push around the partly opened door from floor to ceiling and begin to fill the bedroom beyond. The bedroom was 35.5 m^3 in volume and the more buoyant hot gases coming around the door would produce a 0.6- to 0.9-m-deep layer (from the ceiling) within a few minutes.

The radiant heat from a gas layer in a compartment was established by Quintiere and McCaffrey (Quintiere 1998, 62), where smoke layers with temperatures between 200 and 300°C were found to produce radiant heat fluxes of 3–5 kW/m^2 on the floor of the compartment. Since the radiant heat threshold for second-degree burns is of the order of 4 kW/m^2 (Quintiere 1998, 60), it can be seen that radiant heat from a hot gas layer alone can induce second-degree burns (given 30 s) or more at temperatures of approximately 250°C (far below flame temperatures of 800°C (1472°F)).

In addition, convective heat transfer is occurring as the room fills with hot fire gases and steam and adds to the thermal insult to which the body is subjected, and increases the likelihood of burns to victims in rooms filling with combustion product. This fire is a good illustration of the application of (1) indicators, (2) flame height vs. heat output relationship, and (3) Zukoski's smoke-filling estimations.

This case is courtesy of Halton Regional Police Service, Oakville, Ontario, Canada.

◆ 9.10 SUMMARY AND CONCLUSIONS

This chapter introduced the concept of detailed case histories as a learning tool in applying and understanding the principles of fire engineering analysis. With case histories, many underlying concepts, fire science principles, and new areas of research are uncovered. Controlled tests can be of value in formulating or testing a hypothesis for the origin of a fire or its growth, the behavior of the occupants, or an act or omission that contributed to the fire loss.

While many principles are discussed, the case analyses are not all conclusive; the reader is free to apply additional concepts in reviewing, discussing, or expanding upon the case histories.

■ ■

Problems

9.1. Find a recent fire death case in which "smoking materials" was the identified cause. What data would you collect and what tests would you want to confirm this?

9.2. Examine the photographs in one of the examples. Try to determine additional information from this review that did not appear in the case study.

9.3. Using the equations provided regarding plume height against a wall, determine the maximum-size fire that would produce plume damage as seen in Case Example 1 (Section 9.1). Assuming a fuel mass loss rate of 4 kg/hr and a Δh_{eff} of 32 kJ/g, calculate the duration of the fire.

9.4. Review three of the historic cases covered in the *NFPA Fire Journal* and evaluate how today's knowledge could have helped solve these cases more quickly through testing of alternative hypotheses.

■ ■

Suggested Reading

DeHaan, J. D., S. J. Campbell, and S. Nurbakhsh. 1999. The combustion of animal fat and its implications for the consumption of human bodies in fires. *Science and Justice* 39(1):27–38.

DeHaan, J. D., and S. Nurbakhsh. 2001. Sustained combustion of an animal carcass and its implications for the consumption of human bodies in fires. *Journal of Forensic Science* 46(5):1076–1081.

Holleyhead, R. 1996. Ignition of flammable gases and vapors by cigarettes: A review. *Science and Justice* 36(4):257–266.

Lentini J. J., D. M. Smith and R. W. Henerdson, 1993 "Unconventional Wisdom: The Lessons of Oakland" *Fire and arson investigator.* 43(June) pp. 18–20.

Zeigler, D. L. 1993. The One Meridian Plaza fire—A team response. *Fire and Arson Investigator* 43(September) pp. 38–39.

Future Tools for the Fire Investigator

10 CHAPTER

> *I had come to an entirely erroneous conclusion which shows, my dear Watson, how dangerous it always is to reason from insufficient data.*
>
> Sir Arthur Conan Doyle,
> *"Adventures of the Speckled Band"*

◆ 10.1 INTRODUCTION

Fire investigators, particularly those who operate in rural areas, often transform their vehicles into mobile offices. These vehicles carry all the necessary tools, forms, maps, photographic, and evidence collection equipment needed to document fully a fire scene and compose a draft preliminary investigative case report. Operating in a reactive mode in high-visibility cases, they are expected to deliver their investigative reports and photographs on a timely basis but are hindered by both technical and administrative hurdles.

Based upon lessons learned from the Major Fire Investigations Program, the Federal Emergency Management Agency's (FEMA) U.S. Fire Administration (USFA) is molding new information technologies to improve the quality and productivity of investigations conducted at fire scenes, explosions, disasters, and other major events. The source of many of these lessons learned by the USFA was insights into the on-site field investigations of the World Trade Center (1993) and Oklahoma City bombing cases.

The Arson Intervention and Mitigation Strategy (AIMS 2000) project, under the direction of the USFA, supports local, state, and federal arson investigations through the introduction and dissemination of innovative fire investigation technologies. The purpose of this section is to report on the productivity successes of the AIMS 2000 strategy.

◆ 10.2 THE "TRIP KIT"—TRANSFER OF MILITARY TECHNOLOGY TO FIRE INVESTIGATORS

Facing the concerns of increasing case loads and shrinking budgets, investigators often inquire as to the availability of U.S. military-developed technologies that could increase their efficiency on the fire scene. Military technology transfers, however, are sometimes difficult to sustain, particularly in the areas of training, hardware support, and procurements.

For example, U.S. military special operations personnel during Operation Just Cause and Desert Storm carried portable mobile field offices (the size of commercial aircraft-sized luggage) onto foreign battlegrounds. These transportable units, which typically contained a laptop computer, digital camera, color scanner, and printer gave soldiers the ability to investigate, photograph, and record battlefield targets accurately. This information was quickly assembled in the field and transmitted back to field commanders for review and dissemination.

The operational requirements for conducting and documenting investigations at scenes of major fires and explosions are similar to those of the U.S. military and include a factor requiring a unique balance of quality and productivity. The AIMS 2000 strategy exploited the military solution and resulted in a novel effort to produce a field-deployable "virtual office" dubbed the Transportable Rapid Information Package (TRIP) Kit. Common practices, challenges, and proposed solutions form the functional foundation for the TRIP Kit design, as shown in Figure 10.1 and Table 10.1.

Figure 10.1 ◆ The AIMS 2000 strategy exploited the military solution and resulted in a novel effort to produce a field-deployable "virtual office" dubbed the Transportable Rapid Information Package, codeveloped by the USFA and U.S. TVA Police. *Photo courtesy of FEMA/USFA; www.usfa.fema.gov/dhtml/fire-service/aims2.cfm.*

TABLE 10.1 ◆ **Technological Assessment Matrix**			
Major Goal	*Current Practice*	*Current Challenge*	*TRIP Kit Solution(s)*
Case management	Nonstandard approach to scene	Reliance on *NFPA 921* and *1033*	Template-based scene exams
Incident reporting systems	Complete separate NFIRS and NIBRS forms	Reduce reporting duplication	Collect data, print NFIRS and NIBRS reports
Fire investigation unit management	Cases assigned and closed as needed	Develop case solvability scoring	Incorporate case solvability factors
Litigation and prosecutive support	Cases based on written files	Automate case files	Case-based data search and retrieval
Investigative efficiency and productivity	Case assignments by call-out	Measure efficiency	Time-motion studies

Source: Kenneth J. Kuntz, National Fire Programs Division, USFA.

DESIGN SPECIFICATION

The TRIP Kit design was first based upon a military model, which was later expanded to include the wider functionality needed by fire and arson investigators, as shown in Table 10.2. The TRIP Kit is the product of a national strategy to

- improve the productivity and quality of criminal fire and explosion investigations by standardizing the documentation process,
- use case management to enhance the investigative efficiency and effectiveness of fire investigation unit personnel,
- demonstrate commercial off-the-shelf (COTS) software and hardware,
- increase the communications and data transfer capability at the scene, and
- provide electronic references and computer fire modeling capability.

It is also intended that the system provide arson data through the states to the USFA's National Fire Incident Reporting System (NFIRS) and the FBI's National Incident Based Reporting System (NIBRS). In addition, AIMS 2000 system users would be able to query a wide variety of federal electronic public access information retrieval services currently available via Web sites, e-mail, or a CD-ROM library carried with the TRIP Kit.

Through partnerships agreements, the USFA is completing pilot testing TRIP Kits at the Tennessee State Fire Marshal's Office and U.S. TVA Police to work out operational issues before the national release of the configuration.

PRODUCTIVITY IMPROVEMENT

It has been documented in several authoritative studies that computer technologies can significantly

- enhance the quality and productivity of criminal investigations;
- improve arrest rates, clear unsolved cases, provide linkages with other similar cases, and share time-sensitive information among multiple agencies;
- reduce an agency's operational costs, and
- meet demands from citizenry for increased services.

TABLE 10.2 ◆ TRIP Functional Capabilities

Feature	Description
Basic virtual office unit	Laptop configured with color scanner, printer, and office software tools
Fire reports	Field data collection of information needed for NFIRS, NIBRS, and fire modeling
Interviews	Written field witness interviews using prompted questions
Evidence collection	Photography by digital and videocameras, printing of evidence labels
Fire scene diagramming	Computer-assisted drafting of building floor plans and fire scene
Mapping and GPS	Plotting the fire scene location on a digital map using GPS
CD-ROM library	Retrieval of fire reports, NFPA fire codes, NFIRS data profiles, and NIST models
Network conferencing	Wireless network with other TRIP units with video desktop conferencing
Telecommunications	Cellular and pager communications, faxing of reports from scene

Source: Kenneth J. Kuntz, National Fire Programs Division, USFA.

The collaborative efforts of sharing information and resources in major case investigations are often a measure of success. For example, in a multiagency criminal investigation in the southeastern United States, 150 local and federal investigators worked on a case containing 100,000 pages of information and more than 7000 leads. Using computer technology for document storage, retrieval, and search strategies was an essential tool for successful management of this case (Shill 1998).

A management study explored the positive impact of the use and reliance of computer databases. Investigators reported *informational* as well as *individual performance benefits* (Danziger and Kraemer 1985). The study also reported that investigators who make extensive use of computer technology are rewarded with substantially higher productivity gains compared to investigators who avoid new technology. Investigators who use computers reported assistance and arrests, clearances, and linkages between persons in custody and unsolved cases.

FUNCTIONAL LESSONS LEARNED

Many lessons have been learned during the initial testing phases of TRIP Kit technologies.

GPS and Computer Mapping. Providing the ability to look for maps using a Global Positioning System (GPS), the deployable computer has been helpful in determining exact jurisdictional boundaries in relation to the crime scene. Also, this use provides the ability to integrate digital computer maps into formal reports.

Cases where lightning strikes are suspected or must be ruled out are becoming more common. GPS data collection allows the investigator to determine if a lightning strike was the source of fire ignition by comparing it to information sources from commercial lightning strike databases (Vaisala Inc. 2002).

Video and Teleconferencing.   A unique function of the TRIP Kit is to capture imagery digitally and transfer it electronically by Internet connection. This enables investigators to provide good on-site capture of photos for preliminary reports and provides the ability for remote "photoconferencing." This capability is used by the State of Tennessee to relay photographs of major fires and explosions for use in timely briefings of upper management, who may need the information to allocate resources to continue to support the case (Figure 10.2).

Fire Modeling.   The use of fire modeling is becoming more prevalent and tests show that accurate collection of building measurements is needed for scene diagrams and models. The fire model later adds interpretive value to investigations.

Computer capabilities include access to fire modeling heat release data and other information stored on CD-ROM formats. These sources include the NFPA *Fire Protection Handbook* (NFPA 2000a), NFPA codes and standards such as *NFPA 921* (NFPA 2001), and *The SFPE Handbook of Fire Protection Engineering* (SFPE 2002a).

FIGURE 10.2 ◆ The Special Operations Response Team (SORT) operated by the Tennessee State Fire Marshal's Office uses mobile technologies at scenes of major fires and explosions to assimilate information, document the scenes, and transfer imagery back to their headquarters using cell phone and satellite telephone communications. *Courtesy of D. J. Icove.*

Records Management. The Records Management department, when facing automation, must address a critical "build vs. buy" decision on the acquisition of software to handle a comprehensive departmentwide records management system. Suggested requirements indicated that printed case reports contain color text, department logos shown on reports, scanned photographs added as attachments, and custom field layouts. Statistical information must be collected at the same time as other case data to avoid replication of work.

EXPLORING WIRELESS TECHNOLOGIES

The TVA Police are presently experimenting with low-cost wireless technologies to extend the limits of TRIP Kits, particularly at the scenes of major cases. The concept is that the use of wireless technologies will impact productivity and real-time information availability.

For an example of an application in industry, Microsoft has deployed about 3000 access points at its Redmond, Washington, campus and satellite branch offices. About 7000 employees are using wireless LAN cards in their laptops, a number that is expected to grow to 20,000 by the end of the year 2003. Microsoft believes that the use of wireless technologies yields a 30-min productivity gain per employee per day.

INFORMATION TECHNOLOGY TRAINING IS KEY

The Metropolitan Toronto Police Department reported that information technology has saved the force $21.3 million in cost avoidance and significantly increased productivity (Wolchak et al. 1995). Toronto, like other jurisdictions receiving increasing demands from their citizenry for expanded services, is using information technology to meet public expectations. These demands also stress the already committed austere budget environment.

Internationally, policing strategies have become more receptive to new technology (Icove, Wherry, and Schroeder 1998). The New Zealand police strategic plan focuses on people, business redesign, and technology. Their strategy for technology includes gains in productivity and performance via operational information management systems, particularly those that have effective interaction between the police and the community they protect (Policing New Zealand, 2000; www.crime.co.nz).

The TRIP Kit is a truly mobile office as shown in Figure 10.1. Investigators in the field have at their disposal the tools necessary to document and store time-sensitive information electronically. Experience has shown that capturing digital images of crime scene photographs, video, and audio increases police productivity (Wise 1995). The TRIP Kit combines such institutional experiences from these and other studies that report to increase investigator productivity and efficiency using low-cost computer information technologies.

COMPUTER SKILL ENHANCEMENT FOR INVESTIGATORS

In this project, TRIP Kits were issued to federal and state law enforcement agents whose responsibility included criminal investigations of fires and explosions. In the initial phase of the project, the only knowledge about operating laptop computers came from "home use" as reported by the participants.

The project managers quickly realized that these investigators needed to obtain baseline knowledge in applying computer skills to operate and maintain their TRIP Kits successfully. The final course consisted of 4 days of instruction in using operating system and major office applications. These applications included the electronic mail, word

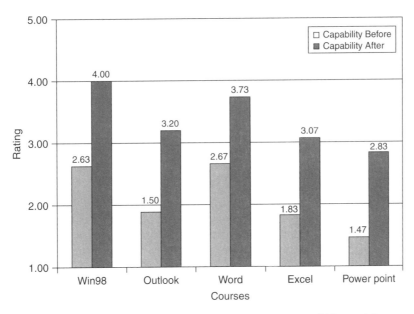

FIGURE 10.3 ◆ Average capability rating per course of 30 participants. *From Blocker and Icove (2001).*

processing, spreadsheet, and presentation software modules. These applications are the core tools used in preparing criminal investigative reports and courtroom presentations.

The investigators also received training in several of the peripherals used in the TRIP Kits, such as digital cameras and a screen-capturing program. They were shown and had the opportunity to integrate digital photos into documents and presentation graphics.

As with the introduction of any innovative technology, the USFA and TVA Police quickly learned that education in information technology is an essential element of success in the testing, evaluation, and deployment of TRIP Kits. With as little as 4 days of instruction, criminal investigators self-reported a 29 to 48 percent improvement in skill levels after they completed a modular seminar concentrating on word processing, spreadsheets, and presentation graphics (Figure 10.3).

A second level of training came with using TRIP Kits integrated into the State of Tennessee and TVA's respective police information management systems (Figure 10.2). The skills obtained during the first phase produced a "level playing field" during the second critical training phase. Additional training is anticipated as these investigators become more accustomed to the technology.

FUTURE CAPABILITIES

The TRIP Kit concept addresses the major contemporary issues of case management, incident reporting systems, unit management, litigation support, and investigator productivity. It will truly be a model investigative tool for the next decade (Madrzykowski 2000).

In summary, there is significant improvement in the overall capability of investigators when exposed to modular training seminars consisting of customized training on pertinent applications. Investigators had more confidence and improved productivity and efficiency, consistent with results of other research studies. Clearly an organized instructional design is needed to undertake comprehensive training.

As the literature demonstrates, computer and information technologies can significantly improve the quality and productivity of criminal investigations, reduce an agency's operational costs when using integrated information management systems, and meet demands for increased services.

Finally, some minimal computer skill information technology training should take place along with delivery of TRIP Kits or similar technologies to fire and arson investigators.

◆ 10.3 PRESENT AND EMERGING TECHNOLOGIES

Several present and emerging technologies have shown promise. Investigators often seek out their own solutions, particularly when approached by universities, government laboratories, and vendors who think they may have an insight into the overall problems.

Reference in this book to any product, process, or service by trade name, trademark, manufacturer, or otherwise does not necessarily constitute its endorsement or recommendation by the authors, their agencies, or the U.S. Government.

INFRARED VIDEO IMAGING

Pioneered by Inframetrics, false-color infrared video recording has been in use in industrial and military applications for more than a decade, having been used to survey remotely and quickly high-tension line insulators for heating from leakage currents, transformers for overheating, chemical delivery pipes, and storage tanks and to measure operating temperatures of surfaces of aircraft or vehicles.

Its applications to fire research would seem self-evident, providing the capability of remotely and accurately measuring the surface temperatures of fuel surfaces or masses of hot gases or of the external surfaces of walls, windows, or other structural elements as they are exposed to fires.

A very early version was used by Helmut Brosz in 1991 in monitoring fire development in compartments (Brosz, Posey, and DeHaan, unpublished). Although DeHaan (1995, 1999a) used Inframetrics equipment to measure the surface temperatures of carpet during the evaporation of volatile fuels, very few papers on the topic have been published in the fire research literature. It can distinguish areas of a surface whose temperatures differ by only 0.25°C in static images (see Figure 10.4). Its applicability to capturing transient events such as the chaotic movement of the high-temperature flames in postflashover fires has been demonstrated (DeHaan 2001).

The high cost of the necessary equipment has been reduced significantly in recent years with the dramatically increased power of low-cost computer components. It is to be hoped that more researchers will use this type of technology to study surface temperatures that presently are impractical to measure accurately using traditional "hard-wire" thermocouple devices. Several versions are made today by FLIR Systems.

TOTAL STATION SURVEYING SYSTEMS

This is a resource that has been widely used by accident investigation teams in large police departments (as well as by commercial surveying companies) for many years in the United States, the United Kingdom, and Australia, yet it is rarely used by fire investigators, probably due to the high initial cost of the systems.

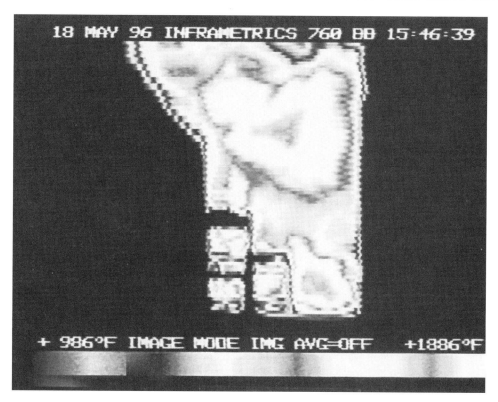

FIGURE 10.4 ◆ DeHaan (2001) used Inframetrics infrared video to study the dynamics of chaotic postflashover fires. *Photo courtesy of John DeHaan.*

As described in *Kirk's Fire Investigation* (DeHaan 2002) and in Chapter 4 of this text, it uses a laser beam from a surveyor's theodolite to measure the distance from the reference point (which in turn is located by GPS data) to any point in view. The data system captures the distance, direction, and elevation of every point at which the laser is aimed and codes it as an edge, corner, point, or feature as instructed by the operator.

The computer data are then downloaded to a PC for translation into a finished diagram with scale dimensions. The newest generation of systems will operate remotely at distances of up to 100 m (330 ft) without a reflector or marker and at much greater distances when a reflector or marker can be positioned to reflect the laser beam back to the system with a higher efficiency. Plotting software is then used to produce accurate, scaled diagrams of the scene (in both plan and elevation views).

The newest versions (such as the Leica FMS307) have a small video screen to display a preview version of the plot to ensure that all important data points are measured. The system can be used indoors or out and can capture critical measurements of a large scene far more accurately than can be achieved with a tape measure or visible-light optical range finder (see Figure 10.5 for a typical system). Such systems deserve more evaluation and use by fire and explosion investigators. They are currently offered by Leica, Topcon, and Sokkia.

LASER SCANNING OF SCENES

A technology firm in Chapel Hill, North Carolina (3rd Tech, Inc., 119 East Franklin Street, Third Floor, Chapel Hill, N.C. 27514; http://www.3rdtech.com/), has developed a

FIGURE 10.5 ◆ The Total Station Surveying System is a resource that has been widely used by accident investigation teams in large police departments for many years in the United States, the United Kingdom, and Australia, yet it is rarely used by fire investigators, probably due to the high initial cost of the systems. *Photo courtesy of Michael J. Capman, by permission.*

laser-scanning device that uses a time-of-flight, 5-mW laser range finder that, when mounted on a photographer's or surveyor's tripod, will scan an entire room with the laser beam, taking 25,000 measurements per second. Figures 10.6 and 10.7 show a data image and a laser scan of the same scene.

Using the same laser interferometry and data analysis presently used in Total Station surveying and measuring systems, but on an automated scanning basis, the device will record the distance, direction, and elevation of all visible surfaces in the room. The data system will then permit reconstruction of the image by virtual imaging, as well as precise measurements of its visual features.

After processing, the linear distance between any two points in the image can be extracted. The scanner can be fitted with a digital camera to capture visual images that can be blended into the digitally mapped data. The system will record at distances of 0.3–12 m (1–40 ft), with an accuracy of 0.008 m (0.3 in.) at 12 m, and over an elevation range of −60° to +90°. A single 360° rotation to scan a room takes approximately 20 min. The system uses a Windows 2000/XP-type PC for data storage and processing and is fully portable.

DIGITAL IMAGING CAPTURE

A recently introduced optical technology combined with digital processing called the iPIX system allows the investigator to capture a 360° view of a scene and produce a fully immersive, navigable (virtual reality) image. It uses a 185° fisheye lens on a digital camera. The camera is rotated horizontally 180° between two photos, and iPIX software seams the two images together, allowing the viewer to "virtually" enter the

FIGURE 10.6 ◆ Image of a simulated crime scene from a laser scan. *Courtesy of 3rd Tech, Inc., Chapel Hill, N. C., by permission; www.3rdtech.com.*

scene and look up, down, to either side, or even directly behind. The iPIX (2003; www.iPIX.com) technology has been used successfully at fire scenes under the AIMS 2000 project.

MRI AND CT IMAGING

Progress in magnetic resonance imaging (MRI) and computerized tomography (CT or multiple-position X-ray imaging) has been very rapid. Recently MRI and multiple-slice CT were demonstrated, in the examination of a very badly charred body, to establish the presence or absence of fractured bones and reconstruction of impact injuries as well as to document the presence of surgical implants and the like (Thali et al. 2002). Such techniques have also been suggested for use in examination of melted or charred artifacts from fire scenes where nondestructive testing for pockets of liquid or similar nonradiopaque inclusions would be desirable. While expensive, such techniques may provide answers where X rays cannot.

POLYNOMIAL TEXTURE MAPPING

A new method for increasing the photorealism of maps and complex surfaces is known as Polynomial Texture Mapping (PTM). This technique could be used to enhance and visualize fire patterns on various surfaces such as wood paneling, gypsum, and concrete.

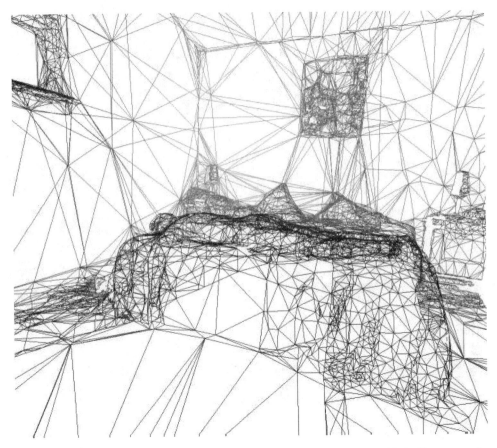

FIGURE 10.7 ◆ Laser scan of the simulated crime scene in Figure 10.6. *Courtesy of 3rd Tech, Inc., Chapel Hill, N. C., by permission; www.3rdtech.com.*

PTM was originally used to enhance the surfaces of hard-to-read ancient clay tablets for interpretation by archaeologists and a pilot system is available through a Hewlett–Packard/Compaq Web site (http://www.hpl.hp.com/ptm/).

THE FIRE RESEARCH LABORATORY

The ATF Fire Research Laboratory in Ammendale, Maryland, has been completed. It will have the physical facilities to observe and measure fires of all sizes, from bench-scale tests to very large fires involving entire rooms, vehicles, and even multiple furnished rooms. It is anticipated that its unique blend of fire engineers, fire and forensic scientists, and public-sector fire investigators will create a wealth of knowledge about the properties of fires involving commonly encountered fuels such as cardboard boxes, liquid fuel spills, furniture, and even vehicles in "real-world" events.

This knowledge is expected to validate or refine many of the existing models but may also refute a few of them (or at least show the limitations of their applicability) with scientifically valid studies with demonstrable uncertainties or error rates. This is exactly the kind of knowledge that the *Daubert/Kumho* decisions have begun to require of all fire investigators. It holds exciting promise for the future of fire investigation as it aims toward meeting the goals set out in 1997 at the planning conference described in the Preface to this book (Dillon, Hill, and Sheppard 2003); (Nelson and Tontarski 1998)

◆ **10.4 SUMMARY AND CONCLUSIONS**

The future progress of fire investigation, then, is really dependent not just on advances in technology, but on the ever-increasing cooperation and communication among the disparate elements of the fire community. The daring training project that came to fruition as InterFire brought together the resources of the International Association of Arson Investigators (IAAI), U.S. Fire Administration (USFA), National Fire Protection Association (NFPA), Bureau of Alcohol, Tobacco and Firearms (ATF), and American Re-Insurance (as well as individual contributors from all over the United States) to produce a revolutionary, interactive CD training guide and "instant" resource library for every fire investigator. This tremendously successful effort demonstrated what could be accomplished by interdisciplinary cooperation.

The contact among fire investigators, engineers, and scientists has become more routine through shared sources of information. The IAAI, ATF, SFPE, and NFPA all host conferences and special training sessions that are intended not for investigators only or researchers only, but for both. Interscience Communications of London hosts major biennial conferences such as InterFlam, Fire and Materials, Human Behavior, and Fire Retardant Materials at which fire investigation is a major topic, along with state-of-the-art academic or industrial research.

Every year the IAAI hosts national and regional training conferences that bring investigators and researchers together in a variety of formats. The International Association of Fire Safety Sciences hosts a triennial international conference that usually includes papers of special interest to the investigative community.

For those unable or unwilling to travel, there are a host of peer-reviewed publications that offer very useful information. These include the *NFPA Journal*, *Fire Technology*, the IAAI's *Fire and Arson Investigator*, *Fire and Materials*, *Fire Safety Journal*, and the *Journal of Forensic Sciences*. While it is not practical for an individual to belong to more than a few professional organizations, these journals and the published proceedings from the conferences mentioned above make it possible to keep up with the ever-increasing wealth of information. It is this information that will make the greatest difference to the quality of fire investigation in the future.

■ ■

Suggested Reading

DeHaan, J. D. 2001. Full-scale compartment fire tests. *CAC News* Second Quarter: 14–21.

FLIR Systems: www.flirthermography.com

IPIX: www.ipix.com

3rd Tech, Inc.: www. 3rdtech.com

Vaisala: www.lightningstorm.com

References

Accu-Line. 2003. Accu-Line Drawing Products, Buildersight Co. www.Accu-line.com

Ahonen, A., M. Kokkala, and H. Weckman. 1984. Burning characteristics of potential ignition sources of room fires. Research Report 285. Espoo, Finland: Technical Research Centre.

Albers, J. C. 2002. Pour pattern or product of combustion? Fire-Arson Investigator, California Conference of Arson Investigators, July, p. 5.

Alpert, R. L. 1972. Calculation of response time of ceiling-mounted fire detectors. *Fire Technology* 9:181–195.

American Psychiatric Association. 1994. *Diagnostic and statistical manual of mental disorders,* 4th ed. (DSM-IV). Washington, D.C.: APA.

Armor Forensics Inc. 2003. 13386 International Parkway, Jacksonville Fla., 32218. www.armorholdings.com

Associated Press. 2001. Shadowy eco-militants stepping up arson campaign. Terrence Petty, AP, June 2.

ASTM. 1999a. *Fire test standards,* 5th ed. West Conshohocken, Pa.: American Society for Testing and Materials Committee E-5 on Fire Standards. ISBN 0-8031-2597-6.

ASTM. 1999b. *E860 Standard practice for examining and testing items that are or may become involved in products liability litigation.* West Conshohocken, Pa.: American Society for Testing and Materials Committee E-30.05 on Forensic Sciences.

ASTM. 2000. ASTM E502-84 *Standard test method for selection and use of ASTM standards for the determination of flash point of chemicals by closed-cup methods.* West Conshohocken, Pa.: American Society for Testing and Materials.

ASTM. 2001. *Forensic science book of standards, Vol. 14.02.* West Conshohocken, Pa.: American Society for Testing and Materials Committee E-30 on Forensic Sciences.

ASTM. 2002. *E1321–97a Standard test method for determining material ignition and flame spread properties.* West Conshohocken, Pa.: American Society for Testing and Materials Subcommittee E-05.22.

ASTM. 2003. *E603-01 Standard guide for room fire experiments.* West Conshohocken, Pa.: American Society for Testing and Materials Subcommittee E05.13.

ASTM. 2003. *E1354-03 Standard test method for heat and visible smoke release rates for materials and products using an oxygen consumption calorimeter.* West Conshohocken, Pa.: American Society for Testing and Materials Subcommittee E05.21.

Ayala, F. J., and B. Black. 1993. Science and the courts. *American Scientist* 81:230–239.

Babrauskas, V. 1980. Estimating room flashover potential. *Fire Technology* 16:94–103, 112.

Babrauskas, V. 1981. Applications of predictive smoke measurements. *Journal of Fire and Flammability* 12:51–64.

Babrauskas, V. 1988. Upholstered furniture room fires—Measurements, comparison with furniture calorimeter data, and flashover predictions. *Journal of Fire Sciences* 2:5–19.

Babrauskas, V. 1996. Fire modeling tools for FSE: Are they good enough? *Journal of Fire Protection Engineering* 8(2):87–96.

Babrauskas, V. 1997. The role of heat release rate in describing fires. *Fire and Arson Investigator* 47(June):54–57.

Babrauskas, V. 1998. Fire safety improvements in the combustion toxicity area: Is there a role of LC50 tests? In *Flame Retardants '98*, pp. 213–224. London: Interscience Communications.

Babrauskas, V. 2002. The heat release rate hazard of Christmas trees. Personal communication.

Babrauskas, V., D. Baroudi, J. Myllymaki, and M. Kokkala. 1997. The cone calorimeter used for predictions of the full-scale burning behaviour of upholstered furniture. *Fire and Materials* 21:95–105.

Babrauskas, V., and S. J. Grayson, eds. 1992. *Heat release in fires*. London: Chapman and Hall. ASIN: 1851667946.

Babrauskas, V., and J. F. Krasny. 1985. *Fire behavior of upholstered furniture*. NBS Monograph 173. Gaithersburg, Md.: U.S. National Bureau of Standards.

Babrauskas, V., and R. D. Peacock. 1992. Heat release rate: The single most important variable in fire hazard. *Fire Safety Journal* 18:255–272.

Babrauskas, V., and W. D. Walton. 1986. A simplified characterization of upholstered furniture heat release rates. *Fire Safety Journal* 11:181–192.

Babrauskas, V., et al. 1982. Upholstered furniture heat release rates measured with a furniture calorimeter. NBSIR 82-2604. Gaithersburg, Md.: National Bureau of Standards.

Bailey, J. W. 1983. Archaeology: Help for arson investigation from an unexpected source. *Fire Chief* (Washington, D.C.) March: 24–27.

Beering, P. S. 1996. *Verdict: Guilty of burning—What prosecutors should know about arson*. Indianapolis, Ind.: Insurance Committee for Arson Control, May.

Berrin, E. R. 1977. Investigative photography. Technology Report No. 77-1. Boston: Society of Fire Protection Engineers.

Blocker, C. M. and D. J. Icove. Improved quality and productivity of criminal investigators through information technology education. TVA University Conference Paper Series. U.S. Tennessee Valley Authority. Knoxville, Tenn., February 2001, pp 30–37.

Bohnert, M., T. Rost, and S. Pollak. 1998. The degree of destruction of human bodies in relation to the duration of the fire. *Forensic Science International* 95:11–21.

Bryan, J. L. 1983. An examination and analysis of the dynamics of the human behavior in the Westchase Hilton Hotel Fire, Houston, Texas, on March 6, 1982. Report prepared for the National Fire Protection Association, March.

Bryan, J. L., and D. J. Icove. 1977. Recent advances in computer-assisted arson investigation. *Fire Journal*, January.

Bukowski, R. W. 1991. Fire models: The future is now! *Fire Journal* 85(2):60–69.

Bukowski, R. W. 1992. Analysis of the Happyland Social Club fire with HAZARD I. *Fire and Arson Investigator* 42(3):36–47.

Bukowski, R. W. 1995a. How to evaluate alternative designs based on fire modeling. *Fire Journal* 89(2):68–74.

Bukowski, R. W. 1995b. Predicting the fire performance of buildings: Establishing appropriate calculation methods for regulatory applications. In *ASIAFLAM'95, International Conference on Fire Science and Engineering, 1st Proceedings,* Kowloon, Hong Kong, March 15–16, pp. 9–18.

Bukowski, R. W. 1996. Modelling a backdraft incident: The 62 Watts Street (New York) fire. *Fire Engineers Journal* 56(185):14–17; *NFPA Journal*, Nov/Dec 1995: 85–89.

Burnette, G. E. 2000. Spoliation of evidence: A fire scene dilemma. *InterFIRE*. www.interfire.org

Burnette, G. E. 2003. Fire scene investigation: The Daubert challenge. Personal communication. (See also www.interfire.org)

Butler, C. P. 1971. Notes on charring rates in wood. Fire Research Note No. 896. London: Department of the Environment and Fire Officers Committee, Joint Fire Research Organization.

Campagnolo, T. 1999. The Fourth Amendment at fire scenes. *Arizona Law Review* 41 (Fall):601.

Carey, N. 2002. Powerful techniques, arc fault mapping. *Fire Prevention* (UK) March:47–50.

Chesbro, K. J. 1994. Taking Daubert's focus seriously: The methodology conclusion distinction. *Cardozo Law Review* (Yeshiva University) 15(6–7):1745–1753.

Clifford, R. C. 2000. *Qualifying and attacking expert witnesses,* pp. 3–64. Costa Mesa, Calif.: James. ISBN 0-938065.

C. M. Blocker and D. J. Icove, "Improved Quality and Productivity of Criminal Investigations through Information Technology Education." TVA University Conference Papers Series, / U. S. Tennessee Valley Authority, Knoxville, Tennessee February 2001, pp. 30–37.

Coulson, S. A., and R. K. Morgan-Smith. 2000. The transfer of petrol on to clothing and shoes while pouring petrol around a room. *Forensic Science International* 112:135–141.

Curtin, D. P. 1999. Curtin's short courses in digital photography. www.shortcourses.com-chapter11.htm

Damant, G., and S. Nurbakhsh. 1994. Christmas trees—What happens when they ignite? *Fire and Materials* 18:9–16.

Danziger, J. N., and K. L. Kraemer. 1985. Computerized data-based systems and productivity among professional workers: The case of detectives. *Public Administration Review* January–February:196–209.

David, J. 1990. *Seven-fatality Christmas tree fire, Canton, Michigan (December 22, 1990).* Emmittsburg, Md.: Federal Emergency Management Agency.

DeHaan, J. D. 1992. Fire—Fatal intensity. *Fire and Arson Investigator* 43(1).

DeHaan, J. D. 1995. The reconstruction of fires involving highly flammable hydrocarbon liquid. Ph.D. diss. University of Strathclyde, Glasglow, Scotland. (Published by University Microfilms International, 1996.)

DeHaan, J. D. 1999a. The Def . . . . *Interflam99.*

DeHaan, J. D., S. J. Campbell, and S. Nurbakhsh. 1999b. The combustion of body fat and its implications for fires involving bodies. Science and Justice, 39:2738.

DeHaan, J. D. 2000. Why is it important to recognize flashover in a room fire? California Institute of Criminalistics. www.interfire.org./res_file/flashovr.htm

DeHaan, J. D. 2001. Full scale compartment fire tests. *CAC News* Second Quarter:14–21.

DeHaan, J.D. 2002. *Kirk's fire investigation,* 5th ed. Upper Saddle River, N.J.: Prentice Hall. ISBN 0-13-060458-5.

Dillon, S. E., S. Hill, and D. Sheppard. 2003. Role of the ATF Fire Research Laboratory in fatal fire investigations. Presentation lecture, American Academy of Forensic Sciences (AAFS) 55th Annual Meeting, Chicago, February 17–22.

Drysdale, D. 1999. *An introduction to fire dynamics,* 2nd ed. New York: John Wiley and Sons. ISBN 0-471-97290-8.

DSM-IV. 1994. *Diagnostic and statistical manual of mental disorders,* 4th ed. Washington, D.C.: American Psychiatric Association.

Eastman Kodak. 1968. Basic police photography. Publication No. M77. Rochester, N.Y.: Eastman Kodak Co.

Eastman Kodak. 1971. Fire and arson photography. Publication No. M67. Rochester, N.Y.: Eastman Kodak Co.

Eos Systems Inc. 2003. 101-1847 West Broadway, Vancouver, British Columbia, Canada V6J 1Y6. www.eossystems.com

Factory Mutual. 1996. Fire protection for offices: From office to ashes in 7 minutes—Can you afford it? Publication P8802 (Rev 4-97). Norwood, Mass.: Factory Mutual Engineering Corp.

FBI. 1996. *Crime in the United States for 1995*. Washington, D.C.: U.S. Department of Justice, Federal Bureau of Investigation.

FBI. 1999. Definitions and guidelines for the use of imaging technologies in the criminal justice system. *Forensic Science Communications* 1(October): No. 3.

Feld, J. M. 2002. The physiology and biochemistry of combustion toxicology. In *Proceedings of Fire Risk and Hazard Assessment Research Applications Symposium,* FP Research Foundation, July 24–26.

FEMA. 1997a. USFA fire burn pattern tests. Emmitsburg, Md.: Federal Emergency Management Agency, U.S. Fire Administration, July 16.

FEMA. 1997b. Arson in the United States. Report FA-174. Emmitsburg, Md.: Federal Emergency Management Agency, U.S. Fire Administration, Tri-Data Corp., August.

FLIR Systems, 16 Esquire Rd., N. Billerica, Mass. 01862. www.flirthermography.com

Fleischmann, D. M., P. J. Pagni, and R. B. Williamson. 1994. Salt water modeling of fire compartment gravity currents. In *Proceedings. 4th International Symposium, International Association for Fire Safety Science,* Ottawa, Ontario, Canada, July 13–17, ed. T. Kashiwagi, pp. 253–264. Boston: IAFSS.

FLIR 2003. FLIR Systems, 16 Esquire Rd., N. Billerica, MA 01862 www.flirthermography.com.

Forney, G. P., and W. F. Moss. 1992. Analyzing and exploiting numerical characteristics of zone fire models. Report NISTIR 4763. Gaithersburg, Md.: National Institute of Standards and Technology, March.

FPRF. 2002. Recommendations of the Research Advisory Council on Post-fire Analysis: A white paper. Quincy, Mass.: Fire Protection Research Foundation, February.

FRCP. 2000. Orders of the Supreme Court of the United States Adopting and Amending Rules, Order of April 17, 2000. Federal Rules of Civil Procedure as amended effective on December 1, 2000. www.gamb.uscourts.gov

Friedman, R. 1992. An international survey of computer models for fire and smoke. *SFPE Journal of Fire Protection Engineering* 4(3):81–92.

FRS. 2002. Fire Research Station, the Research-Based Consultancy and Testing Company of BRE, Garston, Watford, U.K. www.bre.co.uk/frs

Fulghum, D. 1997. ANG pilot: Jet hit by object. *Aviation Week and Space Technology,* March 10.

Geller, J. L. 1992. Arson in review: From profit to pathology. *Journal of Clinical Forensic Psychiatry* 15:623–645.

Geller, J. L., J. Erlen, and R. L. Pinkas. 1986. A historical appraisal of America's experience with pyromania—A diagnosis in search of a disorder. *International Journal of Law and Psychiatry* 9(2):201–229.

Georges, D. E. 1967. The ecology of urban unrest in the city of Newark, New Jersey, during the July 1967 riots. *Journal of Environmental Systems* 5(3):203–228.

Georges, D. E. 1978. The geography of crime and violence: A spatial and ecological perspective. Resource Paper for College Geography No. 78-1. Washington, D.C.: Association of American Geographers.

Goldbaum, L. R., T. Orellano, and E. Dergari. 1976. *Annals of Clinical and Laboratory Science* 6:372.

Grosselin, S. D. 1998. The application of fire dynamics to fire forensics. Master's thesis. Worcester Polytechnic Institute, Fire Protection Engineering, Worcester, Mass.

Hajpal, M. 2002. Changes in sandstones of historical monuments exposed to fire or high temperatures. *Fire Technology* 38:373–382.

Harmon, R. B., R. Sosner, and M. Wiederight. 1985. Women and arson: A demographic study. *Journal of Forensic Sciences* 30(2):467–477.

Harris, R. J. 1983. *The investigation and control of gas explosions in buildings and heating plants,* pp. 96–100. London: British Gas Corp., E&FN Spon.

Hasemi, Y., and M. Nishihata. 1989. Fuel shape effects on the deterministic properties of turbulent diffusion flames. In *Proceedings of the Second International Symposium, International Association of Fire Safety Science,* pp. 275–284. Washington, D.C.: Hemisphere.

Hasemi and Tokunaga. 1984. *Combustion Science Technology* 40.

Health Canada. 1995. *Investigating human exposure to contaminants in the environment: A handbook for exposure calculations.* Ottawa, Ontario, Canada: Health Protection Branch, Health Canada, Ministry of National Health and Welfare. www.hc-sc.gc.ca

Heskestad, G. H. 1982. Engineering relations for fire plumes. SFPE Technology Report 82-8, p. 6. Boston: Society of Fire Protection Engineers.

Heskestad, G. H. 1988. Fire plumes. In *The SFPE handbook of fire protection engineering,* ed. J. DiNenno, sect. 1, chap. 6, pp. 1-107–1-115. Quincy, Mass.: National Fire Protection Association.

Hewitt, Terry-Dawn. A primer on the law of spoilation of evidence in Canada. *Fire and Arson Investigation* Vol. 48, No. 1, September 1997, pp. 17–21.

Holleyhead, R. 1999. Ignition of solid materials and furniture by lighted cigarettes—A review. *Science and Justice* 39(2): 75–102.

Huff, T. G. 1997. *Killing children by fire. Filicide: A preliminary analysis.* Quantico, Va.: Federal Bureau of Investigation.

Hunt, S. 2000. Computer fire models. *Section News NFPA* 1(2):7–9.

Hurd, R. M. 1903. *Principles of city values.* New York: Records and Guide.

Hurley and Monahan. 1969.

Icove, D. J. (1979). Principles of incendiary fire analysis. Unpublished Ph.D. diss. Knoxville: College of Engineering, University of Tennessee.

Icove, D. J. (1995). Fire Scene Reconstruction. First International Symposium on the Forensic Aspects of Arson Investigations, Federal Bureau of Investigation, Fairfax, Va., July 31.

Icove, D. J., E. C. Escowitz, and T. G. Huff. 1983. The geography of violent crime: Serial arsonists. In *Agenda and abstracts, 8th Annual Geographic Resources Analysis Support System (GRASS) Users Conference,* Washington, D.C., Reston, Va., March 14–19, p. 46.

Icove, D. J., J. E. Douglas, G. Gary, T. G. Huff, and P. A. Smerick. 1992. Arson. In *Crime classification manual,* eds. J. E. Douglas, A. W. Burgess, A. G. Burgess, and R. K. Ressler, pp. 165–166. New York: Macmillan.

Icove, D. J., and M. H. Estepp. 1987. Motive-based offender profiles of arson and fire-related crimes. *FBI Law Enforcement Bulletin,* April.

Icove, D. J., and M. M. Gohar. 1980. Fire investigation photography. In *Fire investigation handbook.* Gaithersburg, Md.: National Bureau of Standards.

Icove, D. J., and P. R. Horbert. 1990. Serial arsonists: An introduction. *Police Chief* (Arlington, Va.) December: 46–48.

Icove, D. J., P. E. Keith, and H. L. Shipley. 1981. An analysis of fire bombings in Knoxville, Tennessee. U.S. Fire Administration Grant EMW-R-0599. Knoxville: Knoxville Police Department Arson Task Force, Department of Public Safety.

Icove, D. J., V. B. Wherry, and J. D. Schroeder. 1998. *Combating arson-for-profit: Advanced techniques for investigators.* Columbus, Ohio: Battelle Press.

Inberg, S. H. 1927. Fire tests of office occupancies. *NFPA Quarterly* 20:243.0.

Inciardi, J. A. 1970. The adult firesetter: A typology. *Criminology* 8(August).

iPIX. 2003. Interactive Pictures Corp., 1009 Commerce Park Drive, Suite 400, Oak Ridge, Tenn. 37830. www.ipix.com

Iqbal, N., and M. H. Salley. 2002. Development of a quantitative fire scenario estimating tool for the U.S. Nuclear Regulatory Commission Fire Protection Inspection Program. Washington, D.C.: Fire Protection Engineering and Special Projects Section, Nuclear Regulatory Commission.

Janssens, M. 2000. *Introduction to mathematical fire modeling*, 2nd ed. Lancaster, Pa.: Technomic.

J. D. DeHaan, S. J. Campbell, and S. Nurbakhsh. 1999b. "The Combustion of Body Fat and Its Implications for Fires Involving Bodies," Science and Justice, 1999; 39: 2738.

Jensen, G. 1998. Wayfinding in heavy smoke: Decisive factors and safety products: Findings related to full-scale tests. IGPAS, InterConsult Group ICG. www.interconsult.com, May 1.

Jin, T. 1975. Visibility through fire smoke. Part 5. Allowable smoke density for escape from fire. Report of Fire Research Institute of Japan, No. 42, p. 12.

Karlsson, B., and J. G. Quintiere. 1999. *Enclosure fire dynamics,* p. 64. Boca Raton, Fla.: CRC Press. ISBN 0-8493-1300-7.

Karter, M. J., Jr. and S. Badger. 2002. 2001 Fire loss/large loss. *NFPA Journal* 2001. Quincy, Mass.: National Fire Protection Association, November/December.

K. B. McGrattan, G. P. Forney, J. E. Floyd, S. Itostikka, K. Prasad, "Fire Dynamics Simulator (Version 3)—User's Guide," National Institute of Standards and Technology, Gaithersburg, MD, NISTIR 6784, 2002 Edition.

Khoury, G. A. 2000. Effect of fire on concrete and concrete structures. *Progress in Structural Engineering Materials* 2:429–442.

King, C. G. and J. I. Ebert. 2002. Integrating archaeology and photogrammetry with fire investigation. *Fire Engineering,* Feb. p 79.

Kolczynski, P. J. 2000. *Preparing for trial in federal court,* 2nd ed. Costa Mesa, Calif.: James.

König, J., and L. Walleij. 2000. Timber frame assemblies exposed to standard and parametric fires. Part 2: A design model for standard fire exposure. Report No. 100010001. Tratek. Stockholm: Swedish Institute for Wood Technology Research.

Krasny, J. F., W. J. Parker, and V. Babrauskas. 2001. *Fire behavior of upholstered furniture and mattresses.* New York: William Andrews. ISBN 0-8155-1457-3.

Lawson, J. R., and J. G. Quintiere. 1985. Slide-rule estimates of fire growth. NBSIR 85-3196. Gaithersburg, Md.: National Bureau of Standards.

LeBeau, J. L. 1987. The methods and measures of centrography and the spatial dynamics of rape. *Journal of Quantitative Criminology* 3:125–141.

Lentini, J. J. 1992. Behavior of glass at elevated temperatures. *Journal of Forensic Sciences* 37(5):1358–1362.

Lentini, J. J. 2001. Standards impact: The forensic sciences. *ASTM Standardization News* February: 17–19.

Lentini, J. J., J. A. Dolan, and C. Cherry. 2000. The petroleum-laced background. *Journal of Forensic Sciences* 45(5):968–989.

Levin, B. 1976. Psychological characteristics of firesetters. *Fire Journal* National Fire Protection Association. Quincy, Mass. March:36–41.

Lewis, N. D. C., and H. Yarnell. 1951. *Pathological firesetting (pyromania).* Nervous and Mental Disease Monographs, No. 82. New York: Coolidge Foundation.

Lilley, D. G. 1995. Fire dynamics. American Institute of Aeronautics and Astronautics, AIAA-95-0894, Meeting and Exhibits, Reno, Nev., January 9–12.

Lipson, A. S. 2000. *Is it admissible?* pp. 44-2–44-8. Costa Mesa, Calif.: James Publishing.

MacQueen, J. 1967. Some methods for classification and analysis of multivariate data. In *Proceedings of the 5th Berkeley Symposium of Probability and Statistics*. Berkeley: University of California Press.

Madrzykowski, D. 2000. The future of fire investigation. *Fire Chief* October:44–50.

Madrzykowski, D., and R. L. Vettori. 2000. Simulation of the dynamics of the fire at 3146 Cherry Road, NE, Washington, D.C., May 30, 1999. NISTIR 6510. Gaithersburg, Md.: National Institute of Standards and Technology, Center for Fire Research, April.

Magnusson, S. E., and S. Thelandersson. 1970. Temperature-time curves of complete process of fire development. Theoretical study of wood fuel fires in enclosed spaces. Civil and Building Construction Series No. 65. Stockholm: Acta Polytechnica Scandinavia.

McCaffrey, B. J. 1979. Purely buoyant diffusion flames: Some experimental results. NBSIR 79-1910. Gaithersburg, Md.: National Bureau of Standards.

McCaffrey, B. J., J. G. Quintiere, and M. F. Harkleroad. 1982. Estimating room fire temperatures and the likelihood of flashover using fire test data correlations. *Fire Technology* 17(2):98–119.

McGill, D. 2003. Fire Dynamics Simulator, FDS 683, participants handbook. School of Fire Protection, Seneca College, Toronto, Canada, January.

McGratten, K. B., G. P. Forney, J. E. Floyd, S. Hostikka and K. Prasad. Fire dynamics Simulator (version 3) user's guide. National Institute of Standards and Technology. Gaithersburg, Md. NISTIR 6784, 2002 Edition.

Microsoft. (2002). Visio Crime Scene Template. www.microsoft.com

Milke, J. A., and F. W. Mowrer, 2001. Application of fire behavior and compartment fire models seminar. Tennessee Valley Society of Fire Protection Engineers (TVSFPE), Oak Ridge, September 27–28.

Mitler, H. E. 1991. Mathematical modeling of enclosure fires. NISTIR 90-4294. Gaithersburg, Md.: National Institute of Standards and Technology, May.

Moodie, M., and S. E. Jagger. 1991. The technical investigation of the fire at London's King Cross Underground Station.

Morgan-Smith, R. 2000. Persistence of petrol on clothing. Presented at ANZFSS Symposium of Forensic Sciences, Gold Coast, Queensland, Australia, March.

Mowrer, F. W. 1992. Methods of quantitative fire hazard analysis. TR-100443. Research Project 3000-37. Prepared for Electric Power Research Institute (EPRI) by the Society of Fire Protection Engineers. Boston.

Mudan, K. S., and P. A. Croce. 1995. Fire hazard calculations for large open hydrocarbon fires, p. 3–203. In *The SFPE handbook of fire protection engineering,* 2nd ed. Bethesda, Md.: Society of Fire Protection Engineers.

Myers, R. A. M., and R. A. Cowley. 1979. CO poisoning. *Journal of Combustion Toxicology.* 6:86.

Nelson, G. L. 1998. Carbon monoxide and fire toxicity: A review and analysis of recent work. *Fire Technology* 34(1).

Nelson, H. E. 1987. An engineering analysis of the early stages of fire development—The fire at the Dupont Plaza Hotel and Casino on December 31, 1986, p. 50. Gaithersburg, Md.: National Institute of Standards and Technology, Center for Fire Research, April.

Nelson, H. E. 1989. An engineering view of the fire of May 4, 1988, in the First Interstate Bank Building, Los Angeles, California. NISTIR 89-4061. Gaithersburg, Md.: National Institute of Standards and Technology, Center for Fire Research, March.

Nelson, H. E. 1990. Fire growth analysis of the fire of March 20, 1990, Pulaski Building, 20 Massachusetts Avenue, N.W., Washington, D.C. NISTIR 4489. Gaithersburg, Md.: National Institute of Standards and Technology, Center for Fire Research, December.

Nelson, H. E. 1991. Engineering analysis of the fire development in the Hillhaven Nursing Home Fire, October 5, 1989. NISTIR 4665. Gaithersburg, Md.: National Institute of Standards and Technology, Center for Fire Research, September.

Nelson, H. E. 2002. From Phlogiston to computational fluid dynamics. *Fire Protection Engineering,* Winter: 9–17. www.sfpe.org

Nelson, H. E., and R. E. Tontarski. 1998. Proceedings of the International Conference of Fire Research for Fire Investigators. HAI Report 98-5157-0001, April.

Newman, J. S. 1993. Integrated approach to flammability evaluation of polyurethane wall-ceiling materials. Polyurethane World Congress, October 10–13. Washington, D.C.: Society of the Plastics Industry.

NFPA. 1943. The Cocoanut Grove Night Club fire, Boston, November 28, 1942 (preliminary report). Quincy, Mass.: National Fire Protection Association, January 11.

NFPA. 1997. *NFPA 705—Field flame test for textiles and films, recommended practice.* Quincy, Mass.: National Fire Protection Association.

NFPA. 1998. *NFPA 906—Guide for fire incident field notes.* Quincy, Mass.: National Fire Protection Association.

NFPA. 2000. *NFPA 555—Guide on methods for evaluating potential for room flashover.* Quincy, Mass.: National Fire Protection Association.

NFPA. 2000a. *Fire protection handbook,* 18th ed. Quincy, Mass.: National Fire Protection Association.

NFPA. 2000b. *NFPA 92B—Guide for smoke management systems in malls, atria and large areas.* Quincy, Mass.: National Fire Protection Association.

NFPA. 2001. *NFPA 921—Guide for fire and explosion investigations.* Quincy, Mass.: National Fire Protection Association.

NFPA. 2002. *NFPA 1403—Standard on live fire training evolutions.* Quincy, Mass.: National Fire Protection Association.

NFPA. 2002. *NFPA 72—National fire alarm code,* appendix B. Quincy, Mass.: National Fire Protection Association.

NIJ. 2000. Fire and arson scene evidence: A guide for public safety personnel. NCJ 181584. Washington, D.C.: National Institute of Justice, June.

NIST. 1991. Users' guide to BREAK1, the Berkeley algorithm for breaking window glass in a compartment fire. NIST-GCR-91-596. Gaithersburg, Md.: National Institute of Standards and Technology.

NIST. 1997. Full-scale room burn pattern study. NIJ Report 601-97. Washington, D.C.: National Institute of Justice, December.

Ogle, R. A. 2000. The need for scientific fire investigations. *Fire Protection Engineering* 8(Fall): 4–8.

Orloff, L., A. T. Modak, and R. L. Alpert. 1977. Burning of large-scale vertical surfaces. In *16th International Symposium on Combustion*, Combustion Institute, Pittsburgh, Pa., p. 1345.

Park, R. E., and E. W. Burgess. 1921. *An introduction to the science of sociology.* Chicago: University of Chicago Press.

Paulsen, T. 1994. The effect of escape route information on mobility and wayfinding under smoke-logged conditions. In *Fire Safety Science—Proceedings of the Fourth International Symposium*, International Association of Fire Safety Science, Ottawa, pp. 693–704.

Peacock, R. D., G. P. Forney, P. Reneke, R. Porter, and W. W. Jones. CFAST, the consolidated model of fire growth and smoke transport. National Institute of Standards and Technology. Gaithersburg, Md. NIST Technical note 1299. February 1993.

Peacock, R. D., P. A. Reneke, R. W. Bukowski, and V. Babrauskas. 1999. Defining flashover for fire hazard calculations. *Fire Safety Journal* 32(4):331–345.

Peige, J. D., and C. E. Williams. 1977. *Photography for the fire service.* Oklahoma City: International Fire Service Training Association, Oklahoma State University.

Penney, D. G., ed. 2000. *Carbon monoxide toxicity.* Boca Raton, Fla.: CRC Press.

Perry, R. H., and D. W. Green, eds. 1984. *Perry's chemical engineers' handbook,* 6th ed. New York: McGraw-Hill.

Peterson, J. E., and R. D. Stewart. 1975. Predicting the carboxyhemoglobin levels resulting from carbon monoxide exposure. *Journal of Applied Physiology* 39:633–638.

Policing New Zealand. 2000. *New Zealand police today—Policing 2000*. Video. www.crime.co.nz

Pope, E. J., and O. C. Smith. 2003. Features of preexisting trauma and burned cranial bone. Presentation lecture, American Academy of Forensic Sciences (AAFS) 55th Annual Meeting, Chicago, February 17–22.

Purser, D. 2001. Human tenability. The technical basis for performance based fire regulations, United Engineering Foundation Conference, San Diego, Calif., January 7–11.

Purser, D. 2002. Toxicity assessment of combustion products. In *The SFPE handbook of fire protection engineering,* 3rd ed., pp. 2-83–2-171. Quincy, Mass.: National Fire Protection Association.

Putorti, A. D. 2001. Flammable and combustible liquid spill-burn patterns. NIJ Report 604-00, NCJ 186634. Gaithersburg, Md.: U.S. Department of Justice, National Institute of Justice, National Institute of Standards and Technology, March.

Quintiere, J. G. 1994. A perspective on compartment fire growth. *Combustion Science and Technology* 39:11–54.

Quintiere, J. G. 1997. *Principles of fire behavior.* Albany, N.Y.: Delmar.

Quintiere, J. G., and M. Harkleroad. 1984. New concepts for measuring flame spread properties. NBSIR 84-2943. Gaithersburg, Md.: National Bureau of Standards, November.

R. D. Peacock, G. P. Forney, P. Reneke, R. Porter, and W. W. Jones, "CFAST, the Consolidated Model of Fire Growth and Smoke Transport." National Institute of Standards and Technology, Gaithersburg, MD NIST Technical Note 1299, February 1993.

Reardon, J. J. 2002. The warning signs of a faked death life insurance beneficiaries can't recover without providing due proof of death. *Connecticut Law Tribune* June 10:5.

Rengert, G., and J. Wasilchick. 1990. Space, time, and crime: Ethnographic insights into residential burglary. Grant 88-IJ-CX-0013. Final Report to the U.S. Department of Justice. Philadelphia: Department of Criminal Justice, Temple University.

Robbins, E. S., and L. Robbins. 1967. Arson with a special reference to pyromania. *New York State Journal of Medicine* March.

Rossmo, D. K. 1999. *Geographic profiling.* Boca Raton, Fla.: CRC Press. ISBN 0849381290.

R. Shill (1998). "Beyond Search and Retrieval: Fire Benefits of Document Management" *New York Law Journal* New York Law Publishing Co., March 31, 1998.

Sapp, A. D., G. P. Gary, T. G. Huff, and James. 1993. *Characteristics of arsons aboard naval ships. A monograph.* Quantico, Va.: FBI National Center for the Analysis of Violent Crime, U.S. Department of Justice.

Sapp, A. D., G. P. Gary, T. G. Hugg, D. J. Icove, and P. R. Horbert. 1994. *Motives of serial arsonists: Investigative implications. A monograph.* Quantico, Va.: FBI National Center for the Analysis of Violent Crime, U.S. Department of Justice.

Sapp, A. D., T. G. Huff, G. P. Gary, and D. J. Icove. 1995. *A motive-based offender analysis of serial arsonists.* Monograph. Quantico, Va.: FBI National Center for the Analysis of Violent Crime, U.S. Department of Justice.

Schroeder, R. A., and R. B. Williamson. 2003. Post-fire analysis of construction materials—Gypsum wallboard. Fire and Materials, 8th International Conference, January 28–29. (See also http://www.schroederfire.com/)

SFPE. 1995a. *The SFPE handbook of fire protection engineering,* 2nd ed. Quincy, Mass.: National Fire Protection Association. ISBN: 0-87765-354-2.

SFPE. 1995b. *Fires-T3: A guide for practicing engineers.* Bethesda, Md.: SFPE Task Group on Documentation of Computer Models, Society of Fire Protection Engineers.

SFPE. 1999. *Engineering guide for assessing flame radiation to external targets from pool fires.* Bethesda, Md.: SFPE Task Group on Engineering Practices, Society of Fire Protection Engineers.

SFPE. 2000. *Engineering guide to predicting 1st and 2nd degree skin burns*. Bethesda, Md.: SFPE Task Group on Engineering Practices, Society of Fire Protection Engineers.

SFPE. 2002a. *The SFPE handbook of fire protection engineering,* 3rd ed. Quincy, Mass.: National Fire Protection Association. ISBN: 0-87765-451-4.

SFPE. 2002b. *SFPE engineering guide to piloted ignition of solid materials under radiant exposure.* Bethesda, Md.: SFPE Task Group on Engineering Practices, Society of Fire Protection Engineers.

SFPE. 2002c. *SFPE engineering guide—The evaluation of the computer model DETECT-QS.* Bethesda, Md.: Society of Fire Protection Engineers.

Shields, T. J., G. W. H. Silcock, and M. F. Flood. 2001. Performance of a single glazing assembly exposed to enclosure corner fires of increasing severity. *Fire and Materials* 22:123–152.

Shill, R. 1998. Beyond search and retrieval: five benefits of document management. *New York Law Journal.* New York Law Publishing Co., March 31, 1998.

Shipley, H. L., D. J. Icove, and P. E. Keith. 1981. An analysis of fire bombings in Knoxville, Tennessee. U.S. Fire Administration Grant No. EMW-R-0599. Knoxville Police Arson Task Force, Department of Public Safety, December 17.

Silcock, G. W. H., and T. J. Shields. 2001. Relating char death to fire severity conditions. *Fire and Materials* 25:9–11.

Smith, O. C., and E. J. Pope. 2003. Burning extremities: Patterns of arms, legs, and preexisting trauma. Presentation lecture, American Academy of Forensic Sciences (AAFS) 55th Annual Meeting, Chicago, February 17–22.

Spearpoint, M. J., and J. G. Quintiere. 2001. Predicting the ignition of wood in the cone calorimeter. *Fire Safety Journal* 36(4):391–415.

Stanton, J., and A. Simpson. 2002. Filicide: A review. *International Journal of Law and Psychiatry* 25:1–14.

Steinmetz, R. C. 1966. Current arson problems. *Fire Journal,* September.

Stickevers, J. 1986. Factors to consider when attempting to determine point of origin. *Fire Engineering* January:36–46.

Stoll, A. M., and L. C. Greene. 1959. Relationship between pain and tissue damage due to thermal radiation. *Journal of Applied Physiology* 14(3):373–382.

Stroup, D. W., L. A. DeLauter, J. H. Lee, and G. L. Roadarmel. 1999. Scotch pine Christmas Tree fire tests. Report of Test. FR 4010. Gaithersburg, Md.: National Institute of Standards and Technology, December.

Tassios, T. P. 2002. Monumental fires. *Fire Technology* 38:311–317.

Tewarson, A. 1995. In *The SFPE handbook of fire protection engineering,* 2nd ed., chap. 3. Quincy, Mass.: Society of Fire Protection Engineers.

Thali, M. J., et al. 2002. Charred body: Virtual autopsy with multi-slice computer tomography and magnetic resonance imaging. *Journal of Forensic Sciences* 47(6).

Thomas, P. H. 1971. Rates of spread of some wind-driven fires. *Forestry* 44:155–175.

Tobin, W. A., and K. L. Monson. 1989. Collapsed spring observation in arson investigation: A critical metallurgical evaluation. *Fire Technology* 25(4):317–335.

Tou, J. T., and R. C. Gonzalez. 1974. Pattern recognition principles, pp. 94–97. Reading, Mass.: Addison–Wesley.

Vaisala Inc., 2002 (formerly Global Atmospherics). 2705 E. Medina Rd., Tucson, Ariz. 85706 www.lightningstorm.com

Vandersall, T. A., and J. M. Wiener. 1970. Children who set fires. *Archives of General Psychiatry* 22(January).

Vreeland, R. G., and M. B. Waller. 1978. The psychology of fire setting: A review and appraisal. Grant No. 7-9021. Gaithersburg, Md.: National Bureau of Standards, December.

Wheaton, S. 2001. *Personal accounts: Memoirs of a compulsive firesetter.* Washington, D.C.: American Psychiatric Association. psjournal@psych.org

Wilkinson, P. 2001. Archaeological survey site grids. *Practical Archaeology* 5(Winter):21–27.

Wise, B. 1995. Catching crooks with computers. *American City and County* 110(6):54.

Wolchak, et al. 1995. *Public sector stars,* p. 22. Framingham, Mass.: Computerworld.

Wolford, M. R. 1972. Some attitudinal, psychological and sociological characteristics of incarcerated arsonists. *Fire and Arson Investigator* 22.

Zukoski, E. E. 1978. Development of a stratified ceiling layer in the early stages of a closed-room fire. *Fire and Materials* 2(2).

LEGAL REFERENCES

Abu-Hashish, Eid, and Sheam Abu-Hashish v Scottsdale Insurance Company, Case No. 98 C 4019, 88 F. Supp. 2d 906, 2000 U.S. Dist. LEXIS 3663. Decided March 16, 2000.

Allstate Insurance Company, as Subrogee of Russell Davis v Hugh Cole Builder, Inc. Hugh Cole individually and dba Hugh Cole Builder, Inc., Civil Action No. 98-A-1432-N, 137 F. Supp. 2d 1283, 2001 U.S. Dist. LEXIS 5016. Decided April 12, 2001.

American Family Insurance Group v JVC Americas Corp., Civil Action No. 00-27(DSD-JMM), 2001 U.S. Dist. LEXIS 8001. Decided April 30, 2001.

Booth, Jacob J. and Kathleen Booth v Black and Decker, Inc., Civil Action No. 98-6352, 2001 U.S. Dist. LEXIS 4495. Decided April 12, 2001.

Daubert v Merrell Dow Pharmaceuticals, Inc. (1993), 509 U.S. 579, 113 S. Ct. 2756, 215 L.Ed.2d 469.

Daubert v Merrell Dow Pharmaceuticals, Inc. (Daubert II), 43 F.3d 1311, 1317 (9th Cir. 1995).

Kumho Tire Co. Ltd. v Carmichael, 119 S. Ct. 1167 (1999), U.S. LEXIS 2199 (March 23, 1999).

LaSalle National Bank et al. v Massachusetts Bay Insurance Company et al., Case No. 90 C 2005, 1997 U.S. Dist. LEXIS 5253. Decided April 11, 1997. Docketed April 18, 1997.

Michigan Miller's Mutual Insurance Company v Benfield, 140 F.3d 915 (11th Cir. 1998).

Royal v Daniel Construction. (Royal Insurance Company of America as Subrogee of Patrick and Linda Magee v Joseph Daniel Construction, Inc.), Civil Action No. 0 Civ. 8706 (CM), 208 F. Supp. 2d 423, 2002 U.S. Dist. LEXIS 12397. Decided July 10, 2002.

Pappas, Andronic, et al. v Sony Electronics, Inc. et al., Civil Action No. 96-339J, 136 F. Supp. 2d 413, 2000 U.S. Dist. LEXIS 19531, CCH Prod. Liab. Rep. P15,993. Decided December 27, 2000.

Snodgrass, Teri, Robert L. Baker, Kendall Ellis, Jill P. Fletcher, Judith Shemnitz, Frank Sherron, and Tamaz Tal v Ford Motor Company and United Technologies Automotive, Inc., Civil Action No. 96-1814 (JBS), 2002 U.S. Dist. LEXIS 13421. Decided March 28, 2002.

United States of America v Lawrence R. Black Wolf, Jr., CR 99-30095, 1999 U.S. Dist. LEXIS 20736. Decided December 6, 1999.

United States of America v Ruby Gardner, No. 99-2193, 211 F.3d 1049, 2000 U.S. App. LEXIS 8649, 54 Fed. R. Evid. Serv. (Callaghan) 788.

Glossary

accelerant. A fuel (usually a flammable liquid) that is used to initiate or increase the intensity or speed of spread of fire.

accident. Unplanned or unintentional event; an event occurring without design or intent.

adiabatic. Conditions of equilibrium of temperature and pressure.

adsorption. Trapping of gaseous materials on the surface of a solid substrate.

aliphatic. Hydrocarbons with straight-chain structures of the carbon atoms; *normal* hydrocarbon.

alligatoring. Rectangular patterns of char formed on burned wood.

ambient. Surrounding conditions.

ampacity. Current-carrying capacity of electric conductors (expressed as amperes).

ampere. Quantity of electrical charge passing a point in an electrical circuit per unit time (1 coulomb per second).

annealing. Loss of temper in metal caused by heating.

anoxia. Condition relating to an absence of oxygen.

appliance. Equipment, usually nonindustrial, which is installed or connected as a unit to perform one or more functions such as clothes washing, air-conditioning, food mixing, and cooking. Normally built in standardized sizes and types.

arc. Flow of current across a gap between two conductors, generally producing high temperatures and luminous gases.

arcing through char. Unintended passage of current through a semiconductive degradation product.

area of confusion. Mixture of fire directional indicators in a wildlands fire.

area of origin. The general locale in which a fire was ignited.

aromatic. Hydrocarbon compound whose structure is based on a benzene ring.

arson. The intentional setting of a fire with intent to damage or defraud.

atom. Smallest unit of an element which still retains its properties.

Auger spectroscopy. A means of identifying elements trapped within a substrate by analyzing the electrons emitted during electron microscopy.

autoignition. Ignition by sufficient surrounding temperature in the absence of an external source of ignition; non-piloted ignition.

autoignition temperature. The temperature at which a material will ignite in the absence of any external pilot source of heat; spontaneous ignition temperature.

backdraft. A deflagrative explosion or rapid combustion of gases and smoke from an established fire that has depleted the oxygen content of a structure, most often initiated by introducing oxygen through ventilation or structural failure.

BLEVE. Boiling-liquid, expanding-vapor explosion. A mechanical explosion caused by the heating of a liquid in a sealed vessel to a temperature far above its boiling point.

boiling point. The (pressure-dependent) temperature at which a liquid changes to its gas phase and that transition reaches equilibrium.

branch circuit. The circuit conductors between the outlet(s) and the final over-current device protecting that circuit.

brisance. The amount of shattering effect that can be produced by a high explosive.

Btu. British thermal unit. A standardized measure of heat, it is the heat energy required to raise the temperature of 1 pound of water by 1 degree Fahrenheit.

calcination. Loss of water of crystallization caused by heating.

calorie. The amount of heat necessary to raise the temperature of 1 gram of water by 1 degree Celsius.

char. Carbonaceous remains of burned organic materials.

chromatography. Chemical procedure that allows the separation of compounds based on differences in their chemical affinities for two materials in different physical states, i.e., gas/liquid and liquid/solid.

circuit breaker. A device designed to open a circuit automatically at a predetermined overcurrent without injury to itself when properly applied within its ratings.

clean burn. An area of wall or ceiling where the charred organic residues have been burned away by direct flame contact.

combustible. A material that will ignite and burn when sufficient heat is applied and when an appropriate oxidizer is present.

combustible liquid. A liquid having a flash point at or above 38°C (100°F).

combustion. Oxidation that generates detectable heat and light.

concealed wiring. Wiring rendered inaccessible by the structure or finish of the building. Wiring in covered raceways is considered concealed.

conduction. Process of transferring heat through a material or between materials by direct physical contact.

conductor. Any material capable of permitting the flow of electrons. (1) *Bare.* A conductor having no covering or electrical insulation whatsoever. (2) *Covered.* A conductor encased within a material whose composition or thickness is not considered insulative. (3) *Insulated.* A conductor encased within a material recognized by the electrical code as insulation.

convection. Process of transferring heat by movement of a fluid. In convective flow, the warm fluid becomes less dense than the surrounding fluid and rises, inducing a circulation.

corpus delicti. Literally, the body of the crime. The fundamental facts necessary to prove the commission of a crime.

crazing. Stress cracks in glass as the result of rapid cooling.

dead load. The weight of a structure and any equipment and appliances permanently attached.

deep-seated. Fire that has gained headway and built up sufficient heat in a structure to require great cooling for extinguishment; fire that has burrowed deep into combustible fuels (as opposed to a surface fire); deep charring of structural members.

deflagration. A very rapid oxidation (typically of fuel gas, vapor, or dust) with the evolution of heat and light and the generation of a low-energy pressure wave that can accomplish damage. The reaction proceeds between fuel elements at subsonic speeds.

detonation. An extremely rapid reaction that generates very high temperatures and an intense pressure/shock wave that produces violently disruptive effects. It propagates through the material at supersonic speeds.

device. Any chemical or mechanical contrivance or means used to start a fire or explosion.

diatomic. Molecules consisting of two atoms of an element.

diffusion flame. Combustion supported by fuel and oxygen molecules diffusing into one another.

direct attack. Application of hose streams or other extinguishing agents directly on a fire, rather than attempting extinguishment by generating steam within a structure.

dropdown. The collapse of burning material in a room that induces separate, low-level ignition; falldown.

duty. Conditions of use in electrical service: (1) *Continuous duty.* Operation at substantially constant load for an indefinitely long time. (2) *Intermittent duty.* Operation for alternate intervals of (a) load/no load, (b) load/rest, or (c) load/no load/rest. (3) *Periodic duty.* Intermittent operation in which the load conditions are regularly recurrent.

(4) *Short-time duty*. Operation at substantially constant load for a short and definitely specified time. (5) *Varying duty*. Operation at loads and for intervals of time, both of which are subject to wide variation.

endothermic. Absorbing heat during a chemical reaction.

entrain. The mixing of two or more fluids as a result of laminar flow or movement.

eutectic. An alloy of two materials having special physical or chemical properties, typically having the lowest melting point of any combination of the two.

exothermic. Generating or giving off heat during a chemical reaction.

explosion. The sudden conversion of potential energy (chemical or mechanical) into kinetic energy with the production of heat, gases, and mechanical pressure.

explosion-proof apparatus. Apparatus enclosed in a case that is capable of withstanding an explosion of a specified gas or vapor within it or preventing the ignition of a specific gas or vapor surrounding it by sparks, flashes, or explosion of fuels within and that operates at such an external temperature that a surrounding flammable atmosphere will not be ignited by it.

explosive. Any material that can undergo a sudden conversion of physical form to a gas with a release of energy.

explosive limits (flammability limits). The lower and upper concentrations of an air/gas or air/vapor mixture in which combustion or deflagration will be supported.

exposure. Property that may be endangered by radiant heat from a fire in another structure or an outside fire. Generally, property within 40 feet is considered an exposure risk, but larger fires can endanger property much farther away.

fire. Rapid oxidation with the evolution of heat and light; self-sustaining (glowing or flaming) combustion.

fire behavior. The manner in which a fuel ignites, flame develops, and fire spreads. Unusual fire behavior may reveal the presence of added fuel or accelerants.

fire load. The total amount of fuel that might be involved in a fire, as measured by the amount of heat that would evolve from its combustion (expressed as units of heat).

fire point. See flame point.

fire-resistive. A structure or assembly of materials built to provide a predetermined degree of fire resistance as defined in building or fire prevention codes (calling for 1-, 2-, or 4-hour fire resistance).

firestorm. Overwhelming progression of fire through structures or wildlands caused by a combination of convective and radiative processes.

fire wall. A solid wall of masonry or other noncombustible material capable of preventing passage of fire for a prescribed time (usually extends through the roof with parapets).

flame. A luminous cloud of burning gases.

flameover. The flaming ignition of the hot gas layer in a developing compartment fire.

flame point (fire point). Temperature at which a flame is sustained by evaporation or pyrolysis of a fuel.

flame resistant. Material or surface that does not maintain or propagate a flame once an outside source of flame has been removed.

flame spread. The rate at which flames extend across the surface of a material (usually under specific conditions).

flammable (same as *inflammable*). A combustible material that ignites easily, burns intensely, or has a rapid rate of flame spread.

flammable liquid. A liquid having a flash point below 38°C (100°F).

flashback. The ignition of a gas or vapor from an ignition source back to a fuel source (often seen with flammable liquids).

flashover. The final stage of the process of fire growth; when all combustible fuels within a compartment are ignited, the room is said to have undergone flashover.

flash point. Temperature at which an ignitable vapor is first produced by a material.

fragmentation. The fast-moving solid pieces created by an explosion. Primary fragmentation is that of the explosive container itself; secondary fragmentation is that of the target shattered by an explosion.

fuel load. All combustibles in a fire area, whether part of the structure, finish, or furnishings.

fully involved. The entire area of a fire building so involved with heat, smoke, and flame that entry is not possible until some measure of control has been obtained with hose stream attacks.

ghost marks. Stained outlines of floor tiles produced by the dissolution and combustion of tile adhesive.

glowing combustion. The rapid oxidation of a solid fuel directly with atmospheric oxygen creating light and heat in the absence of flames.

ground. A conducting connection, whether intentional or accidental, between electrical circuit, equipment, and the earth, or to some conducting body that serves in place of the earth.

ground fault. An interruption of the normal ground return path of electricity in a structure that leads to unintended current flows.

heat flux. The rate at which heat energy is transferred to a surface per unit time per unit area.

heat release rate. The rate at which heat is generated by a source, usually measured in watts, joules/sec, or Btu/sec.

heat transfer. Spread of thermal energy by convection, conduction, or radiation.

high explosive. Any material designed to function by, and capable of, detonation.

hydrocarbon. Chemical compound containing only hydrogen and carbon.

hypoxia. Condition relating to low concentrations of oxygen.

ignitable liquid. A liquid fuel classified by NFPA as either a flammable liquid or a combustible liquid.

ignition energy. The quantity of energy that must be transferred into a fuel/oxidizer combination to trigger a self-sustaining combustion.

ignition temperature (same as *autoignition*). The minimum temperature to which a substance must be heated in air to ignite

independently of the heating source (i.e., in the absence of any other ignition source; nonpiloted ignition).

incendiary fire. A deliberately set fire.

incipient. Beginning stages of a fire.

indirect application. A method of firefighting by applying water fog into heated atmospheres to obtain heat absorption and smothering action by generating steam.

inorganic. Containing elements other than carbon, oxygen, and hydrogen.

isochar. Line drawn on a scene diagram that connects points of similar char depth on wood surfaces.

low explosive. Any material designed to function by deflagration.

Molotov cocktail. A breakable container filled with flammable liquid, usually thrown. It may be ignited by a flaming wick or by chemical means.

nonflammable. Material that will not burn under most conditions.

normal hydrocarbons (*n*-hydrocarbons). Hydrocarbons having straightchain structures with no side branching; aliphatics.

ohm. Unit of measurement for electrical resistance.

olefinic. Hydrocarbons containing double carbon–carbon bonds (denoted $C=C$); nonsaturated; alkenes.

open wiring. Uninsulated conductors or insulated conductors without grounded metallic sheaths or shields, installed above ground but not inside enclosures or appliances.

organic. Compounds based on carbon.

outlet. A point on the wiring system at which current is taken to supply equipment or appliances by means of receptacles or direct connections.

overcurrent. Current flow above the intended or design current.

overhaul. The firefighting operation of eliminating hidden flames, glowing embers, or sparks that may rekindle the fire, usually accompanied by the removal of structural contents.

oxidation. Combination of an element with oxygen; chemical conversion involving a loss of electrons.

paraffinic. Hydrocarbon compounds involving no double or triple C—C bonds; alkanes; saturated aliphatic hydrocarbons.

piloted ignition temperature. The minimum temperature at which a fuel will sustain a flame when exposed to a pilot source.

plume. The convective column of hot gases generated by a flame, may include both flames and nonflaming products.

point of origin. The specific location at which a fire was ignited.

pyrolysis. The chemical decomposition of substances through the action of heat, in the absence of oxygen.

pyromania. Uncontrollable psychological impulse to start fires.

pyrophoric. Capable of oxidizing on exposure to atmospheric oxygen at normal temperatures.

raceway. Any channel for holding wires, cables, or busbars that is designed expressly for, and is used solely for, this purpose.

radiation. Transfer of heat by electromagnetic waves.

receptacle. A contact device installed as the outlet for the connection of an appliance by means of a plug.

rekindle. Reignition of a fire due to latent heat, sparks, or embers.

resistance. Opposition to the passage of an electrical current.

rollover. See flameover.

salvage. Procedures to reduce incidental losses from smoke, water, and weather following fires, generally removal or covering of contents.

saturated. Hydrocarbons that have no double or triple C—C bonds.

seat of explosion. The area of most intense physical damage caused by high explosive pressures and shock waves in the vicinity of a solid or liquid explosive.

seat of fire. Area where main body of fire is located, as determined by outward movement of heat, flames, and smoke.

self-heating. An exothermic chemical or biological process that can generate enough heat to become an ignition source; spontaneous ignition.

self-ignition. See *Autoignition*.

service. The conductors and equipment for delivering electricity from the supply system to the equipment of the premises served.

service conductors. The supply conductors that extend from the street main or transformer to the equipment of the premises served.

service drop. The overhead service conductors from the pole, transformer, or other aerial support to the service entrance equipment on the structure, including any splices.

service equipment. The necessary hardware that constitutes the main control and means of cutoff of the electrical supply, usually consisting of circuit breakers, fuses, or switches located in a panel box near the point of entrance of supply conductors.

set. Device or contrivance used to ignite an incendiary fire.

short-circuit. Direct contact between a current-carrying conductor and another conductor.

smoldering combustion. The direct combination of a solid fuel with atmospheric oxygen to generate heat in the absence of gaseous flames. *See Glowing combustion.*

soot. The carbon-based solid residue created by incomplete combustion of carbon-based fuels.

spall. Crumbling or fracturing of concrete or brick surface as a result of exposure to thermal or mechanical stress.

spark. Superheated, incandescent particle.

spoliation. Destruction, loss, or alteration of material that is or can be evidence at a judicial proceeding.

spontaneous ignition. Chemical or biological process that generates sufficient heat to ignite the reacting materials. See *Self-heating*.

stoichiometry. Balance of chemical reactants and products.

suspicious. Fire cause has not been determined, but there are indications that the fire was deliberately set and all accidental fire causes have been eliminated.

thermal inertia. A value calculated from the thermal conductivity, density, and heat capacity that relates to the ignitability of that material.

thermal protector. An inherent device against overheating that is responsive to temperature or current and will protect the equipment against overheating due to overload or failure to start.

thermoplastics. Organic materials that can be melted and resolidified without chemical degradation.

thermosetting. Materials that, once solidified, will not melt but will chemically degrade as they are heated.

torch. A professional fire setter.

trailers. Long trails of fast-burning materials used to spread a fire throughout a structure.

vapor. The gaseous phase of a liquid or solid that can be returned to that phase by the application of pressure.

vapor density. The ratio of the weight of a given volume of gas or vapor to that of an equal volume of air.

vented. The fire has extended outside the structure or compartment by destroying the windows or burning an opening in the roof or walls.

ventilation. A technique for opening a burning building to allow the escape of heated gases and smoke to prevent explosive concentrations (smoke explosions or backdrafts) and to allow the advancement of hose lines into the structure.

volatile. A liquid having a low boiling point; one that is readily evaporated into the vapor state.

volt. The basic unit of electromotive force.

voltage, nominal. A value assigned to a circuit or system for the purpose of conveniently designating its voltage class (120/240, 480Y/277, 600, etc.).

watt. Unit of heat release, 1 watt = 1 joule per second; unit of power or work (in electrical circuits, equivalent to voltage multiplied by amperes).

Mathematics Refresher

C.1 FRACTIONAL POWERS

When a number (or value) is followed by a superscript number or fraction, such as X^n, it means that the number (X) is raised to a power (n). If the power is 2, the number is squared (multiplied by itself) ($3^2 = 3 \times 3 = 9$). If the power is 3, the number is cubed (multiplied by itself and then by itself again) ($3^3 = 3 \times 3 \times 3 = 27$).

The number n can be a fraction, such as $\frac{1}{2}$.

$$X^{1/2} \quad \text{or} \quad X^{0.5} = \sqrt{X} \qquad \text{(the square root of } X\text{).} \tag{C.1}$$

The number n can be any value. $n = 2/5, 5/2,$ and $3/2$ are common calculations in fire dynamics. Such values must be calculated using a "scientific notation" pocket calculator with a y^x or y^n function (where $y = X$, and x [or n] = the power to which y is raised). If n is a negative number, it denotes $1/X^n$ (X^n is put in the denominator of a fraction).

C.2 LOGARITHMS

Logarithms are numerical equivalents calculated by raising 10 to a power where log_{10} = the value to which the number 10 must be raised to be equal.

The

$$log_{10} \quad \text{of} \quad 100 = 2 \tag{C.2}$$

because

$$10^2 = 100. \tag{C.3}$$

Also,

$$log_{10} \quad \text{of} \quad 10 = 1 \tag{C.4}$$

because

$$10^1 = 10. \tag{C.5}$$

This value can be found using a pocket calculator, tables, or a slide rule. For instance,

$$log_{10} 3.5 = 0.54, \tag{C.6}$$

while

$$log_{10} 2330 = 3.3673. \tag{C.7}$$

The inverse function is sometimes called the antilog:

$$antilog_{10}(2) = 10^2 = 100$$

When the base unit e is used ($e = 2.71$), the value is called the natural log. For example,

$$log_e 5 = \ln 5 = 1.6094, \tag{C.8}$$

while

$$\ln 10 = 2.3025. \tag{C.9}$$

The antilog e^x is the value calculated by raising e to that power,

$$e^5 = 148.413 \tag{C.10}$$

and

$$e^{10} = 22,026.5 \tag{C.11}$$

Figure 6.1 uses a modified "power of 10" notation as shorthand for very large or very small numbers. For example,

$$3.40\text{E} - 05 = 3.4 \times 10^{-5} = 0.000034,$$
$$2.88\text{E} + 00 = 2.88 \times 10^{0} = 2.88,$$
$$5.00\text{E} + 02 = 5.0 \times 10^{2} = 500.$$

C.3   DIMENSIONAL ANALYSIS

A useful concept for checking calculations is called dimensional analysis. If one keeps track of the units used for variables and constants and applies the rules of canceling out units when they appear in both the numerator and the denominator of a function, for instance, one can verify that the correct relationship was used.

For example
$$\dot{Q} = m'' A \Delta H_c, \tag{C.12}$$

where

m'' =   mass flux ($kg/m^2 s$),

A =   area of burning surface (m^2), and

ΔH_c =   heat of combustion (kJ/kg),

multiplying

$$\left(\frac{kg}{m^2 s}\right)(m^2)\left(\frac{kJ}{kg}\right) = \frac{kJ}{s} = kW, \tag{C.13}$$

which is the correct unit for $\dot{Q}$.

For conduction calculations,

$$\dot{Q} = k\frac{T_2 - T_1}{l}A, \tag{C.14}$$

where
k = $\dfrac{W}{m} \cdot K,$

l =   length (m),

$T_2 - T_1 = \Delta T(K),$ and

A =   area (m^2).

Solving for

$$\dot{Q} = \frac{\left(\dfrac{W}{m \cdot K}\right)(K)(m^2)}{m} = W, \tag{C.15}$$

which is the correct unit for $\dot{Q}$.

Index